# No Standard Oil

# No Standard Oil

*Managing Abundant Petroleum in a Warming World*

—◆◦◆—

## DEBORAH GORDON

OXFORD
UNIVERSITY PRESS

# OXFORD
## UNIVERSITY PRESS

Oxford University Press is a department of the University of Oxford. It furthers
the University's objective of excellence in research, scholarship, and education
by publishing worldwide. Oxford is a registered trade mark of Oxford University
Press in the UK and certain other countries.

Published in the United States of America by Oxford University Press
198 Madison Avenue, New York, NY 10016, United States of America.

Library of Congress Cataloging-in-Publication Data
Names: Gordon, Deborah, author.
Title: No standard oil : managing abundant petroleum in a
warming world / Deborah Gordon.
Description: New York : Oxford University Press, 2022. |
Series: Carnegie endowment for intl peace series | Includes index.
Identifiers: LCCN 2021034530 (print) | LCCN 2021034531 (ebook) |
ISBN 9780190069476 (hardback) | ISBN 9780190069490 (epub) |
ISBN 9780190069506
Subjects: LCSH: Petroleum—Environmental aspects. | Climate change.
Classification: LCC QH545.O5 G68 2021 (print) | LCC QH545.O5 (ebook) |
DDC 363.738/2—dc23
LC record available at https://lccn.loc.gov/2021034530
LC ebook record available at https://lccn.loc.gov/2021034531

DOI: 10.1093/oso/9780190069476.001.0001

1 3 5 7 9 8 6 4 2

Printed by Sheridan Books, Inc., United States of America

*To my husband, Eric, and our sons, Michael and Josh*

# Contents

# *Foreword*

IN A QUIET, workmanlike way, this book makes a revolutionary argument. If it becomes widely understood and applied, it could spell the difference between success and dismal failure in keeping climate change from greenhouse gas emissions to less than 2 degrees Celsius by 2050, the official international target for avoiding catastrophic damage from a rapidly changing climate.

Deborah Gordon was trained as a chemical engineer and began her career working in the oil industry—for Chevron. She later worked in state government, academia, think tanks, and advocacy groups. Her transformative recognition that vast amounts of greenhouse emissions are hiding in plain sight—not recognized, not counted, and not addressed—came from this rare variety of experience and of approaches to problem solving packed in one person. She is relentlessly driven by data, a technical expert who is equally experienced in policymaking, and she is constitutionally unable to see any of the key actors in this global challenge as the bad guys. At one time or another in her career they have all been colleagues and friends.

All of this came together over the past half-dozen years, as she realized that what we used to think of as simply oil—such a dependable thing that John D. Rockefeller named his company Standard Oil for a reason—is actually now a heterogeneous mix of very different chemical entities ranging from heavy sludge to viscous liquid to volatile gas. From there it was a natural step to wonder how greatly these very different entities differed in their greenhouse gas emissions. To answer this question, from her base at the Carnegie Endowment for International Peace, a global think tank, Gordon assembled a team of experts at far-flung universities who together built and integrated a suite of models. With this unprecedented tool in hand, they were naturally led to consider how such varied resources differed not only in their inherent carbon content but also in the emissions intensity of their extraction, processing, refining, transportation, byproducts, and leakages as well as their eventual end use. From this work came the Oil Climate Index, later expanded to include natural gas, and now known as the OCI+.

The OCI+ transforms one's understanding of the steps that will have to be taken to control climate change. It focuses attention on the many kinds of unconventional oil and gas being produced with new technologies like horizontal drilling and hydraulic fracturing. It demands that advocates for slowing climate change devote their attention not just to a chemical fantasy called "oil" and to its eventual end use—say, gasoline in cars—but to every step in the supply chain in between. The OCI+ revealed that emissions intensities vary by a factor of ten from producing and processing different types of oil and gas and that over their full lifecycle including end use, barrels of otherwise equivalent oil and gas can differ by a factor of three or more. Because these differences are now largely ignored, attractive opportunities to control emissions are being "squandered," as Gordon terms it. Collecting enough accurate data to fully populate and make use of the OCI+ will be a difficult undertaking requiring governmental action, but if accomplished, it promises a huge—transformative—payoff, allowing policies to be targeted to the lowest-cost, highest-impact steps to control emissions.

Having fully described the current state of oil production and the OCI+, this book spells out a second, equally profound, data-based insight: namely that the supply side is just as important for controlling greenhouse gas emissions as are demand-side policies. In the short term—and the thirty years between now and 2050 is the short term—it is more important. Oil and gas account for 54 percent of global primary energy supply. Moreover, petroleum products are embedded throughout the economy from food companies, pharmaceutical firms, big-box retail, airplanes, plastics—indeed, in just about every industry one can name. So in the crucial few decades ahead, we need to "shift our mindset to assume that oil and gas are here to stay, [so that] we can actually minimize their supply-side emissions now instead of imagining that they will go away." Divestment strategies, which largely shift resources and emissions from one company to another, will not meet the need. Even combined with aggressive demand-side policies, they will not get the world's emissions to where they need to be. Rather, governments, academics, and civil society advocates for slowing greenhouse emissions must all recognize that in the near term oil and gas will be a major part of global emissions and that the massive industry that produces them has to be a constructive participant in the global effort.

Relentlessly practical and knowing from direct experience the particular strengths of each set of players, in the closing chapters of the book Gordon lays out an extensive menu of partial solutions to climate change. There is no silver bullet. Instead, what she calls "2 percent solutions," layered together, can achieve what needs to be done.

Few books offer a genuinely new way of looking at the world. This one offers two: one grounded in the technical knowledge of a chemical engineer, the other, equally important, based on the experience gained from having addressed climate policy from the vantage of producer, policymaker, analyst, and advocate. Its insights deserve the closest attention by all those concerned with addressing the existential challenge of the changing global climate.

Jessica T. Mathews
Distinguished Fellow, Carnegie Endowment for International Peace

# *Acknowledgments*

UNDERSTANDING THE PROPERTIES of oil and gas—and analyzing how the production, processing, refining, and end uses of these resources affect the environment—has been my life's passion. My interest in the oil and gas sector began in the early 1980s, when I worked in a hydrogen catalysis lab during my undergraduate years at the University of Colorado and subsequently as a gas field summer roustabout with Conoco and then as a chemical engineer with Chevron. Later, my graduate work at the Goldman School of Public Policy at UC Berkeley and my professional experiences at the Lawrence Berkeley National Laboratory, Union of Concerned Scientists, and Yale School of the Environment afforded me the opportunity to study and develop climate-related transportation policies. But it was not until I joined the Carnegie Endowment for International Peace that I was able to connect my knowledge about oil and gas properties to the study of climate change. This book grew directly out of my energy and climate research while working at Carnegie between 2010 and 2019. This research has followed me to RMI (formerly Rocky Mountain Institute), where I currently lead the Oil and Gas Solutions Initiative in the Climate Intelligence Program. I wish to express my appreciation to Jules Kortenhorst and my RMI colleagues for their superb collaboration. I also would like to thank Edward Steinfeld and my colleagues at the Watson Institute for International and Public Affairs at Brown University, where I am a Senior Fellow and have the opportunity to help mentor the next generation of students.

Many people made important contributions to this project. First and foremost, I would like to thank my Oil Climate Index Plus Gas (OCI+) research partners, Adam Brandt, Joule Bergerson, and Jonathan Koomey, as well as their amazing cadre of past and present graduate students, including Mohammad Masnadi, Hassan El-Houjeiri, Jeff Rutherford, Jacob Englander, Yuchi Sun, Zhan Zhang, Kavan Motazedi, and Liang Jing. This book would not have been completed without their stellar research, keen insights, and enduring friendship.

The launch of the OCI+ benefited greatly from the support of Carnegie's former presidents, Jessica Mathews and William J. Burns. Numerous former Carnegie Junior Fellows and Energy and Climate Program staff were instrumental in this project, including Frances Reuland, Sam Wojcicki, Smriti Kumble, Jeffrey Feldman, Eugene Tan, David Livingston, Wang Tao, Matt Ferchen, Florencia Franzini, Shin-pei Tsay, Rachel Flaherman, Madeleine Bronstein, Anisha Mehta, Hannah Donart, Kate Garner, Yevgen Sautin, Sabine Jonhson-Reiser, Adnan Vatensever, Yuhan Zhang, Tom Carothers, and the late David Burwell. While they are too numerous to name, I am indebted to my many other former Carnegie colleagues and board members for their support and encouragement. Chapter 3 draws directly from the *Know Your Oil* report that was published by Carnegie in 2015. The OCI+ web tool exceeded expectations and was built (and is being expanded) by Drew Bollinger, Ian Schuler, and their crew at Development Seed.

I owe a debt of gratitude to Ryan DeVries for his meticulous review of the manuscript and his masterful editing skills. Thank you to Devyn Collado-Nicol, Stephen Ziman, Julia Benz, Hilary Levy Friedman, Janet Peargin, Caroline Blanck, and Abigail Lambert for reading drafts and providing feedback and encouragement. Fran Reuland created two superb websites, one for the Oil Climate Project and another for this book. Dina Cappiello, Christian Roselund, Nick Steel, and Jennifer Stokes provided communications, marketing, and development support. For hosting events, participating on panels, convening and attending meetings, and having conversations that sharpened my thinking, thank you to Madhav Acharya, Rachel Adams-Heard, Nate Aden, Ann Alexander, Robert Armstrong, Paul Balaran, Jonathan Banks, Alex Barba, Morgan Bazilian, Johan Bergenas, Heidi Binko, Angela Blanchard, Peter Blanck, Bill Bradley, Sian Bradley, Daniel Bresette, Bruce Cain, Edil Sepulveda Carlo, Tom Carver, Marion Chertow, Jeff Colgan, Lisa Couch, the late John Courtis, Jon Creyts, Cynthia Cummis, Daniel Cusworth, Aaron Davitt, Phil DeCola, Simon Dietz, Kathryn Dunkelman, Riley Duren, Jane Durkin, Chris Elvidge, Stephanie Epner, Dan Esty, Harvey Fineberg, Chaz Freeman, Dan Gardiner, Andrew Glickman, former V.P. Al Gore, Tim Gould, Jeremy Grantham, William Hafker, Leslie Harroun, Hal Harvey, Ned Harvey, Karl Hausker, Sean Hecht, Wolfgang Heidug, John Holdren, Taku Ide, Benjamin Israel, Lucinda Jackson, Daniel Jacob, Sarah Jordaan, Maciej Kolaczkowski, Kevin Knobloch, Jules Kortenhorst, Glada Lahn, Judy Lai-Norling, Sarah Landislaw, Rick Lattanzio, Andrew Leach, Henry Lee, Syed Munir Khasru, Reid Lifset, TJ Loudermilk, Amy Luers, Loreana Marciante, Jeremy Martin, Kevin Massy, Gavin McCormick, Christophe McGlade, Nathan Meehan, Patty Monahan, Kathy Mulvey, Robert Murphy, Moises Naim, Mary Nichols, Susann Nordrum, Cathy Paglia, Margarita Parra, Edward Parson, Arvind Ravikumar, Rafael Reif, Timmons Roberts, Philippe Roos, Joseph Ryan,

Wendy Schiller, Jigar Shah, Drew Shindell, Laura Singer, Sarah Smith, Irena Spazzapan, Dan Sperling, Lekha Sridhar, Andrew Stevenson, Benjamin Storrow, Georges Tijbosch, Dan Tuden, Kelly Vaughn, Gabriela Volpato, Carol Werner, Becky White, Catherine Witherspoon, Heidi Yamaguchi, Durwood Zaelke, and so many others that I do not have the space to name.

I wish to thank the following philanthropic institutions for providing generous financial support to create and expand the OCI+ model and write this book: Grantham Foundation, Alfred P. Sloan Foundation, Oak Foundation, Hewlett Foundation, ClimateWorks Foundation, Energy Foundation, Pisces Foundation, Rockefeller Family Fund, and Blue Moon Fund. High Tide Foundation, Lyda Hill Philanthropies, and many others continued to support this work when I joined RMI.

To my lifelong friends Barbara Bernstein, Karen Bloomfield, Ayelet Harnof, Beth Johnke, Wendy Kislik-Blanck, Michele Kupfer, Lisa Lawrence, Jill Rutter, Cheryl Wetmore-Simpson, and Shari Weinberger, thank you for patiently listening to me muse about oil and climate change over the years. I also wish to acknowledge my professional mentors, including Eugene Bardach, the late Hank Dittmar, John Falconer, Emil Frankel, Kurt Gottfried, the late Henry Kendall, Jessica Mathews, the late Arthur Rosenfeld, and Kenneth Train. I am extremely grateful that David McBride, Oxford University Press editor in chief, believed in this book project from the start. Thank you to Holly Mitchell, Helen Nicholson, and others at the press who ushered this project to fruition and to the anonymous reviewers for the press whose suggestions greatly improved the manuscript.

Finally, I wish to thank my husband, Eric; my sons, Michael and Josh; my parents, Coco and the late Mike Gordon; my siblings, Jan and Rob; and my extended family members Derek, Anne, Bernie, Dave, and Dayna for their love and support. You are my rocks.

Deborah Gordon
Providence, RI
August, 2021

# *Abbreviations*

| | |
|---|---|
| ALEC | American Legislative Exchange Council |
| APEC | Asia-Pacific Economic Cooperation |
| API gravity | American Petroleum Institute gravity |
| AR | assessment report (produced by the IPCC) |
| ARCO | Atlantic Richfield Company |
| ARPA-C | Advanced Research Projects Agency - Climate |
| ARPA-E | Advanced Research Projects Agency - Energy |
| BOE | barrel of oil equivalent |
| BP | formerly known as the British Petroleum Company |
| Carbon Mapper | an public-private-nonprofit consortium locating methane and $CO_2$ emissions from air and space |
| CAFE | corporate average fuel economy |
| CCAC | Climate and Clean Air Coalition |
| CCL | Citizens Climate Lobby |
| CCS | carbon capture and storage (or CCUS is when captured carbon is utilized and not stored) |
| CDR | carbon dioxide removal |
| CEO | chief executive officer |
| Climate TRACE | a emissions tracking project involving several NGOs, including RMI, and the use of artificial intelligence |
| CMS | Carbon Monitoring System |
| CNPC | China National Petroleum Corporation |
| $CO_2$ | carbon dioxide |
| $CO_2e$ | carbon dioxide equivalent |
| COP | Conference of the Parties |
| CTL(s) | coal-to-liquids |
| DRIVE+ | a feebates program started in the state of California |
| EDF | Environmental Defense Fund |
| EIA | (US) Energy Information Administration |
| EITI | Extractive Industries Transparency Initiative |
| EOR | enhanced oil recovery |
| EPA | Environmental Protection Agency |
| ESG | environmental, social, and governance (factors for evaluating corporate performance) |
| EU | European Union |
| EV(s) | electric vehicles |
| FPSO | floating production storage and offloading |
| GHG(s) | greenhouse gas(es) |
| GTL(s) | gas-to-liquids |
| GWP(s) | global warming potential(s) |
| GWP* | proposed replacement for GWP |

| | |
|---|---|
| IEA | International Energy Agency |
| INOCs | international national oil companies |
| IOCs | international oil companies |
| IPCC | Intergovernmental Panel on Climate Change |
| IPO | initial public offering |
| LCA | lifecycle assessment |
| LCFS | low-carbon fuel standard |
| LiDAR | light detection and ranging |
| LNG | liquefied natural gas |
| LPG | liquefied petroleum gas |
| MiQ | a standard to reduce methane emissions in the oil and gas sector through certified differentiated gas |
| MMS | US Minerals Management Service |
| MRV | monitoring, reporting, and verifying (GHG emissions) |
| NAS | National Academy of Sciences |
| NASA | National Aeronautics and Space Administration |
| NDCs | nationally determined contributions |
| NET(s) | net-zero emissions technologies |
| NGL(s) | natural gas liquids |
| NGO(s) | nongovernmental organization(s) |
| NOAA | National Oceanic and Atmospheric Administration |
| NOC(s) | national oil companies |
| OCI+ | Oil Climate Index + Gas |
| OGCI | Oil and Gas Climate Initiative |
| OPEC | Organization of Petroleum Exporting Countries |
| OPEM | Oil Products Emissions Module |
| OPGEE | Oil Production Greenhouse Gas Emissions Estimator |
| PM | particulate matter |
| POFP | photochemical ozone-forming potential |
| PRELIM | Petroleum Refinery Lifecycle Inventory Model |
| R&D | research and development |
| RDD&D | research, development, demonstration, and deployment |
| Saudi Aramco | Saudi Arabia Oil Company |
| SCO | synthetic crude oil |
| SDGs | Sustainable Development Goals |
| SEC | US Securities and Exchange Commissions |
| SINOPEC | China Petroleum and Chemical Corporation |
| SIQ | SystemIQ |
| SLCPs | short-lived climate pollutants |
| SMR | steam methane reforming |
| SOE(s) | state-owned enterprises |

| | |
|---|---|
| TAN | total acid number |
| TCFD | Task Force on Climate-Related Financial Disclosures |
| TCI | the name of a UK-based hedge fund |
| TPI | Transition Pathway Initiative |
| TROPOMI | Tropospheric Monitoring Instrument |
| UAE | United Arab Emirates |
| UK | United Kingdom |
| UNFCCC | United Nations Framework Convention on Climate Change |
| USGS | US Geological Survey |
| VIIRS | Visible Infrared Imaging Radiometer |
| VOCs | volatile organic compounds |
| ZEV | zero-emission vehicle (mandate) |

# *Introduction*

## THE UNEXPECTED PITFALLS OF CONTENDING
## WITH OIL AND GAS

IT'S THE SPRING of 1984. I am a chemical engineer for Chevron, testifying at a public hearing in Santa Barbara, California, in front of the county's powerful Air Pollution Control Board. Chevron is seeking to develop one of the largest-ever western US oil leases off California's central coast.[1] Back-to-back oil crises during the mid- to late 1970s—and unpleasant memories of long lines at gasoline pumps—are fueling exploration.[2] The national search for oil is on.

Without an air permit, Chevron's massive project cannot proceed. But the air district's control officer, John English, has serious reservations about how Chevron's operations will increase local smog levels. My job is to state how the company can effectively reduce its emissions.

But more oil development is decidedly not what Santa Barbara County's elected officials want or need.[3] A dozen platforms already dot the Southern California coast. Oily blobs of tar naturally seep from the ocean floor, washing ashore and semi-permanently tattooing the feet of those who casually stroll in the sand. Fifteen years earlier, a well blow-out caused the largest oil spill in California's history, coating Santa Barbara's coast and killing thousands of birds and marine wildlife.[4] Public outrage extended beyond the state's borders, propelling America's modern environmental movement and the regulatory frameworks that followed.[5]

Regardless, the Reagan administration is now expediting offshore oil leasing, citing vast energy supplies and promises of US energy security.[6] Although the president's vacation home is located only a few miles away, other famous Santa Barbara residents adamantly oppose the development of this new oil field.[7] Rock stars Jackson Browne and Bonnie Raitt perform free concerts to raise money for a local public interest environmental group, Get Oil Out![8] But even the rich and the famous cannot stem the flow of oil, and the coveted air permit for Chevron's Point Arguello–Gaviota project is granted.

To win approval, Chevron makes a commitment to install and demonstrate (before it is required by law) a novel environmental technology to reduce smog-forming air pollutants from the project's large bank of engines.[9] This, along with a promise to abide by roughly 160 other environmental permit conditions, secured a regulatory green light.[10] By 1987, construction is complete: three platforms, an onshore oil and gas processing plant and flare, massive storage tanks, expanding marine terminal, and over thirty miles[11] of specially insulated pipelines.

## More Questions Than Answers

I left Chevron in 1987 and headed to grad school at the University of California, Berkeley. Five years as an engineer in the oil industry had piqued my interest in public policy. I wanted to step back from the conflicts over the merits of a single energy project's development plans and ask deeper questions. If oil is abundant, why are we proceeding as if it is scarce? Are the differences between hydrocarbon resources greater than their similarities? Why are we overlooking the wide-ranging climate impacts of oil and gas resources? How can we get stakeholders to work together? Is there a better course for energy and environmental decision-making in the United States and the world?

While I was no longer around to observe events in Santa Barbara firsthand, the saga continued. Drilling at Point Arguello began in 1989. Then, just as production peaked in 1993, oil prices tumbled.[12] Five years later, oil prices hit rock bottom at $17 per barrel in real terms—their lowest level since World War II.[13] Chevron decided to shutter the Gaviota plant and reinject its gas to produce more oil, which was heavy like tar and contained poisonous hydrogen sulfide gas and heavy metals. Within a year, the majority stake in Point Arguello was sold to little-known Plains Resources and then, in 2013, was acquired by the copper conglomerate Freeport-McMoRan.[14]

In 2015, an interconnected pipeline installed in 1987 ruptured due to severe corrosion, spilling an estimated 100,000 barrels of viscous oil,[15] again marring Santa Barbara's beaches and killing scores of marine wildlife.[16] The Point Arguello–Gaviota pipeline was ordered to be completely emptied of all its oil and gas contents after surveys revealed corrosion characteristics similar to those that caused the spill.[17] A moratorium was placed on offshore oil production. The pipeline operator, Plains All American, was found guilty in 2019 and ordered to pay over $3 million in fines and penalties. Freeport-McMoRan announced plans to start decommissioning Point Arguello's three platforms in 2020.[18]

In retrospect, the project I testified about in 1984, asserting that Chevron was capable of managing all risks, ultimately failed on every dimension. This failure did not result from a lack of trying. It was due to a lack of full understanding. Chevron could not manage what they did not know.

Everyone's perception of reality was distorted by the conventional thinking of the mid-1970s to early-1980s. Oil was thought to be running out, and this elevated the project's status as a national priority and masked its many challenges. Chevron's engineers assumed they could easily convert Point Arguello's unconventional oil into gasoline. Regulators focused on reducing local smog, yet climate pollutants were entirely ignored. And a pitched battle between government, industry, and civil society was assumed to be the only certain way forward. All these assumptions turned out to be wrong or misguided.

## *Shouldering Climate Risks*

The oil and gas industry has confronted many challenges over its century-plus history. It has weathered economic risks (an inevitability of making investments based on very long-term and uncertain price forecasts), political risks (such as losing control of oil reserves due to domestically driven resource nationalization), and operational risks (like delving deeper and faster into more extreme recesses of the earth in pursuit of hydrocarbons), as well as environmental risks (such as mitigating the local impacts of water and air pollution). Another hidden danger, reputational risk—how companies are viewed in relation to their competitors— is also always lurking and can oust corporate heads and threaten the survival of a company.

But the risks posed by climate change are different.[19] Climate change cuts across all of these risks: economic, political, operational, environmental, and reputational. It is an existential threat to oil companies' profit-making model and strikes at the heart of the petroleum industry's very being. With damages that escalate dramatically over time, the intergenerational costs of global warming are at odds with the long-term returns needed to pay for major capital investments up front.[20]

At the same time that John D. Rockefeller was growing his Standard Oil empire, the Swedish scientist Svante Arrhenius theorized that the build-up of heat-absorbing gases in the atmosphere would increase Earth's temperature.[21] By the early 1900s, fossil fuel combustion was labeled as a main culprit in global warming.[22] Carbon dioxide ($CO_2$) measurements have been taken at Mauna Loa, Hawaii, since they were first recorded in 1958 at 313 parts per million (ppm).[23] As of June 2021, atmospheric $CO_2$ concentrations had risen to 419 ppm.[24] And other petroleum-related greenhouse gas (GHG) emissions, such as methane, are also breaking records at 1,889 parts per billion in March 2021, up 16 percent since July 1983.[25]

## Oil Reality Check

The myths and half-truths about oil and gas and their climate implications muddle both public and private decision-making. Exposing them can realign markets, steer policymaking, guide civil society, and facilitate a durable, low-carbon energy transition. Conversely, glossing over facts or avoiding the truth can set in motion wasteful efforts that impose lasting damages to the local community, global environment, and industry reputations. Using outdated assumptions and methods that ignore external social and environmental impacts (negative externalities) forces citizens to bear the costs rather than companies themselves. These realities, exposed through the lens of my Point Arguello experience, are explored in great depth in this book. They are more pressing today than when I experienced them decades ago.

### Reality #1: Oil and Gas Resources Are Physically Abundant

Point Arguello did not stop producing for lack of oil.[26] This supergiant field was mired in local politics and unfavorable economics. Since then, however, technological innovations and oil production breakthroughs like hydraulic fracturing, horizontal drilling, and enhanced recovery methods have been unearthing unconventional oil and gas supplies with novel properties (compared to conventional resources), which require new methods to be extracted and processed into the standard petroleum products we consume like gasoline for our cars, propane for our grills, asphalt for our roads, sulfur in our medicine, and petrochemicals in our clothes.[27] Massive amounts of conventional and unconventional oil and gas may ultimately be recoverable: trillions of barrels of oil and quadrillions of cubic feet of gas.[28] The full bounty of oil and gas exceeds totals we can safely combust. Since oil and gas will not run out, we need to wrestle anew with the climate impacts of abundant fossil fuels.[29]

### Reality #2: Oil and Gas Resources Are Chemically Diverse

Oil and gas may appear to be deceptively simple, consisting of mostly two basic elements—hydrogen and carbon. But their recipes are not. Conventional oil and gas are pooled beneath the surface, freely flow when tapped, and have the consistency of maple syrup (oil) and air (gas). Unconventional resources, however, are trapped, do not readily flow, and have consistencies ranging from chunky peanut butter to nail polish remover.[30] They also can have unusual compositions. Unconventional oils may be heavier (more solidified and containing excess carbon) or extremely light (more gaseous and containing excess hydrogen). Unconventional gases can contain excess carbon (making them quasi-liquid or high in $CO_2$) and other unusual contaminants.

Point Arguello's oil was unconventional. It was produced from a shale formation (Monterey shale) and was viscous like oil sands, and its gas contained deadly contaminants, like hydrogen sulfide. Its extraction required hot water and chemical additives.[31] Nearby refineries required costly retrofits to process its oil. Safety precautions were enacted to decontaminate its gas, and pipelines had to be insulated to make it flow.[32] Treating heterogeneous oil (or gas) resources as if they are homogeneous introduces significant economic, safety, and environmental risks.

## Reality #3: Oil and Gas Have Wide-Ranging Lifecycle Climate Impacts

Regulatory authorities threw just about every rule in the book at Chevron's Point Arguello–Gaviota project, except climate change.[33] Broad awareness of this global disaster in the making was emerging with the formation of the Intergovernmental Panel on Climate Change (IPCC) in 1988. Yet, during years of permitting, the inert gas ($CO_2$) was never mentioned, while the powerful climate-warming gas, methane, was expressly disregarded because it was not considered a reactive organic compound that caused smog.[34] Other climate-forcing gases were ignored, including the black carbon emitted from burning the project's heavy-oil byproducts. All in all, producing, processing, and shipping a barrel of Point Arguello crude likely has over twice the climate impact as the oil California now imports from Saudi Arabia does.[35] Failing to account for these material differences in lifecycle emissions between the production, refining, and end uses of various oil and gas resources has us flying blind in the battle to combat climate change.

## Reality #4: Companies Come and Go, but Oil and Gas Endure

While Chevon spearheaded this multi-billion-dollar project, some two dozen industry partners large and small were involved, and only a few of these companies exist today.[36] Even Big Oil—the consortium of Western multinational oil companies—remains in flux. The seven major international oil companies known as the Seven Sisters once controlled over 80 percent of the world's petroleum reserves.[37] Today, their fortunes have reversed. National oil companies

(NOCs)—state-owned enterprises (SOEs) in the Middle East, Africa, Eastern Europe, and Latin America—are now estimated to control more than 85 percent of the global oil and gas market.[38] And new actors continue to gain footholds, including independent producers and refiners, shippers, traders, limited partnerships, and financial firms.

The shifting fortunes and changing roster of the oil and gas industries' major players are an important consideration. The diverse and globalized ecosystem of oil and gas enterprises is often called on to take full responsibility for its role in contributing to climate change. But it is not clear exactly who is in charge, both because the market is highly decentralized and because the biggest players of the past and the present are not the same. Even if that were not the case, oil and gas assets can change hands many times over a project's lifetime, so it is not always evident how responsibility for climate-related externalities should be apportioned to the various firms involved in a given project over its lifetime. In addition to the example of Point Arguello, with its revolving door of partners, consider the difficulty allocating lifetime GHGs for the Martinez refinery (the third largest in California) where nine different companies (large and small) have come and gone since its operations commenced in 1913.[39] For these reasons, it is not enough and not always feasible to pursue individual companies when it is the oil and gas resources themselves that endure and continue to damage the climate.

## Reality #5: Governments Face Conflicting Incentives

Chevron followed government regulators' marching orders on a litany of requirements. The firm was expected to protect monarch butterflies, kelp beds, fishing sites, and raptor roosts; safeguard Chumash Native American sacred sites; secure coastal access for public hiking; relocate a local public school; placate armed ranchers along the pipeline route; avoid interfering with nearby Vandenberg Air Force Base; continuously monitor for releases of deadly hydrogen sulfide; and on and on. The project involved at least five federal, state, and local government agencies, each with a designated charge.[40]

But were these government regulators actually *in* charge? Or were the budget officials—the state controller, governor, and finance directors—actually pulling the strings? California stood to make a lot of money on Point Arguello. The lease granting drilling rights generated over $1 billion in today's dollars,[41] and millions more in annual royalties were split between California and the US federal government.[42] The national, state, and local government actors involved in overseeing such projects may impose specific protections, but they are often pitted against one another and even themselves—working at cross-purposes with larger financial (and political) aims. These tensions are large enough in countries like the United States, but they are even greater in other countries like Iraq, Russia, and Azerbaijan, where oil and gas revenues make up a major share of national budgets.[43]

### Reality #6: Civil Society Demonizes Oil and Gas but Can't Stop Using Them Either

I was loath to discuss my work in Santa Barbara because it led to intense arguments. So I engaged local residents, posing as a grad student from the University of California, Santa Barbara. I asked them what their biggest concerns were and how they squared those worries with the fact that their every car trip, plane ride, hot shower, makeup application, medical treatment, and house insulation job was made possible by oil and gas. Oil and gas were designated public enemies to be vanquished. Was this the case because they felt their own backyards were at stake?

Outspoken critics of oil and gas never offered up personal austerity measures, which highlights an inconvenient truth: the world is hooked on oil and gas, and cutting back on them will not be easy. Virtually every drop in a barrel of oil is exploited for one purpose or another. Every molecule of hydrocarbons is turned into something that someone, somewhere will buy—or else it is leaked into the air and water or pumped back into the ground. Environmentally conscious civil society is missing an honest conversation about oil and gas and how to actually dial down their climate impacts as we chart the long, winding road to net-zero GHG emissions in the petroleum sector.[44]

## *An Oil Brain Trust Reunites*

As 2020 approached, more of my former Chevron colleagues opted to retire. A light bulb went off in my head: why not reunite those of us who permitted Point Arguello? I imagined that, together, Big Oil retirees with their vast know-how could help the industry move beyond business as usual and do business better. A dozen of us initially connected. We agreed that the oil industry was remiss in its duty to deal with climate change head on. As the earth's temperature reached new highs, Chevron (and its competitors) were not acting urgently enough to address a colossal problem of their own making.

For three years, we connected by phone, sharing ideas. We cultivated and produced a technical document that identified what we saw as low-hanging fruit—numerous swift actions to leverage assets while improving efficiency, reducing GHG emissions, and cutting costs. A couple of us even published a "how to" climate plan for petroleum companies.[45] In February 2020, nine of us planned a reunion (seven attended in person). It was amazing to see old work friends and meet others who had picked up the torch after I left Chevron. With collectively some 300 years on the job, our technical and policy wisdom was palpable.

The next day we had an audience with Chevron management. We knew the company could not divulge corporate intel to us, even if we once served in its ranks. At most, we were now only shareholders. (For my part, I am divested.) But we wanted Chevron to hear from us so we could offer up technical solutions and share win-win opportunities to decarbonize their operations. The four guiding principles we discussed apply well beyond Chevron to the oil and gas industry writ large. These recommendations underpin a trailblazing way forward to actually reduce the GHG footprint of the oil and gas industry by leveraging our expertise of the industry itself and working in tandem with established players.

First, oil and gas companies should increase transparency by publicly releasing and routinely updating credible corporate climate plans using open-source lifecycle assessments to evaluate their GHGs.[46] Second, companies should advocate for GHG pricing on actual, measured, auditable data for all climate pollutants. Third, companies should accept climate responsibility by exiting industry associations that challenge climate science and by rewarding employees for excellent climate performance. And fourth, companies should position themselves on a climate path below 2 degrees Celsius by adopting science-based mitigation targets[47] and investing in research, development, demonstration, and deployment (RDD&D) on low-GHG technologies and pathways.[48]

Some of us left the meeting more upbeat than others. But we all agreed that real, concerted change takes perseverance on the part of the industry, government, and civil society. Measurable and durable action is urgently needed. The oil sector is at a turning point and the next decade will make or break its future. The transition from Big Oil (and Big Gas) to Big (Low-Carbon) Energy starts by reducing the climate intensity of the industry's own operations.

## The Problems, the Players, and the Pathways

This story has three parts. The first section details overlooked aspects of the problems that have spawned climate change. Chapters 1 and 2 detail deeply rooted, mistaken patterns of conventional thinking about fossil fuels and the market failures that stand in the way of progress: massive externalities, incomplete knowledge, oligopolistic behavior, and intergenerational inequities. Chapters 3 and 4 introduce and apply a more complete and accurate open-source framework for assessing the climate impact of the world's diverse array of hydrocarbons—the Oil Climate Index + Gas (OCI+).[49]

For all the awareness, optimism, and determination efforts like the Paris Agreement and the Green New Deal have inspired, even the most well-intentioned policies are often hampered by mistaken assumptions about the nature of the problem, misconceptions that could easily reduce the effectiveness of proposed solutions. This struggle is not a new one. After fifty years of pursuing a coveted, wholesale, low-carbon future powered by renewable energy, the global economy is no closer to reducing its dependence on petroleum than it was a half-century ago. Oil and gas continue to supply the majority of global primary energy (54 percent, compared to renewables at 5 percent) and emit the majority of energy-related GHG emissions.[50] After all, oil and gas remain highly abundant. And due to their heterogeneous composition, some hydrocarbon resources have far greater impacts on the climate than others. By mistakenly treating diverse oils and gases as homogeneous resources, the world misses real opportunities to reduce emissions now, as the planet dangerously warms.

The main characters of this energy-focused climate saga—industry, government, and civil society—are featured in the book's second act in chapters 5 through 7. Assorted industry players are as diverse as the oil and gas resources they peddle. They impose sizeable climate footprints even before their products are sold.[51] Despite accidents, attacks, and epidemics, the market directs an ever-changing guard of companies to churn out millions of barrels of oil, gas, and assorted petroleum products a day. Governments are charged with providing public oversight on this well-oiled machine that is changing the climate. While no single government actor has figured out how to stop oil and gas from warming the planet, a varied collection of actors are trying different approaches. And a hodgepodge of civil society actors—nongovernmental organizations (NGOs), philanthropies, academia, think tanks, and the public—are doggedly finding different ways to balance private interests and the public good in a largely uncoordinated fashion.

Those of us with a wealth of experience and expertise on the inner workings of the oil and gas industry can offer vital lessons on how these different actors can overcome these climate change coordination challenges and work better in tandem. Chapter 8 details climate pathways forward, or solutions for the tightly linked oil and gas supply chain, a system that is geared for just-in-time delivery of assorted petroleum products that compose daily life. Unfortunately, there is no silver bullet. Successfully shrinking the climate footprint of the oil and gas sector will take the accumulated results of countless carefully crafted 2 percent solutions. Different actors can fashion different climate tools for the oil and gas sector to monitor, report, and verify GHG emissions levels; forge voluntary agreements and binding regulations; use financial incentives and disincentives;

impose sanctions and prohibitions; and foster research, development, and technology transfer.

Greater transparency would help guide strategy development so that industry, governments, and civil society actors can pinpoint opportune targets and play a constructive role. This need gave rise to the OCI+. This first-of-its-kind open-source model quantifies and ranks oil and gas resources by their lifecycle GHGs, identifies where in the supply chain emissions are produced, and informs the design of climate mitigation efforts.

Reducing demand for oil and gas alone will not prevent the climate devastation upon us. We all witnessed how sheltering in place and the initial economic shutdowns during the early months of the coronavirus pandemic in 2020 eliminated daily commutes and canceled flights, throttling the global oil industry.[52] While a pandemic temporarily depresses demands for gasoline, diesel, and jet fuel, atmospheric levels of climate-warming gases are cumulative and will not decrease until more emissions are removed than go into the air year after year for decades to come.[53]

Supply-side mitigation of oil and gas GHG emissions is imperative to putting the world on a path toward staying below global temperature rises of 2 degrees Celsius. The endgame involves turning the hydrocarbon (oil and gas) sector into a renewable hydrogen enterprise—keeping the carbon underground ex ante. In a world where ongoing threats of new infectious diseases are exacerbated by fossil fuel–driven climate change, the petroleum sector needs a low-carbon business plan.[54] Capturing and sequestering carbon or employing other climate engineering fixes after oil and gas have been extracted and GHGs emitted will not be enough.

The world cannot wait for oil and gas resources to run out—that simply will not happen. We need to acknowledge that hydrocarbons are extremely complex and mutable. And oil and gas are causing irreparable damage to the global climate. We are long overdue for a clean energy transition. But hydrocarbons have become so pervasive in so many disparate corners of the global economy that no single group—companies, governments, or activists—can root them out alone. Together, the industry, governments, and civil society possess complementary forms of ingenuity needed to develop supply-side oil and gas climate solutions.

Thinking back over the past three-plus decades since my Chevron testimony, I still do not have all the answers. But I know where to look to uncover the unknowns. This book is a shining beacon lighting the way.

# PART I

## *The Problem*

# *I*

# *Beyond Peak Oil and Gas*

*The world will run out of oil in 10 years.*[1]

—US BUREAU OF MINES (1914)

*The world will run out of oil in 13 years.*[2]

—US DEPARTMENT OF THE INTERIOR (1939 AND 1950)

*The world will run out of oil in 2011.*[3]

—JIMMY CARTER, presidential debate (1976)

*The decline of oil and gas will affect the world population more than climate change.*[4]

—UNIVERSITY OF UPPSALA, Sweden (2003)

*The global supply of oil is expected to meet the world's demand through 2050.*[5]

—US ENERGY INFORMATION ADMINISTRATION (2019)

OIL AND GAS were never expected to last forever.[6] For over a century, predictions that oil and gas would become scarce have stoked fear and distracted industry, have confused policymakers, and have vexed civil society.[7] Yet oil and gas keep flowing. What accounts for this vast discrepancy between expected supply and actual supply?

It turns out that hydrocarbon resources are *not* in short supply. According to analysis, at least 100 years of petroleum resources exist, and even longer with continued innovations in technology.[8] A 2006 study estimated that the world has used up only about 5 percent of known, technically recoverable oil reserves.[9] In 2018, energy experts reconfirmed these findings.[10] The sheer abundance of oil and gas may be surprising to many. But it is a fundamental fact of the energy landscape: the world must confront oil and gas supplies to tackle climate change. These resources cannot simply be wished away.

Oil and gas are not merely plentiful—they are also widespread. While some nations contain more hydrocarbon resources than others, oil and gas (often buried together) reside nearly everywhere on Earth, underwater and on land alike. Hydrocarbon building blocks are also present in outer space on other celestial bodies. Space missions have identified oil and gas resources off-planet that could be accessed someday.[11]

Abundance does not guarantee that all these resources can be readily recovered, however. Scientists and engineers are continually probing how much oil and gas are present, where these resources are buried, and what it will take to extract them and turn them into different commodities.

This chapter explores the ample stores of hydrocarbons located worldwide that are poised to meet the world's rising demand for energy resources and other commodities over time and into the future. The world's hydrocarbons are highly abundant, geographically dispersed, and quite varied in their composition and (consequently) their effects on the climate. A complex value chain links oil, gas, and other widely consumed petroleum products, for many of which there are currently few substitutes. The task of protecting the climate cannot afford to wait for the unlikely event that the mere market forces of supply of or demand for petroleum products peak (and then begin to abate) on their own. Instead, we must prepare for a future in which oil and gas are and remain abundant. This task begins with understanding the fundamental characteristics underpinning oil and gas resources to better grasp how to reduce the petroleum sector's greenhouse gas (GHG) emissions in the short term and set the stage for a clean energy transition.

## An Unreachable Peak

In 1956, a Shell geophysicist, M. King Hubbert, published a paper claiming that oil production increases exponentially in the early stages, reaches a peak when half of the hydrocarbons in a field are extracted, and then falls into a terminal decline.[12] Hubbert's work theorized "peak supply," the point at which the global output of oil and gas reaches its maximum levels and their rate of production permanently decreases.[13]

Hubbert predicted that US oil production would peak in 1965.[14] It did not. In fact, that still has not happened. US oil and gas production stood at all-time highs at the end of 2019, prior to the global coronavirus pandemic.[15] So too did global production levels.[16]

Still, the search for oil and gas resources continues. In recent years, a new concept of "peak demand" has gained traction.[17] Some experts project that the rapid adoption of electric vehicles (EVs), new mobility trends such as Uber and other ridesharing platforms, vehicle automation, higher oil and gas prices, and dramatic growth in renewable power will soon replace oil and gas. They suggest that the connection between economic growth and oil use is breaking down due to reduced petroleum consumption in wealthy, developed nations; more fuel-efficient motor vehicles; declining prices for alternative fuels; and accelerating urbanization by migrating populations around the world.[18]

Unlike Mount Everest's lofty summit, however, there is no clear-cut sign of whether or when we will reach peak oil and gas supply or demand. This is because peak scenarios fail to consider the dynamic response of oil and gas markets to changing economic, technological, and social conditions. Like the accounting of petroleum reserves discussed in the next section, the answer to how much oil and gas remains is: it depends largely on variables that cannot be easily modeled like industrial innovation and consumer behavior.

Peak assumptions about oil markets were tested in 2020 when the COVID-19 pandemic dialed down the global economy and put the brakes on petroleum consumption around the globe. At the same time, producers refused to reduce supplies in line with shrinking demand, turning oil markets upside down. By early 2021, however, oil and gas flows were rebounding to their all-time high.[19] Even historic, unparalleled events like the pandemic that temporarily snuff out oil and gas demand are unlikely to shutter the petroleum industry altogether.[20] As long as any demand for some combination of petroleum products exists—such as jet fuel, roadways, roof shingles, fertilizer, synthetic fabrics, and machine lubricants—oil and gas will continue to be supplied.

## *How to Measure Buried Troves of Hydrocarbons*

Quantifying the remaining volume of hydrocarbons at any point in time is difficult.[21] Changing economic, technological, geopolitical, geological, and policy conditions affect the accounting of petroleum reserves and the prospects for recovering them. There are different measures for assessing oil and gas reserve volumes, each with its own set of assumptions that change over time and necessitate ongoing evaluations. Uncertainty resides with the type of reservoir, transformations of hydrocarbons within the reservoir, and quantity and quality of available data.

Overall volumes consider original oil and gas in place, estimating the total amount of these resources buried in the earth. These estimates can then be broken out into cumulative production to date, proved reserves, remaining reserves, technically recoverable oil and gas, economically recoverable oil and gas, and remaining oil and gas in place. Figure 1.1 illustrates these various reserve accounting categories.

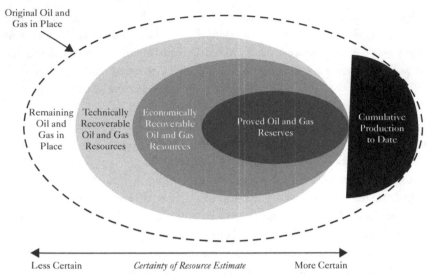

**FIGURE I.I**  Oil and Gas Resource Categories

*Notes:* Resource categories are not drawn to scale relative to the actual amount of each resource category. Gas resources include natural gas and natural gas liquids.

*Source:* US Energy Information Administration, "Oil and Natural Gas Resource Categories Reflect Varying Degrees of Certainty," July 17, 2014. https://www.eia.gov/todayinenergy/detail. php?id=17151.

The Society of Petroleum Engineers goes into great detail reporting reserve figures.[22] Others, like the US Energy Information Administration (EIA), for example, do not report economically recoverable resources because they are tied to a specific set of price and cost assumptions that change rapidly.[23] Yet other estimates may bundle together figures for technically and economically recoverable reserves,[24] or they may use categories such as proven, possible, and probable reserves to estimate the chances of recovery based on accessibility, legality, and other assumptions.[25] Since economic and technological conditions can change markedly over time, updating estimated oil and gas volumes helps guide short-term market and policymaking decision-making. This information is also useful for calculating potential GHG emissions from the future development of oil and gas resources.

## Proved Reserves

Proved reserves are resources from confirmed oil and gas reservoirs that are known to be recoverable under existing operating and economic conditions. The volume of these reserves is smaller than that of other future oil and gas resources because proved reserves are estimated based on current market conditions, including the prevailing price of oil and gas. As oil and gas prices rise and operating costs fall, volumes in the economically recoverable category are redesignated as proved reserves. Conversely, the lower oil and gas prices go, the fewer proved reserves the world has. As such, proved reserves are estimated at a point in time and are constantly shifting. For example, in a single year between 2014 and 2015, the EIA estimated that US annual proved reserves of oil and gas fell 12 and 17 percent, respectively, due to the decline in resource prices.[26]

Since the valuations and share prices of publicly traded oil and gas companies are partly based on their proved reserves, disclosing these assets is required. In the United States, the US Securities and Exchange Commission has standardized the metric of reserve valuation.[27] Other countries—such as Canada, China, Norway, and Russia—have their own definitions. And, even when countries do disclose their proved reserves, there may be too little data transparency to validate these figures, especially in countries that have nationalized their oil and gas resources. This makes it difficult for the market (and investors) to value those state-owned enterprises that control the majority of the world's oil and gas. This helps explain why Saudi Aramco, for example, had a difficult time ascertaining its market value (from a low of $1.1 trillion up to $2 trillion) when it launched an initial public offering of its stock in 2019.[28]

## Economically Recoverable Resources

If there is money to be made or there are government rents to be collected, hydrocarbon resources will be recovered. Whereas proved reserves are calculated by the prevailing price of oil and gas, the volume of economically recoverable oil and gas depends, at any point in time, on a host of economic and other factors, including projected global pricing dynamics, national growth trends, regional conditions, geopolitics, cartel decisions, arbitrage, pandemics, and other market imperfections and societal conditions. Lower prices and higher costs lead to lower estimates of economically recoverable resources. For example, far less Canadian oil sands reserves were economically recoverable in 2020 during the COVID-19 pandemic due to low oil prices and high production costs, increasing the volume of technically recoverable reserves for potential future development.[29]

## Technically Recoverable Resources

Rather than considering the price of oil and gas, the determination of technically recoverable reserves takes into account current technology, industry practice, and geologic knowledge. Improving industry practices and new technologies can change such assessments. The recent history of hydraulic fracturing, for example, illustrates how industry breakthroughs can vastly increase technically recoverable oil and gas volumes. A secondary category, unproved technically recoverable resources, is calculated by taking the total technically recoverable resources and subtracting the proved oil and gas reserves. In other words, if oil and gas resources are known to be present, it is usually just a matter of time before techniques are developed to access them. Both industry and governments drive change to obtain more oil and gas to replace spent reserves. Technological advances bring new societal impacts like climate change that must be anticipated and resolved.

## Remaining Oil and Gas in Place

Assessing the volume of oil and gas still trapped within rocks is difficult and speculative. Estimates are based more on assumptions than facts. Imagine trying to assess the amount of water sitting in thousands of enormous, slowly draining bathtubs over the course of a century. This amorphous figure would represent the amount of hydrocarbon resources remaining in place. Physical limitations both geologically and technologically prevent 100 percent of the original oil and gas in place from being recovered. But future innovation keeps upping recovery rates of remaining oil and gas in place. The greatest uncertainties lie and the largest resource volumes reside in this final category.

California showcases an extreme example of shifting reserves between different categories over time. The massive and highly complex Monterey formation spans the state, subject to varying geologic and seismic conditions. In 2011, the EIA estimated that the Monterey formation could hold up to 24 billion barrels of oil, with technically recoverable reserves of 14 billion barrels.[30] But a big brouhaha ensued in 2014 when the EIA slashed its estimate by 96 percent, to a mere 600 million barrels.[31] Experts generally agree that billions of barrels of oil are in place in California's Monterey. But technically recoverable reserves estimates are a moving target that vacillates by an order of magnitude depending largely on drilling and enhanced recovery innovations. Economically recoverable reserves are even more fraught, depending on fluctuating oil prices, uncertain future demand, and even California's strict environmental regulatory environment.

## *Resources by the Barrelful*

If current projections are accurate, there are many trillions of barrels of oil and gas in place worldwide. A fraction of these resources is technically recoverable today. More will be accessible in the future due to technological advances. At current consumption rates, hydrocarbons in place are projected to last some 500 more years.[32] But, if the past is any indication, we will likely have discovered more hydrocarbons in place by then.

As of 2018, the world's total proved reserves of oil stood at 1.7 trillion barrels.[33] Remaining technically recoverable oil resources are currently estimated as high as 10 trillion barrels.[34] Proved gas reserves stood at 215 trillion cubic meters, or 1.3 trillion barrels of oil equivalent (BOE).[35] Remaining technically recoverable gas resources are estimated at 800 trillion cubic meters (nearly 5 trillion BOE).[36] To put these reserves in perspective, global consumption amounted to roughly 90 mbpd of oil and 4,500 bcm a year of gas.[37] As such, if another drop was not found, today's proved oil and gas reserves alone could supply the world through at least 2070.

Most of these supplies are not conventional hydrocarbons, however. Unconventional resources that make up the lion's share are likely undercounted, mostly because they are not considered technically or economically recoverable at present. However, it may also be the case that some forms of unconventional oil and gas remain largely unknown and therefore undercounted. These unconventional hydrocarbons will require significant modifications to existing oil and gas value chains, as discussed in the next chapter.

## Barrels Beyond Earth

Petroleum is also abundant in the wider solar system from planets like Saturn and Mars to comets and space dust.

Methane rivers flow into a methane sea on Saturn. The National Aeronautics and Space Administration's (NASA) *Cassini* spacecraft, which orbited Saturn nearly 300 times over twenty years, recorded Saturn's bounteous stores of hydrocarbons.[38] Saturn's mammoth moon, Titan, has an atmosphere that contains 5 percent methane.[39] Where all this methane comes from remains a mystery, but its presence is clear in Titan's deep, hydrocarbon lakes. Titan is one of the "most Earth-like worlds" in the solar system, with a hydrologic cycle similar to Earth's.[40] Instead of evaporating and raining water, however, Titan's atmosphere circulates methane and ethane. These hydrocarbons are gases on Earth. But Titan is so cold that natural gas behaves like liquid gasoline. Similarly, Mars is thought to be a source of hydrocarbons. Space exploration of Mars has identified "oil-like extractable biomarkers that closely resemble terrestrial hydrocarbon source rock kerogen and bitumen usually observed in shale and carbonates."[41]

In addition to planetary sources, comets and space dust throughout the galaxy could be mined for petroleum inputs. Asteroids are made of rock, water, and metals that could be mined for hydrocarbon resources, including platinum, rhodium, and other rare-earth metals.[42] Many of these serve as catalysts that are essential in oil and gas operations and wider energy industry applications.[43] While the emissions impacts of gathering hydrocarbons in space are unknown and could be considerable, perhaps some methods of extracting space-based oil and gas resources could be viable—such as gathering space dust or collecting methane raindrops—using negligible energy if routine space missions (powered by reusable methane-fueled rockets) get underway.[44]

## Mapping Oil and Gas

The US Geological Survey (USGS) is an authority figure that not only issues reports on reserves but also maps out the locations of oil and gas deposits.[45] Its ongoing geologic studies update the country's understanding of the "quantity, quality, and geologic distribution" of global oil and gas resources.[46] The agency has identified more than 170 basins of conventional and unconventional hydrocarbon resources. In 2019, the USGS mapped undiscovered oil and gas resources in the continental United States.[47]

Oil and gas resources are widespread across the globe. Every continent possesses hydrocarbons—even Antarctica,[48] where a current treaty bans resource extraction until the middle of the twenty-first century (and then only if a majority of parties agree and if there is a suitable regulatory system in place).[49]

## X Marks the Spot

Pinpointing where oil and gas resources lie is not a simple task. Neither is parsing how the USGS classifies them with its confusing forest of terminology.

It is helpful to start broadly with the survey terms used to designate the largest, continent-spanning geological units and zoom in from there. The total petroleum system is an assessment of already-discovered and yet-to-be-discovered pockets of petroleum residing in source rocks worldwide.[50] From there, the USGS divides the globe into geologic provinces, areas made up of thousands of square kilometers that contain geologic resources.[51] A single geologic province may contain several formations, the unique name that classifies rocks with similar characteristics in a specific location. When two or more formations exist together, they are considered a group.[52] Zooming in still farther, a play is an oil and/or gas reservoir—an accumulation of hydrocarbons that share similar geologic properties.[53] In most cases (but not always), a play is limited to a single formation. In the United States alone, over 100,000 oil and gas reservoirs had been documented by the late twentieth century.[54]

From there, the classifications become even more granular. A field encompasses connected production areas that contain adjacent pools of oil and gas.[55] Although the number of fields is hard to ascertain and not routinely updated, there are reportedly more than 65,000 oil and gas fields of all sizes worldwide as of 2011.[56] While that figure may seem quite high, some 9,000 fields worldwide currently account for approximately 98 percent of all current oil production.[57] A smaller number of gas fields dominate global production and are concentrated in the United States, Russia, Iran, Canada, China, and Qatar.[58]

Numerous wells typically operate in a given field at the same time. Millions of oil and gas wells have been drilled around the globe, with nearly 1 million currently active in the United States alone.[59] Many more wells have been drilled, though some of them either are temporarily inactive due to economic factors or other operational reasons or have been abandoned.

Looking beyond existing wells, as long as there is ongoing demand for oil and gas, ample hydrocarbon resources remain in place for development for generations to come.

## New Finds

Over the past decade, numerous nations have announced new discoveries of large oil and gas deposits. Equatorial Guinea, Suriname, Guyana, Cyprus, and the Arctic are just a few of the live oil prospects.[60] In 2018 alone, nearly 10 billion barrels of new oil and gas deposits were discovered cumulatively worldwide on every continent, and a decade earlier during the run-up in prices, some 90 billion barrels were found.[61] Shale basins in Southeast Asia have abundant, untapped gas resources. Meanwhile, countries like Malaysia, China, India, Thailand, and Pakistan have been assessed for significant future oil and gas production volumes.[62] And new oil and gas resources were identified in the famed Permian Basin in Texas and New Mexico, possibly denoting the largest pool of hydrocarbons on Earth.[63]

Discovering oil and gas is an ongoing quest. The USGS conducts assessments, studying a dozen different regions and nations each year.[64] All this probing is bound to lead to new finds and increased estimates of worldwide oil and gas deposits. As these various numbers suggest, the planet's hydrocarbon holdings are vast and continue to expand: constraints on supply are not poised to curtail the use of oil and gas worldwide anytime soon.

Despite the global prevalence of oil and gas resources and continued reserve growth, not all regions have local abundance. Western Europe, Japan, and New England, for example, are minimally endowed. However, even those places lacking oil and gas reserves, such as Singapore, can still have ample petroleum product supplies if they are home to oil refineries and gas processing plants.

## Durable Demands

With the world's supply of hydrocarbons showing no signs of abating, worldwide demand for oil and gas has climbed steadily for decades. Figure 1.2 depicts the quadrupling of global consumption since 1965, a spike that has not been deterred by global recessions followed by price hikes on fuel. At the end of 2019, global demand for oil reached 101 million barrels per day, the majority as crude oil along with small amounts of natural gas liquids (NGLs), biofuels, and refinery gains.[65]

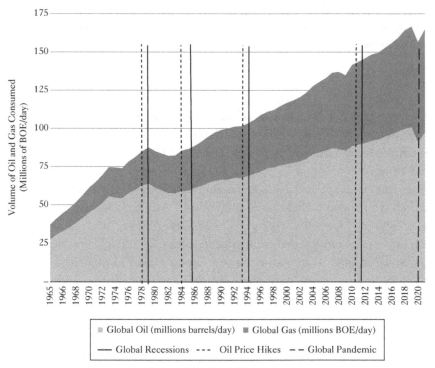

**FIGURE 1.2** Global Oil and Gas Consumption (1965–2021)

*Note:* To convert billion cubic meters (bcm) gas to million barrels of oil equivalent (BOE), multiply by 6.1.

*Sources:* Author's calculations using data from BP Statistical Review, https://www.bp.com/en/global/corporate/energy-economics/statistical-review-of-world-energy.html. And with additional inputs from the International Energy Agency, US Energy Information Administration (EIA), Macrotrends, Investopedia, Statista, and SP Global Platts. 2021 Projected by EIA (oil) and SP Global Platts (gas).

Global demand for natural gas stood at nearly 4 trillion cubic meters in 2019, up 30 percent since 2009.[66]

The COVID-19 pandemic temporarily curtailed demand through 2020, though this trend is not likely to continue for a long period. Short-term forecasts on the economic recovery rate remain uncertain, but long-term oil and gas demand began trending upward again in early 2021 and are projected to grow through 2022.[67] Already, China is driving a rebound in global demand for oil, as the country's levels of demand have recovered to more than 90 percent of those seen before the pandemic—trends that could be mirrored elsewhere as vaccines are distributed.[68]

In 2018, the International Energy Agency (IEA) projected that, by 2023, oil demand will reach nearly 105 million barrels per day, bolstered by the growth of the petrochemicals industry in the United States and China and strong overall economic signals. By 2022, annual demand for natural gas is projected to pass current consumption levels, with China and other emerging Asian countries accounting for the majority of the growth, with demand shifting from power generation to the industrial sector, and with gas trade spurred by expanding liquefied natural gas (LNG) export capacity.[69]

As long as global demand for oil and gas continues to grow, it will take durable changes to supply-side oil and gas emissions to lessen the effects of climate change. And even if oil and gas consumption is ultimately curtailed, the petroleum value chain will also need to be decarbonized.

## *The Petroleum Value Chain*

The oil and gas that flow out of the ground are essentially useless to consumers.[70] You cannot pump crude into your car's gas tank without killing the engine or wash your hair with gas condensates without harming yourself. But once petroleum is processed and refined into consumer products, the usefulness of oil and gas becomes apparent. This petroleum value chain undoubtedly generates massive profits for petroleum companies, but the public also benefits greatly—much more so than most people acknowledge. Modern society runs on oil and gas—and not just for fueling our vehicles and generating electricity. Our roads, tires, buildings, medicine, clothing, and much more are made from hydrocarbons.

Turning oil and gas into myriad petroleum products is a big undertaking. In industry lingo, this colossal task entails upstream extraction and processing, midstream refining and shipping, and downstream marketing and end use. (The industry classifies refining as downstream. Since the refinement process is quite distinct from end-use consumption, however, I separate it out here.)

Market forces largely beyond the industry's control affect petroleum product demands that do not necessarily rise and fall in sync with each other. For example, during the COVID-19 pandemic, demands for jet fuel and gasoline plummeted, while diesel fuel remained relatively steady and plastics for protective personal gear soared. But the petroleum value chain as a whole is not readily adaptable. Infrastructure in place to produce, process, refine, and ship oil and gas takes decades and costs billions. In other words, it is not enough to simply turn off oil and gas production to meet climate goals.

## Drill, Baby, Drill

At the start of the extraction process, hydrocarbon resources are initially probed using seismic sensors that render detailed images of what lies below the earth's surface.[71] The right to explore land holdings for oil and gas deposits is granted through leases, partnerships, or acquisitions, depending on whether oil and gas rights in a given place are privatized or state owned.[72] The US government, for example, oversees oil and gas exploration and drilling in the Outer Continental Shelf (which consists of areas that are beyond three miles offshore),[73] in national parks, and on federal lands.[74] When private landowners hold mineral rights, they can control whether or not exploration takes place on their property.[75] In other countries, special rules can apply when exploratory drilling is conducted to assess oil and gas resource prospects.

Companies often bid against one another to obtain a lease. Through this process, financial terms—such as rent payments, cost sharing, and royalty rates—and the leasing period are set. With a lease agreement in hand, companies drill exploratory wells to test the quantity and quality of the land's hydrocarbon resources. If their quantity and quality are deemed sufficient, the project gets a green light.

At that point, engineers put pen to paper, modeling systems, designing equipment, and planning the necessary facilities. Depending on the location in question, multiple local, state, federal, and international government approvals may be required, each with its own conditions and associated costs. With official consents granted, oil and gas field equipment can be ordered, constructed, transported, and installed. This litany of equipment and infrastructure includes drilling rigs, pipelines, terminals, processing plants, and more that may be sourced from around the globe. Once everything is assembled and field tested, drilling and extraction operations can commence.

For over a century, vertical wells have been drilled essentially straight down from the surface of the earth into the targeted oil and gas formations. (That practice is changing with the advent of horizontal drilling and hydraulic fracturing, as discussed in chapter 2). With a vertical well, the drill bit is removed and completion techniques prepare the well for production. A cement casing typically supports the hole out of which various multiple-diameter pipes can exit. Once the source rock is punctured, oil and gas enter the wellbore and flow to the surface.

## Field Notes

At first glance, the production phase, when oil and gas finally emerge, may appear to be the final stage of the petroleum value chain. But drilling and extraction are just the beginning. After oil and gas are extracted, it takes significant field work—inputting considerable energy with resulting emissions—to turn hydrocarbons into petroleum products.

When most people think of hydrocarbons, they have certain images of oil (smooth, thick, and black) and of natural gas (invisible and burning with a perfect blue flame). In reality, they certainly do not come out of the ground that way: getting crude and its associated hydrocarbon compounds into a usable state is quite a sordid affair.

When encountered in their natural state, oil and gas are typically mixed together at random ratios. Strange hydrocarbon species can be mixed in that swing between vapor and liquid phases, depending on operating conditions like temperatures and pressures. These liquid gases and vaporous oils—condensates and NGLs—can be readily removed and turned into petrochemical products that make their way into our daily lives as plastics, barbeque fuel, detergents, tires, refrigerants, and thousands of other commodities. (The growing advent of the multiplicity of hydrocarbon products is discussed in chapter 2.)

Unwanted contaminants can accompany extracted resources, including dirty water, tainted air, smelly sulfur compounds, heavy metals, volatile and sometimes toxic compounds, caustic acids, remnants of drilling fluids and other injected substances, invisible pockets of methane, potent traces of carbon dioxide ($CO_2$), and more. Solids may also be present: mud, rock, salt, and sand. Removing these impurities starts with surface processing to separate liquids from gases, drain water, and discard solids.

These undesirable byproducts can easily leak if operators are not careful to handle them responsibly. Some byproducts, like hydrogen sulfide, cause instant death. Others, like benzene, toluene, and xylene, are known or suspected carcinogens. Heavy metals such as vanadium and nickel can leach into water and cause organ damage. Diesel exhaust from onsite engines emits airborne toxins that cause heart and lung damage. Air emissions from wells and equipment form smog and generate GHGs that pollute regional air and the global climate. All told, oil and gas production facilities present risks for workers and other nearby residents and wildlife.

## Exit Gas

After surface processing, raw gas and crude oil typically part company, breaking the oil and gas value chain into two distinct branches.

Before they diverge, however, processed gas and intermediary oil products may be used in petroleum sector operations. Gas and condensates can be reinjected into formations to extract more resources; used to generate onsite heat, steam, and power; dissected to manufacture hydrogen; inserted into pipelines to clean them; or blended with heavy oils to make them flow. Up to 82 percent of the heat needed for processing in the petroleum supply chain currently comes from products and sidecuts of the petroleum industry itself.[76] These quantifiable, but often overlooked, intermediary petroleum uses are crucial for calculating the total lifecycle climate impacts from the petroleum sector, as discussed in chapter 4.

Natural gas that is not needed upstream for drilling, extraction, and processing can be marketed to end users for power, heat, steam, and hydrogen production. It also can serve as a feedstock to make petrochemicals. Or it can be sold to other industrial users. Gas travels to these different users through regional pipelines as compressed natural gas or by oceangoing vessels as LNG.

As of 2015, there were some 1,500 gas processing plants worldwide in over sixty countries.[77] The United States led the pack by processing a reported 525 billion cubic meters of gas, while Iran came next, followed by Saudi Arabia and the United Kingdom.[78] Together, these three countries had enough capacity in 2015 to handle half as much gas as America did. Since then, the United States has surged further ahead thanks to the production of shale gas using unconventional methods, discussed in chapter 2.

## The World's Biggest Kitchens

Unlike gas, condensates, and NGLs, the oil supply chain follows its own unique path to commercialization. Crude oil cannot be directly marketed without a critical additional step—refining—which rearranges hydrocarbons into a broad slate of valuable petroleum products.

Instead of the edible foodstuff used in a kitchen, the ingredients fed into an oil refinery include hydrogen, carbon, oxygen, and various impurities. Refinery equipment—effectively the stoves, refrigerators, pressure cookers, mixers, bowls, spatulas, and knives—heats, cleaves, blends, and reconfigures crude oil into all sorts of consumable petroleum products.[79] Table 1.1 provides an overview of the

**Table 1.1  Converting Hydrocarbons into Petroleum Products**

| Refinery Fraction | Boiling Range (°F) | Intermediary Products | End Use Products | Commercial Uses | Typical Refinery Yields[a] |
|---|---|---|---|---|---|
| **Lighter Products** | <85° | Ethane, propane, liquefied petroleum gas (LPG), butane | Petrochemicals, LPG, solvents | Plastics, cosmetics, solvents, stoves | 12% |
| *(Lower Boiling Points)* | 85–185° | Gasoline components | Gasoline | Cars | 26% |
| | 185–350° | Naphtha | Gasoline | Lighter fluid, metal cleaner | 6% |
| | 350–450° | Kerosene | Jet fuel | Aircraft | 8% |
| | 450–650° | Distillate | Diesel fuels, heating oil | Trucks, trains, generators, home heating | 28% |
| | 650–1,050° | Heavy gas oil | Bunker fuels, fuel oils | Ships, furnaces | 7% |
| *(Higher Boiling Points)* | >1,050° | Residual fuels | Petroleum coke, tar, asphalt, wax | Roads, industrial boilers, roofs, power plants | 13% |
| **Heavier Products** | | | | | |

[a] These typical refinery yield values represent today's generic approximations. The variation in actual refinery yields depends on the oil feedstocks used, the refinery equipment installed, and current demand.

*Sources:* International Energy Agency (IEA), "Oil 2018: Analysis and Forecasts to 2023," 2018, https://www.iea.org/reports/oil-2018; estimated fractions calculated using Deborah Gordon, Brown University Watson Institute, Oil Climate Index Plus Gas, https://dxgordon.github.io/OCIPlus/#; Deborah Gordon, "Carbon Contained in Global Oils," Carnegie Endowment for International Peace, December 2012, https://carnegieendowment.org/files/global_oils.pdf.

basic distillation of hydrocarbon fractions that result at different temperatures to produce products with different commercial uses and the estimated corresponding refinery yields. (Major refining processes are detailed later in figure 3.4 in chapter 3).

Today's 621 global refineries are spread across 120 nations, with 29 in the United States alone.[80] Several countries that produce little or no crude nevertheless refine oil, including Japan (with 22 refineries), Egypt (with 8), or Peru (with 7).[81] Together, the world's refineries reportedly processed nearly 92 million barrels per day of crude in January 2019.[82] The world's smallest facility, in Nigeria, handles a mere 1,000 barrels per day while the largest is located in India with a 1.2 million bpd capacity.[83]

The world's refineries are currently configured in one of four ways. Hydroskimming refineries, the most basic kind, heat crude oil to separate it into refined products without additional transformations. While these simple refineries turn lighter crude oil into gasoline and remove unwanted sulfur, the amount of product they yield depends entirely on the crudes fed into them. They do not perform optimally on heavier crude slates. Roughly 7 percent of the world's crude is processed in hydroskimming refineries, located predominantly in Russia and Italy.[84]

Adding another level of complexity, medium-conversion refineries combine hydroskimming process units with additional equipment to crack (break the bonds of) and rearrange heavier oils that contain more carbon. These midlevel refineries are equipped to process lighter, higher-value petroleum products, such as gasoline and petrochemical feedstocks. But they also yield heavier, lower-value heavy products, including heavy residual fuels and asphalt. Medium-conversion refineries comprise 44 percent of today's crude throughput and predominate in Singapore, Iran, Thailand, France, the United Kingdom, and Canada.[85]

The most complex deep-conversion refineries come in two types: coking and hydrocracking. Both utilize medium-conversion process units and incorporate equipment that applies high temperatures to crack heavier residual oil with long carbon chains into lighter, high-quality products. Coking refineries, which account for 45 percent of current refining capacity worldwide, wring the carbon out of the heaviest crudes and produce the coal-like byproduct petroleum coke.[86] These refineries' ability to process unconventional, low-quality crudes and produce customized yields of gasoline and diesel have historically made these high-cost refineries attractive. Coking refineries are common in the United States, China, India, Brazil, and Venezuela.

That being said, future refineries will have to take into account new and different oil supplies as well as changing consumer demands. This task will not be easy.

Many of these so-called kitchens are older and cannot easily adapt to changing ingredients or varying consumption patterns.

As for changing oil inputs, in recent years, refining has had to shift from being "pear shaped" (whereby heavy oil accounts for a growing share of input crude) to an "hourglass" shape (with growing proportions of light and heavy oils relative to conventional medium crudes).[87] These so-called dumbbell crudes are an odd mixture of light, sweet shale oil and condensates from the United States coupled with heavy, sour crudes from Canada, Mexico, Venezuela, and elsewhere.[88]

Dumbbell crudes are technically challenging to refine. It is far easier for refineries—at least as they are currently configured—to break down heavier oils into desired petroleum products than to recombine lighter oils to turn them into products.

Economically, refiners operate on razor-thin margins. Costly renovations are skirted. Past investments are usually viewed as sunk costs that are deeply discounted. In the United States and Europe, environmental permitting all but prohibits refineries from being built or expanded, which is why new refineries are springing up in Asia and the Middle East. All refiners aim to cook up the right petroleum product slate on a daily basis that is readily consumed. Past product consumption has historically guided refiners' ongoing quest for optimal output combinations for the supply and product lines produced using crude. But what if future demands for petroleum shift markedly from the current refinery yields shown in Table 1.1? These questions plague the industry, for example, when considering a wholesale future shift to electric vehicles that could displace gasoline—a major refinery commodity. If this comes to pass, however, it could be strategically coupled with policymaking to shake up the oil industry and decarbonize the refining sector.

## Beyond Fueling Planes, Trains, and Automobiles

The various commercial uses the oil and gas industry has found for petroleum products is simply staggering. Fictional US vice president Selina Meyer (Julia Louis-Dreyfus) found this out the hard way on HBO's hit show *Veep*. In its first episode, the plastics lobby on Capitol Hill and its oil-toting benefactors thwarted her attempt to replace Washington, DC's plastic utensils with a more ecofriendly alternative.

Petroleum inundates and abets our daily lives from morning to night. We wake up to petrochemicals, brushing our teeth with toothpaste and washing our hair with shampoo. We eat our midday snack grown with fertilizer and packed in plastic bags. At dusk, we drive home on tires made of rubber and over roads paved with asphalt. Before bed, we put on our petroleum-based polyester pajamas

### Table 1.2  Petroleum Products in Daily Life

| Relevant Industry (Common petroleum products) | Consumer Market (Petroleum in common products) |
|---|---|
| *Consumer Fuels* | **Health and Beauty** |
| Gasoline | Cosmetics |
| Diesel | Shampoo |
| Heating oil | Soap |
| Propane | Bandages |
| Kerosene | Petroleum jelly |
| LPG | Vitamin capsules |
| Natural gas | Medicines |
| *Commercial Fuels* | **Personal Items** |
| Bunker fuel | Clothes |
| Jet fuel | Eyeglasses and contact lenses |
| Petroleum coke | Dentures |
| Fuel oils | Toys |
| | Crayons |
| *Infrastructure* | **Electronics** |
| Asphalt | Computers and smartphones |
| Tar | Television sets |
| *Industrial Inputs and Other Uses* | **Household Goods** |
| Petrochemicals | Cleaning products |
| Sulfur | Trash bags |
| Paraffin wax | Candles |
| Lubricants | Paint |
| Tires | Roofing and insulation |
| Solvents | Carpet |
| Fertilizer | Upholstery |

LPG, liquefied petroleum gas.

*Source:* Adapted from Deborah Gordon and Madhav Acharya, "Oil Shake Up," Carnegie Endowment for International Peace, April 2018, https://carnegieendowment.org/files/Gordon_DrivingChange_Article_April2018_final.pdf.

and take medicine made from petrochemicals. Table 1.2 showcases a few of the estimated 6,000 consumer products that are made from petroleum.[89]

Despite our ubiquitous daily connections to oil and gas, most people think that their only contact with petroleum is during weekly fill-ups at the local gas station.[90] While gasoline is a byproduct of petroleum refining, it only accounts for one in four barrels of the world's refined oil and none of its gas.[91] That means that three-quarters of the planet's barrels of oil and essentially all its gas provide goods and services to every other economic sector—industrial, commercial, agricultural, residential, power generation, and nonpassenger transport—for which statistics are not clearly disaggregated.[92] And while renewable energy sources are making considerable gains worldwide, most of the stuff made from oil and gas cannot be made from solar and wind energy. It is no wonder, then, why rooting oil and gas out of the global economy is a far trickier task than the average person realizes.

## Searching for Cleaner Substitutes

Getting off oil will be a hard and slow endeavor given the current state of infrastructure in the world and a lack of readily available product replacements. In theory, the petroleum industry could find alternatives for portions of its product slate—petrochemical feedstock, gasoline, and diesel. But replacing the nearly 1 billion cars on the road worldwide with EVs will not be a simple task.[93] The IEA forecasts that alternative fuels could be swapped in for a scant 500,000 barrels a day of gasoline and diesel by 2023.[94] This is a drop in the bucket (0.5 percent of current oil demand) that mirrors past shifts away from large volumes of oil in the electric power sector, which was slashed by 60 percent between 1978 and 1982 after the second oil crisis in the United States.[95] Such large bites out of our oil diet have prevented US oil consumption from ultimately rising.

But the problem is even more complex than that. Even if EVs powered by renewable energy were to propagate worldwide, it takes oil to manufacture, install, and maintain wind turbines and solar panels. The same is true of the cars themselves too. The millions of EVs envisioned to help reduce demand for oil still use rubber tires, drive on asphalt roads, and contain many plastic parts (such as seats, dashboards, and bumpers) that are made from oil and gas.

Diesel and kerosene, on the other hand, have relative monopolies in road freight and aviation. Jet fuel, fuel oil, asphalt, lubricants, waxes, tar, and sulfur have no ready substitutes. Bio–jet fuel manufactured from renewable feedstocks such as sugar, corn, or forest wastes can be made in small amounts, but it will take many decades before its production equals the volumes currently consumed and before airports and aircraft are retrofitted for its universal, safe use. The prospect that liquid fuels will be obtained directly from renewable energy (the so-called solar fuels that apply photosynthesis principles using sunlight, water, $CO_2$, and nitrogen from the air to produce fuels), while promising, is even more distant.[96]

The same is true of a product as humble—and cheap—as asphalt. Solar road-ways, recycled tires and printer toners, and low-GHG cement notwithstanding, the residual solids left over from refining heavy oils will continue to provide pave-ment on which cleaner cars are driven for countless miles.

And then there is sulfur—a chemical element that serves as the backbone of many industries. Sulfur is the key ingredient in sulfuric acid, a component used in fertilizer, food production, paint, paper, detergents, medicines, cosmetics, leather, tires, plastics, dyes, construction materials, sugar, steel, and water treat-ment.[97] Sulfur is used in batteries that store electricity. Since the early 1900s, the main source of sulfur is processing sour oils and gas. The only other sources of sulfur are volcanoes and pyrite mines.[98] So when EV battery developers and bil-lionaires hail sulfur-flow batteries as a breakthrough technology for renewable energy storage, it is troubling that a key ingredient (sulfur) is a byproduct of the very fossil fuels (oil and gas) that it seeks to replace.[99]

To sum up these various obstacles and complications to curtailing humanity's carbon footprint, Table 1.3 charts cleaner substitutes for oil and gas commodities, listed from lowest to highest in terms of technical, market, and social barriers to entry. These disconcerting findings show that public awareness of the polluting effects of gasoline-powered automobiles hardly scratches the surface of the eco-nomic and societal changes that will be needed to roll back the pervasive presence of oil and gas in our daily lives and the global economy. And in many cases, the world is even further away from finding solutions than it is in the case of EVs.

## Market Forces

For now, market price is the best metric for gauging a peak in petroleum supply or demand. Various market factors influence petroleum prices, including the bal-ance in oil and gas trade and exogenous forces from financial markets.[100]

There is no consensus among experts, however, on the efficacy of oil and gas markets. Some experts view oil and gas as durable assets that the market sets a price for based on the available supply. They subscribe to the expectations of future traders to generate changes in price and inventories.[101] Others believe that resources are not durable and that the market price is determined by ongoing imbalances between oil supply and demand. According to this logic, when supply tightens, prices rise. Higher prices dial down demand and unlock new technolo-gies that increase supply and usher new unconventional hydrocarbons into the market. Any resulting oversupply lowers prices, which in turn moderates produc-tion and reaccelerates demand. Round and round the oil and gas market goes.

In 2020, oil and gas market forces were tested by a global pandemic.[102] Quarantines and travel bans squashed petroleum demands and prices fell.

Table 1.3  The Challenges of Finding Cleaner Substitutes for
Petroleum Products

| Current Petroleum Products | Cleaner Substitutes (ranked lower to higher by barriers to entry)[a] |
|---|---|
| Natural gas (power plants and industry inputs) | Renewable energy sources (wind, solar, geothermal, and others) |
| Gasoline (cars) | EVs (recharged with renewable power) |
| Diesel (trucks, two-wheelers, and stationary engines) | Biodiesel and EVs (made from/recharged with renewables) |
| Fuel oils and distillates (engines, boilers, home heating) | Direct from natural gas (no methane leakage) |
| LPG: propane and butane (cooling, heating, and vehicles) | Bio-based LPG (organic materials)[b] |
| Ethane and naphtha (plastics, solvents, and fuel additives) | Bio-based petrochemical feedstocks (organic materials)[b] |
| Residual fuels and bunker fuels (ships, power plants, and industry) | LNG (no methane leakage) |
| Asphalt (roadway paving and roofs) | Low-GHG concrete |
| Hydrogen (refining, fertilizer, pharmaceuticals, polymers, transportation, power, and renewable storage) | Renewable electrolysis of water |
| Jet fuel: kerosene (airplanes) | Bio-based alternatives (organic materials)[b] |
| Lubricants (transportation/industrial applications and metalworking) | |
| Waxes (foods, cosmetics, coatings, and 3D printing) | |
| Tar (paints and sealers) | |
| Sulfur (rubber, gunpowder, fungicide, and medicines) | Surface mining from volcanoes |

(*Continued*)

**Table 1.3  Continued**

ª The barriers to entry are lower at the top and increase on down to the bottom of the table. Barriers include cost, technical hurdles, and social roadblocks.

ᵇ Organic materials include fats, grains, plants, grasses, trees, cellulosic waste streams, and algae. *Sources:* Table adapted from Deborah Gordon and Madhav Acharya, "Oil Shake Up," Carnegie Endowment for International Peace, April 2018, https://carnegieendowment.org/files/Gordon_DrivingChange_Article_April2018_final.pdf. The information on bio-based feedstocks came from the following source: "Biobased Industrial Products: Priorities for Research and Commercialization," National Research Council (US) Committee on Biobased Industrial Products (Washington, DC: National Academies Press, 2000), https://pubmed.ncbi.nlm.nih.gov/25121336/.

The normal uptick in consumption from low prices did not follow. Experts expect demand to recover once a growing share of the population is vaccinated and it is safe to travel and resume social activities. However, the rate at which oil and gas markets recover remains highly uncertain. This raises questions for future market forces as climate change unleashes new pathogens and other forms of economic and social disruption. It also highlights the perils of innovation—essential keys to unlock climate solutions—and why market forces alone cannot be relied on to successfully move oil and gas solutions to the fore.[103]

## Future Price Uncertainty

Although oil and gas markets are closely tracked, perennial uncertainty in forward-looking supply and demand hampers price projections.[104] Over the long run, future price increases depend on growing demand in emerging economies and growing supply from unconventional deposits around the world. But in the short term, disruptive forces in supply or demand—from political instability to global pandemics—can unexpectedly shift prices.

Spot markets for oil, gas, and petroleum products involve the near-term delivery of a single, large-volume sale delivered to a specific location, via a given mode of transport, at a designated time.[105] While spot markets account for a small portion of overall oil and gas trade, they play an outsized role in setting market prices for those buying and selling petroleum commodities under longer-term contracts.[106]

Numerous agencies, nongovernmental organizations (NGOs), and companies publish oil and gas price forecasts. These predictions are rarely one-off estimates. Instead, these forecasts are often updated, and projects are raised or lowered, according to changing circumstances. Figure 1.3 offers a timeline of successive forecasts over nearly two decades plotted alongside posted market prices.

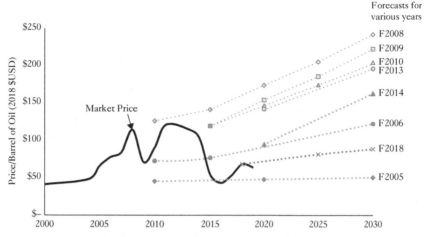

**FIGURE 1.3** Actual and Forecasted Oil Prices (2005–2018)

*Notes:* Dashed lines (e.g., F2008) plot forecasted oil prices made in that year. The dollar values are listed in 2018 dollars. Links for the International Energy Agency's (IEA) *World Energy Outlooks* are available at the IEA webstore. In 2015, the IEA stopped publishing oil price forecasts in the *World Energy Outlook*, so other additional sources were used for price and inflation data.

*Sources:* International Energy Agency, Oil Price Forecasts, *World Energy Outlooks 2005–2014*, https://www.iea.org/topics/world-energy-outlook; International Energy Agency, "World Energy Model," https://www.iea.org/reports/world-energy-model/macro-drivers; CPI Inflation Calculator, https://www.in2013dollars.com/

While this is only one example, it not only illustrates the difficulty of forecasting oil market prices but also exhibits the tendency of experts and their models to overvalue future oil prices, a habit that can prop up oil in the short term when markets decline and sustain oil in the long term. The same price forecasting concerns are expected in gas markets as global trade expands and gas markets grow larger and more dynamic.

## Oil Versus Gas Markets

Even though these hydrocarbons currently satisfy largely different demands (as shown in Table 1.3), oil and gas are often extracted together. The fact that their supply is linked but their demands are not can affect their respective market dynamics. Prior to 2000, crude oil and natural gas prices hovered around parity (1:1), based on their heating values. Figure 1.4 shows that, in recent years, crude oil has captured up to five times more market value than the equivalent energy unit of natural gas. Fracking in the United States may explain this price decoupling because oil is a global commodity but gas is largely domestic. Since oil has a greater market upside than gas, producers extracting these resources together

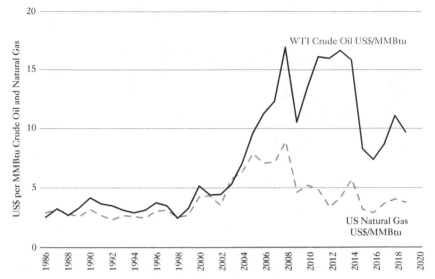

**FIGURE I.4** Comparing Prices of US Oil and Natural Gas (1986–2019)

*Notes:* This figure assumes that 1 thousand cubic feet of gas (Mcf) equals 1.036 million British Thermal Units (MMBtu), while 0.17 barrel crude equals 1 MMBtu.

*Sources:* US Energy Information Administration, "Price of U.S. Natural Gas Exports," https://www.eia.gov/dnav/ng/hist/n9130us3a.htm; and US Energy Information Administration, "Cushing, OK WTI Oil Spot Price," https://www.eia.gov/dnav/pet/hist/LeafHandler.ashx?n=PET&s=RWTC&f=A

have an economic incentive to maximize oil extraction and dispose of unwanted gas, releasing methane into the atmosphere.

## *Market Failures*

Ideally, markets are expected to sustain desirable production and consumption activities and prevent undesirable consequences and inefficient outcomes.[107] This is not the case, however, in the oil and gas (and other) markets. Negative externalities can have harmful spillover effects, like climate change, when the consequences of activities affect others but are not reflected in market prices. Insufficient information results when buyers, sellers, or both are less than certain (or unequally knowledgeable) about the qualities of what is being bought and sold.[108] Free competition is constrained when firms collude and when there are barriers to enter and exit markets. And intergenerational equity bestows the rights of tomorrow's citizens (even those who are not yet born) to inherit a safe environment and conserved natural resources.[109]

A variety of actors recognize that the oil and gas sector suffers from market failures, especially when it comes to climate change. The National Petroleum Council referenced problems of imperfect information in 2011, when it reported that critical aspects of unconventional oils are often not well understood, including the GHG emission intensities of various operations.[110] The premier of Alberta, Canada, highlighted environmental externalities in 2012, when she called for transparency, verifiable social responsibility, and comprehensive standards for oil sands.[111] The Shell Oil Company's chief executive officer (CEO) alluded to issues of market control in 2020 when he professed that his company must figure out the right bets to take amid society's growing concerns about climate change.[112] And the founder of the investment giant BlackRock addressed intergenerational equity when he called out climate change as *the* decisive factor in companies' long-term economic prospects.[113]

Addressing these market failures calls for tailored responses. Imperfect information about oil and gas climate risks and GHG mitigation necessitates potential calls for robust, ongoing, public data transparency to build knowledge and identify gaps to scrutinize industry trends, revise assumptions, improve calculations, and develop solutions. Negative externalities, such as climate change, respond to financial incentives and disincentives, nonbinding agreements, binding regulations, and prohibitions.

Market control (like that which Saudi Arabia demonstrated when it took on Russia in 2020 to return balance in crude markets during the coronavirus pandemic) requires greater oversight through monitoring and reporting followed by financial disincentives and sanctions on efforts to hamper competition.[114] And intergenerational equity necessitates a rallying call for durable actions now that will protect the climate for the well-being of generations to come.

Each of these tools is discussed in the chapters that follow. And their applications are highlighted in the pathway laid out in the book's concluding chapter.

## *The Best Barrels*

A decade ago, strong growth prospects in emerging nations signaled increasing demand for oil and gas. The search for new supplies led to new developments of novel hydrocarbon resources.

But what exactly is this newfound abundance composed of? Are the *best* barrels—those that are most accessible, cheapest, and least damaging—behind us?

Oil and gas are highly heterogeneous—they come in many varieties and are used for numerous purposes. Some resources are easier to manage than others. The easier ones require less effort and fewer inputs, they are more energy efficient, and they are more easily turned into high-value petroleum products.[115] Even after benchmark prices are posted, oil and gas may be discounted according to several factors, including the share of associated gas in a given type of oil, the share of liquids in a given type of gas, an oil's weight (how heavy it is), its sulfur content, whether it has high levels of impurities, and its proximity to trading hubs, refineries, and processing facilities. One thing is certain, however: oil and gas resources' diverse climate impacts are not currently factored into their prices.

As long as it is profitable, the market does not care whether oil and gas are conventional or unconventional. But the changing nature of oil and gas, and the varying impacts that various types of hydrocarbons have on the climate, presents an opportunity to reduce emissions along the entire value chain. The next chapter offers details on what new forms oil and gas are taking, so that these resources can be better managed in the future.

# 2

## *The Overlooked Perils of Heterogeneous Oil and Gas*

THE QUEST TO understand the different types of oil and gas that populate the earth goes all the way back to the industry's early days. Oil was first used in the late 1840s to illuminate buildings in Halifax, Canada.[1] Within a decade, the first modern-day oil well was drilled in northwestern Pennsylvania.[2] Although the two kinds of hydrocarbons these two places produced were readily converted into kerosene-based lamp oils, they had little else in common. Pennsylvania's rock oil was light, with more hydrogen than carbon.[3] The Canadian variety came from heavy bitumen with more carbon than hydrogen.

A few years later, John D. Rockefeller launched Standard Oil, giving birth to the oil industry. The company coined its name as a nod toward the "standard quality of product that the consumer could depend [on]."[4] This branding created the impression that everything associated with oil—its extraction, processing, and shipping—was standard. That impression is false. Hydrocarbons are not all the same: rather, they are incredibly diverse.

By the end of the nineteenth century, the vast Midway-Sunset Oil Field near Bakersfield, California, had turned into a gusher that has since produced billions of barrels from tens of thousands of oil wells.[5] In 1901, oil was struck at Spindletop in Beaumont, Texas—tripling US production in a flash.[6] Geologists at the time noted marked differences between the complex, thick (unconventional) oil from California and the (conventional) crude from Texas.[7]

The varying compositions and settings of conventional and unconventional oils and gas can make for marked differences in their respective carbon footprints. The Oil Climate Index + Gas (OCI+) estimates that the supply-side (production, processing, refining, and shipping) greenhouse gas (GHG) emissions of otherwise-equivalent barrels of oil and gas can vary by up to a factor of ten.[8] Assessing the climate impacts of differing kinds of oil and gas depends largely on the techniques used to supply differing resources. Conventional oil and gas readily flow and are relatively easy to separate, resulting in lower supply-side climate risks. Thus, the majority of conventional oil and gas GHG emissions stem from the end uses of their petroleum products, such as natural gas, gasoline, diesel, and residual fuels.

Unconventional hydrocarbons, however, do not readily flow and are harder to produce, process, refine, and ship. These resources tend to be wedged in rocks, buried deep, bound to sand, solidified, acidic, or frozen. Some unconventional hydrocarbon resources yield byproducts, like petcoke, with high climate impacts of their own. As such, unconventional oil and gas currently have higher climate risks than their conventional counterparts.

Today, resource flows of conventional hydrocarbons are double those of unconventional resources.[9] Looking to the future, however, nearly three times more technically recoverable unconventional resources exist worldwide compared to conventional oil and gas.[10]

This chapter compares volumes of different oil and gas resources. Next, the wide array of these hydrocarbons is surveyed, their various stages of development are discussed, and their different climate risks are gauged. Innovations in the oil and gas sector are enhancing the technically and economically recoverable prospects of unconventional resources. The better their supply chains are understood, the better the chances of successfully mitigating GHG emissions from unconventional oil and gas. The chapter concludes with a side-by-side assessment of GHG emissions from global supplies of different conventional and unconventional oil and gas resources.

## Distinguishing Conventional from Unconventional Resources

Named for its Latin roots, *petra* (meaning rock) and *oleum* (meaning oil), petroleum is formed over time as dead organisms decompose underground, are washed into waterways, and are crushed under tons of rock and sediment. The unique combination of organic matter and geologic conditions in one place or another yields vastly different forms of conventional and unconventional hydrocarbons.[11]

Although hydrogen and carbon are the two main chemical ingredients of conventional and unconventional hydrocarbon resources, contaminants migrate in as they are formed and extracted. Carbon dioxide ($CO_2$), oxygen, and nitrogen (air); heavy metals; water; acids; carcinogens; radioactive elements; and other impurities may be present, and some are more difficult to deal with and have higher GHG impacts than others. For example, the wastewater that is produced along with oil requires extra energy and emits more GHG emissions for it to be pumped, reclaimed, and recycled. But some contaminants return profits. Sulfur is used in the petrochemicals industry, and lithium recovered from wastewater can make rechargeable batteries.[12]

Conventional petroleum takes many physical forms. It can be a liquid with a consistency of water or molasses, it can be gaseous and resemble anything from dry wind to dense fog, or it can be a hybrid of these two phases like a fizzy soda. Conventional resources are typically graded on their weight, sulfur content, and heating value.

Oil density—its weight compared to the same volume of water—is measured in degrees of American Petroleum Institute (API) *gravity.* Conventional oil's gravity typically ranges from 22 degrees API to 32 degrees API.[13] Gas density (measured relative to air) ranges from 0.7 to 0.9 kilograms per cubic meter.[14] Oil and gas that are high in sulfur (over 0.5 percent content by weight) are classified as sour, while low-sulfur varieties (less than 0.5 percent) are considered sweet. Conventional gas heating values (the amount of energy it contains) range from 950 to 1,150 British thermal units (Btu) per cubic foot, depending on its liquid content. Unfortunately, these characteristics cannot be used to quantify the life-cycle GHG emissions of oil and gas because they lack sufficient details on their compositions of the hydrocarbons in question and the processes employed to exploit them.

While conventional oil and gas are loosely defined as those that easily flow, unconventional resources can have too much (or too little) carbon or hydrogen, can contain too many impurities, can be situated in an unusual location, or can be difficult to extract or process for other reasons.[15] In general, the characteristics, locations, volumes, and lifespans of unconventional oil and gas are less well understood and more complex than those of their conventional cousins, as are their technical challenges, economic prospects, and environmental tradeoffs.

The bulk of total GHG emissions from conventional resources (on the order of 90 percent of them[16]) stem from the end uses of their petroleum products. In the short term, the best way to reduce the climate risks of these hydrocarbons is to fine-tune their industrial processes and end-use efficiencies. In the long term, breakthroughs that sequester carbon and methane and produce only hydrogen hold promise in emitting net-zero GHGs.

Unconventional hydrocarbons, however, call for careful management when it comes to climate change. A significant share of their GHGs are emitted during extraction, processing, refining, and shipping—in other words, before their end uses even enter the equation. This difference shifts the climate burden unconventional resources pose from consumers to the petroleum industry. As unconventional stocks compose an increasing share of the oil and gas supply over time, knowing where in the supply chain GHGs are emitted and how best to reduce these emissions will be increasingly important.

While these general industry trends are important, some concrete numbers can be even more illuminating. As of 2019, an estimated 1.4 trillion barrels of conventional oil and 0.7 trillion barrels of oil equivalent (BOE) of conventional gas have been commercially produced since the 1850s.[17] Conventional oil output has averaged 67 million barrels per day—fueling two-thirds of the global oil supply—and the rest has come from unconventional sources, including tight oil, oil sands, condensates, and biofuels.[18] Conventional gas has accounted for 80 percent of worldwide production; the rest has been unconventional.[19]

As conventional oil and gas stores deplete, unconventional hydrocarbon development is on the rise. Figure 2.1 outlines the major classes of diverse petroleum resources.

## Muddled Resource Definitions

Given the differing characteristics and compositions that oil and gas can take, it is not surprising that defining them has not been a straightforward task. Confusion over the definition of oil dates back to the 1970s when US legislation broadly stated that "oil means oil of any kind or in any form."[20] At the time, unconventional oil from fracking, bitumen, or synthetic fuel production had not yet appeared on the scene. In 1980, however, when new oils emerged,[21] Congress specifically exempted all "unconventional" light tight shale oil, oil sands, coal-to-liquids (CTLs), and biomass from the legal definition of oil.[22] This artificial divide carved out in the legal definitions between conventional and unconventional hydrocarbon resources set a precedent that has long hampered the effectiveness of environmental oversight of oil and gas.[23]

A generation later, unconventional oil and gas development took off. In 2010, as oil prices rose, shale gas and light tight oil were extracted by injecting substances into formations, fracturing formations, propping open rocks, and employing horizontal drilling in the United States.[24] In Canada, increasing volumes of extra-heavy oil sands were mined or made to flow by injecting steam.[25]

Advances in technology enabled a widening array of hydrocarbon deposits in once-unreachable places to be extracted and processed into petroleum products.[26] A new assortment of resources emerged in a sector that had long been stuck in familiar patterns.[27] With this change came new uncertainty and risks, especially for the global climate. This sea change for the oil and gas industry underscores the importance of understanding unconventional resources, as detailed in the remainder of this chapter.

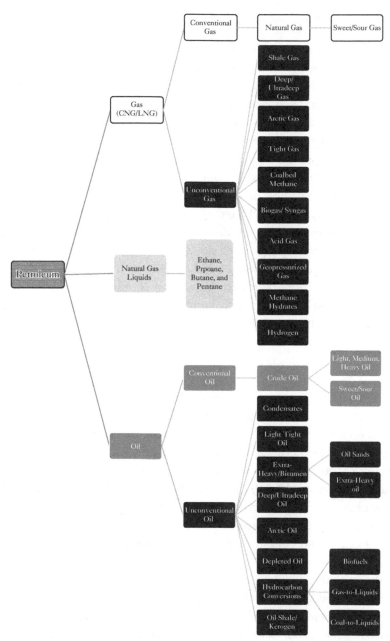

**FIGURE 2.1** Current Classifications of Conventional versus Unconventional Oil and Gas

*Notes:* Conventional gas is depicted in white squares, conventional oil and natural gas liquids are shown in gray squares, and unconventional oil and gas are listed in black squares. CNG, compressed natural gas; LNG, liquefied natural gas.

*Sources:* Author's depiction with data from the following sources: Deborah Gordon, "Understanding Unconventional Oil," 2013, https://carnegieendowment.org/files/unconventional_oil.pdf; and International Energy Agency, *World Energy Outlook 2017*, 2018, figure C.1, https://www.iea.org/reports/world-energy-outlook-2017.

## Understanding Unconventional Gas

There is an ample supply of unconventional gas to meet all imaginable levels of future demand, whether or not the global climate can bear it. Table 2.1 compares the current levels of production and the remaining recoverable reserves of nine unconventional gas resources to those of conventional gas. Since 2000, unconventional gas production has more than quadrupled in volume.[28] The standout, shale gas, increased by over a factor of twenty.[29] Smaller volumes of other unconventional gases have also gained footholds. Of the remaining technically recoverable gas sources, stores of unconventional varieties are collectively several multiples greater than those of conventional gas.

A brief description of each kind of unconventional gas resource follows in the order presented in Table 2.1. These types of gas resources include shale gas, deepwater and ultradeep gas, Arctic gas, tight gas, coalbed methane, biogas and syngas, sour gas, geopressurized gas, and methane hydrates. In addition to presenting some brief background information on these subtypes and assessing their future production prospects, the goal is to convey the current understanding of potential GHG emissions from these burgeoning unconventional gas resources.

## Shale Gas: Trapped in Rocks

There is gas lodged in rock fissures that span thousands of miles on every continent. These continuous shale gas resources require special drilling techniques to fracture (or frack) rocks and recover the gas trapped in them.[30] (Oil and condensates also can be present along with the gas.) Shale gas contains mostly methane—an extremely potent, short-lived, leak-prone GHG—along with other volatile organic compounds (VOCs). When burned (or flared), methane and other hydrocarbons are converted mainly to $CO_2$, depending on the equipment used.

Fracking is not new. Prospectors in the mid-nineteenth century fractured rocks using explosives mostly in oil fields.[31] Gas was produced with the oil, but the lack of pipelines meant that the produced gas was wasted and simply released into the atmosphere.

It took another century to find ways to successfully frack gas. In the late 1940s, the first commercial hydraulic fracturing applications were used in Kansas, Oklahoma, and Texas. Within a decade, petroleum service companies were acquiring licenses to frack gas as well as oil.[32] High energy prices in the 1970s spurred fracking development in gas-rich regions of Colorado, Wyoming, and New Mexico. But the game changer came in the early 2000s, when independent producers coupled fracking with horizontal drilling in East Texas, enabling access

## Table 2.1 Gas Production and Recoverable Resources (2019)

| Gas Resource Type | Current Production (billion cubic meters [bcm]/year) | Remaining Technically Recoverable Resources (trillion m$^3$) |
|---|---|---|
| Conventional Gas[a] | 2,238 | 432 |
| Unconventional Gas Sources | | |
|    Shale Gas | 719 | 233 |
|    Deepwater and Ultradeep Gas | 420 | 19 |
|    Arctic Gas | 300 | 55 |
|    Tight Gas | 285 | 82 |
|    Coalbed Methane | 82 | 50 |
|    Biogas and Syngas | 35 | 1 |
|    Acid Gas[b] | 10 | 4 |
|    Geopressurized Gas[c] | — | 850 |
|    Methane Hydrates[d] | — | 750 |
| Subtotal Unconventional Gas | 1,851 | 2,030 |
| All Gas (Total Estimated Volumes)[e] | 4,089 | 2,471 |

[a] To avoid double-counting, deepwater and ultradeep, Arctic, and acid gas volumes are subtracted from the International Energy Agency's (IEA) conventional gas production volume.

[b] Acid gas is also referred to as "sour" gas because it often contains high sulfur along with the $CO_2$; this value assumes 10 billion metric tonnes/year of "other" gas produced is acid gas.

[c] Geopressurized zones estimated at 8.5 quadrillion cubic meters worldwide with 10% technically recoverable, of which 2.5% is estimated to be produced over the next twenty years.

[d] Methane hydrates are not yet technically recoverable, but some 30 quintillion cubic meters are in place with 2.5 percent deemed technically recoverable, of which 1 percent is estimated to be produced over the next twenty years.

[e] Current gas consumption is estimated at 62 million BOE/day. Double-counting may occur due to resource overlap (including for deep resources that are located in the Arctic, geopressurized zones, or methane hydrates, for instance).

*Notes:* For the gas conversion figures, the calculations assume 164.3 billion cubic meters per BOE. All numbers rounded to the nearest whole number.

*Sources:* International Energy Agency, *World Energy Outlook 2018*, Tables 3.1, 3.5, 4.1 (Gas and Oil Production), https://www.iea.org/reports/world-energy-outlook-2018; International Energy Agency, *World Energy Outlook 2017*, Table 8.2 (Technically Recoverable Gas Resources), https://www.iea.org/reports/world-energy-outlook-2017; James Speight, "Unconventional Gas," 2019, https://www.elsevier.com/books/shale-oil-and-gas-production-processes/speight/978-0-12-813315-6; Statista, "Global Proven Coal Reserves," 2019, https://www.statista.com/statistics/265450/global-proved-reserves-of-coal/; Energy Information Administration, "World Shale Resource Assessments," September 24, 2015 (Tight Oil and Shale Gas), http://www.ieee.es/Galerias/fichero/OtrasPublicaciones/Internacional/2015/EIA_World_Shale_Resource_Assessments_24sept2015.pdf; Wang Hongjun et al., "Assessment of Global Unconventional Oil and Gas Resources," *Petroleum Exploration and Development* 43, no. 6 (December 2016), https://www.sciencedirect.com/science/article/pii/S1876380416301112; Statista, "Production of Biogas Worldwide From 2000 to 2017", https://www.statista.com/statistics/481791/biogas-production-worldwide/; World Biogas Association; and UN Environment Programme, "Frozen Heat: A Global Outlook on Methane Gas Hydrate," 2014, https://sustainabledevelopment.un.org/index.php?page=view&type=400&nr=1990&menu=35.

to continuous shale plays.[33] Through 2007, rising natural gas prices prompted fracking in numerous US states.[34]

The climate footprint of shale gas is largely due to methane leakage throughout the supply chain, from drilling, production, gathering, boosting, shipping, processing, storage, and end-use distribution systems. Undersized and leaky systems with old or faulty equipment have high GHG emissions that can go undetected.[35] Maintenance like well workovers and accidents like blowouts can also generate high volumes of GHGs. Quantifying how much methane is emitted from routine leakage depends largely on operational factors, while how much accidental leakages produce depends on duration and shale gas composition, data that are currently sparse or missing altogether.

## Ultradeep Gas: Buried Way Down

Ultradeep gas wells are drilled one-third of the way through the earth's crust.[36] By comparison, conventional hydrocarbon resources lie hundreds to a few thousand feet below the surface. The difficulties of going deeper—many miles (tens of thousands of feet) underground and underwater—have not thwarted the industry. The depth of a well only carries a relative penalty compared to shallower wells. For example, drilling Russia's ultradeep Chayvo field is estimated to have five times more GHGs per barrel than Angola's shallow Takula field.[37] However, in absolute terms, Chayvo's drilling emissions only contribute an estimated 5 percent to this asset's GHGs and less than 1 percent to total lifecycle GHGs that include end uses.[38] The vast pressures that ultradeep hydrocarbons are subject to make them flow. This translates into average GHG emissions from extraction, but methane can still leak through the rest of the supply chain and present climate risks.

The deepest wells are rarely drilled in a straight line. Instead, they are channeled from left to right, like a skier traversing back and forth down a steep slope. As extraction techniques improve extended-reach drilling, accessing even deeper hydrocarbon resources will become technically feasible. There are existing fields in Russia (the Barents Sea), Mozambique, and Egypt,[39] and new discoveries have been posted in surprising places, including Israel, Tanzania, Mauritania, Senegal, Cyprus, and India.[40]

Gas wells accounted for over half of new deepwater gas volume brought online between 1990 and 2019.[41] However, deepwater gas development has proven less commercially viable than its oil counterpart (discussed later) due to low natural gas prices and high development costs.[42]

## Arctic Gas: Journeying Far North

Above the Arctic Circle lies a plethora of hydrocarbons—gas, oil, and natural gas liquids (NGLs)—that are poorly explored and their possession and access are heavily disputed. Cold, dark, dangerous, and remote conditions hinder the development of these unconventional resources. Six nations—Canada, Denmark (via Greenland), Iceland, Norway, Russia, and the United States—each have a jurisdictional claim over some of these hard-to-reach hydrocarbons, which are spread over seven geologic basins the size of Africa.[43] The Arctic's polar ecosystem is extremely fragile and highly susceptible to warming from the GHG emissions released there.[44]

The Arctic is reported to contain nearly seven times more gas than oil.[45] Limited gas development in parts of Alaska and Siberia dates back to the 1960s. As of 2018, only Norway reported marketable Arctic gas production of 122 billion cubic meters (bcm), while the United States reinjects the bulk of the Arctic gas it produces due to the lack of takeaway capacity, and Russia is just mounting its plans for major Arctic gas developments.[46]

Numerous additional projects have been proposed in subsequent years, many of which have been canceled and not reached fruition. High costs and long lead times hamper Arctic projects, which are shelved when prices fall.[47] Still, in 2020, Russia released its master plan for the Arctic, featuring major new gas and oil drilling and petrochemical processing.[48] As of 2018, GHGs emitted in the Arctic are expected to have an outsized impact both locally and on global weather patterns. A change in the Gulf Stream—the strong current that carries warm water up the East Coast of the United States and onward to Canada and Western Europe—could lead to frequent blasts of Arctic winds in the Northern Hemisphere.[49] Shrinking ice sheets will contribute to rising sea levels and greater ocean warming. And a dangerous feedback loop would enhance global warming as the methane and $CO_2$ frozen beneath Arctic permafrost thaw and enter the atmosphere as regional temperatures rise.

## Tight Gas: Scattered About

Unlike deposits of shale gas trapped between layers of shale, tight gas is dispersed in silt and sand in between hard rocks. Deposited some 250 million years ago, tight gas has been compacted, cemented, and recrystallized, reducing its ability to permeate the rock and readily flow.[50] The large energy inputs needed to force gas through unconnected passages and the potential to leak methane in the process both contribute to the climate impact of tight gas deposits.

Tight gas production requires numerous wells. Detailed seismic data help guide and deviate drilling along various angles that traverse as much of a reservoir as possible. The more inroads that are made, the more gas that is recovered. In addition to unconventional drilling techniques, tight gas often calls for artificial stimulation to promote flow. This technique entails fracturing the rock or acidizing it to dissolve sediment. The goal is to re-establish the original fissures that were present before the source rock was compacted. Challenges arise because every tight formation has its own unique characteristics and engineers continuously are testing new methods. The volumes of GHG emissions stemming from these novel techniques remain highly uncertain.

## Coalbed Methane: Nestled in Coal

Coal, a solid hydrocarbon composed of up to 90 percent carbon, also contains trapped methane, called coalbed methane or coal seam gas.[51] To tap into complex coalbed methane systems, it is necessary to pump water out of coal seams to lower underground pressure, detach the gas from solid surfaces, and help get it flowing.[52] Sizable energy inputs from pumping and reclaiming polluted water plus flared gas and methane leakage contribute to the climate risks of coalbed methane. Roughly 10 percent of global methane emissions are currently estimated from coal mines—and remote sensors are finding methane super emitters.[53]

Coal degasification dates back to late nineteenth-century Europe, where miners removed gas to mitigate explosions. China commercialized the extraction of coalbed methane in the 1950s, and the United States stimulated its development in the 1980s. Countries with large stores of coal—like Australia, India, Russia, and Indonesia—are at various stages of producing these gas resources. Production doubled worldwide between 2000 and 2017, accounting for 2 percent of gas supplied globally.[54] Coalbed methane reserve estimates account for 6 percent of remaining technically recoverable natural gas.[55]

## Biogas: From Recycled Feedstocks

Biogas is produced from organic matter in swamps and lakes, dams, landfills, industrial wastewater, manure ponds, food waste, and the contents of animals' digestive tracts. Biogas can also be manufactured in tanks (anaerobic digesters) that simulate biological processes. Bacteria in the absence of oxygen forms mostly GHGs like methane and $CO_2$. The climate impacts of biogas are wide-ranging and may result in net-negative GHGs, if more carbon is sequestered than emitted. Estimates of the GHG produced by biogas deserve more study and range from negative 85 to positive 251 grams of carbon dioxide equivalent ($CO_2e$) per kilowatt hour, depending largely on methane leakage.[56]

In 2018, 350 US landfills, 44 biogas sewage and industrial wastewater treatment facilities, and 29 large dairies generated 0.3 percent of the country's total utility-scale electricity.[57] In the European Union, biogas plays an even bigger energy role with over 17,000 plants generating electricity and heat.[58]

Biogas can be reformed into synthesis gas (syngas), a mixture of carbon monoxide and hydrogen.[59] Syngas cannot be directly consumed. It is used to generate steam or electricity or as a feedstock for petroleum fuels and chemicals. Many industrial products use syngas, including fertilizer, ammonia, sulfuric acid, and methanol. Researchers are adjusting syngas production methods to recycle increasing volumes of carbon captured from flue gas and directly from the air.[60] Like biogas, the climate impacts of syngas vary widely and need to be assessed case by case.

## Acid Gas: Chock Full of Carbon Dioxide

Acid gas is naturally high in $CO_2$. It is often confused with sour gas because acid gas often contains high sulfur levels. Because acid gas has elevated carbon dioxide concentrations of 15 to 80 percent, its development depends on finding a use for the $CO_2$ that is stripped off.[61] One such use is injecting $CO_2$ to enhanced oil recovery (EOR), a topic discussed more later. Another purpose is to get credit for carbon capture and sequestration (CCS). The most damaging route is to vent excess $CO_2$. The climate risks of acid gas vary widely depending on whether its associated $CO_2$ is used for EOR, is captured and sequestered underground permanently, or is released into the atmosphere.

Acid gas resources have been mapped globally, with large fields in Australia, parts of Southeast Asia, the US state of Wyoming, the Middle East, and North Africa. The latter three places are oil rich, so acid gas production there is typically used for EOR. In Australia, an acid gas project is utilizing CCS.[62] Other acid gas fields in Russia and New Zealand, meanwhile, are known to vent their $CO_2$, so they have four times the upstream GHG emission intensity posed by Australia's operational CCS and twice the upstream GHG emission intensity posed by the United Arab Emirates (UAE) where EOR is employed.[63] The wide-ranging GHG emissions of acid gas raise questions about companies' motivation for developing acid gas fields. If CCS is not coupled with acid gas, its development has very large climate footprints.

## Geopressurized Gas: Under Immense Stress

Hydrocarbons buried underground are subject to pressures that vary with depth, but formations under unusually high pressure (for their depth) are considered geopressurized zones. The gas resources situated in such zones are located up to 25,000 feet below the surface around geologic faults, making attempts to estimate their reserves highly uncertain. Although complex underground conditions can complicate extraction and elevate GHG emissions, geopressurized zones are thought to hold the world's single largest gas reserves.[64] More needs to be uncovered about these unconventional gases and the processes that would be employed to evaluate whether and how much they could elevate GHG levels.

## Methane Hydrates: Frozen Slush

Methane hydrates are naturally occurring cages of ice that contain methane. They carpet the world's oceans at depths over 1,500 feet, stacked hundreds of feet thick and extending horizontally over long distances.[65] Methane hydrates also collect under Arctic permafrost. These hydrocarbons exist undisturbed in stasis, subject to enormous levels of pressure and frigid temperatures. When brought to the surface, 1 cubic foot of methane hydrate releases 164 cubic feet of natural gas, risking a massive methane release.[66] Climate impacts of methane hydrates are highly uncertain.[67] Altering their steady state could perpetuate a dangerous warming cycle with higher ocean temperatures that melt more methane hydrates.[68]

In the late 1970s, drilling expeditions confirmed abundant amounts of methane hydrates around the globe on every continent.[69] Experimental extraction methods include surface dredging, sea mining, alleviating pressure to break down hydrates, and carbon dioxide injection.[70]

Notably, even before production is commercialized, methane hydrates can cause damage. During the Deepwater Horizon explosion and oil spill in 2010, for example, initial efforts to plug the well were thwarted because exploding methane hydrates formed an icy plug and clogged the gear deployed to collect the leaking oil.[71] Moreover, thawing hydrates explosively release gas in the Siberian tundra and at the bottom of the Barents Sea, forming giant craters as large as 3,000 feet wide and nearly 100 feet deep.[72] Since methane hydrates act like glue that hold their surroundings in place, even routine oil and gas extraction in hydrate-rich regions can cause landslides.[73]

While their impacts differ, the general trend toward more unconventional gas resources means that unless the outsized climate impact of such resources is properly tabulated, the industry is at risk of chronically undercounting their climate impact throughout the process from production to processing, shipping, and end uses.

## *Understanding Unconventional Oil*

Like in the case of gas, there is more than enough unconventional oil to last centuries and lead to major climate damage. Table 2.2 compares the current

## Table 2.2 Oil Production and Recoverable Resources (2019)

| Petroleum Resource Type | Current Production (million barrels/day) | Remaining Technically Recoverable Resources (billion barrels) |
| --- | --- | --- |
| Conventional Oil[a] | 57 | 2,245 |
| Unconventional Oil Sources[b] | | |
|     Condensates | 18 | 435 |
|     Light Tight Oil | 8 | 420 |
|     Extra-Heavy Oil and Bitumen | 4 | 1,880 |
|     Deepwater and Ultradeep Oil | 4 | 98 |
|     Arctic Oil | 3 | 134 |
|     Depleted Oil | 2 | 1,650 |
|     Biofuels[c] | 1 | 225 |
|     CTLs/GTLs[d] | 1 | 112 |
|     Oil Shale (Kerogen) | <0.1 | 2,100 |
| Subtotal (Unconventional Oil) | 41 | 7,000 |
| All Liquids (Total Estimated Volumes) | 98 | 8,730 |

[a] To avoid double-counting, deepwater, ultradeep, Arctic, and enhanced oil recovery (EOR) oil volumes are subtracted from the International Energy Agency's conventional oil volume.

[b] Recoverable reserves are derived from original oil in place, including tight oil = 8 trillion bbl; bitumen = 3–5 trillion bbl; and heavy oil = 3–5 trillion bbl.

[c] Biofuels depend on conversion technologies from a wide range of feedstocks.

[d] Coal-to-liquids (CTLs) depend on recoverable coal resources, while gas-to-liquids (GTLs) depend on the volume of natural gas production devoted to liquids conversion.

*Notes:* All numbers rounded to the nearest whole number.

*Sources:* International Energy Agency, *World Energy Outlook 2018*, Tables 3.1, 3.5, 4.1 (Gas and Oil Production), https://www.iea.org/reports/world-energy-outlook-2018; *Journal of Petroleum Technology*, December 2019; Gongcheng Zhang et al., "Giant Discoveries of Oil and Gas Fields in Global Deepwaters in the Past 40 Years and the Prospect for Exploration," *Journal of Natural Gas Geoscience*, February 2019, https://www.sciencedirect.com/science/article/pii/S2468256X19300033; International Energy Agency, *World Energy Outlook 2012*, Table 3.3 (Technically Recoverable Oil Resources), https://www.iea.org/reports/world-energy-outlook-2012; Brandt et al., "Climate-wise Choices in a World of Oil Abundance," April 5, 2018 (Oil in Place), https://iopscience.iop.org/article/10.1088/1748-9326/aaae76; US Energy Information Administration, "World Shale Resource Assessments," September 24, 2015 (Tight Oil and Shale Gas), http://www.ieee.es/Galerias/fichero/OtrasPublicaciones/Internacional/2015/EIA_World_Shale_Resource_Assessments_24sept2015.pdf; Wang Hongjun et al., "Assessment of Global Unconventional Oil and Gas Resources," *Petroleum Exploration and Development* 43, no. 6 (December 2016), https://www.sciencedirect.com/science/article/pii/S1876380416301112.

production of and the remaining recoverable reserves of nine unconventional oil resources to those of conventional oil. Over the past two decades, the production of unconventional oil nearly tripled.[74] For some subtypes of unconventional oil, the spike in production was even steeper. The supply of tight oil rocketed up from zero to 11 million barrels a day,[75] while the supply of NGLs nearly doubled, the bitumen supply quadrupled, and biofuels grew by an order of magnitude as well.[76] By comparison, conventional oil production remained essentially flat.

Three times as much unconventional oil is technically recoverable than conventional oil (see Table 2.2). Kerogen oil shale tops the list, followed closely by the other heaviest hydrocarbons. CTLs and biofuels remain wild cards with technically recoverable resources that stem from huge underlying resource bases.

Like with unconventional gas, there is a greater potential supply of unconventional oil than the global climate can safely accommodate. As a rule of thumb, the lighter the oil, the more methane factors into climate risks, and the heavier the oil, the more $CO_2$ and unburned (black) carbon can elevate GHG emission levels.

A brief description of each unconventional oil resource follows in the order presented in Table 2.2. These subtypes of unconventional oil include condensates, light tight oil, extra-heavy oil and bitumen, ultradeep oil, Arctic oil, heavy depleted oil, biofuels, CTLs and gas-to-liquids (GTLs), and oil shale (or kerogen). In addition to outlining background information on them and their production prospects, the industry's current understanding of their climate impact is summarized.

## Condensates: Shifting States

Condensates are light hydrocarbons that readily shift from gas to liquid, depending on temperature and pressure conditions. These unstable, flammable mixtures are named for their ability to condense into liquid form at the surface.[77] Their lifecycle climate impacts have been poorly studied. But each of their hydrocarbon components are GHGs with varying potency, and they also form ozone, a GHG that also deteriorates local air quality.

All condensates contain NGLs—ethane, propane, butane, pentane (natural gasoline), and hexane—but no two are alike. Those containing more pentane and hexane are volatile liquids, with wide-ranging API gravities between 50 and 120 degrees.[78] NGLs with more ethane and propane are misty, wet gases. Table 2.3 details examples of the end uses of condensates. Their GHG emissions depend largely on how condensates are processed, leakage throughout the supply chain, and whether their end use involves combustion or not.

Table 2.3  Value Propositions for Condensates

| Condensate Compounds | Chemical Formula | Applications and End Uses | Primary Sectors |
|---|---|---|---|
| Ethane | $C_2H_6$ | Petrochemical feedstocks, refinery fuel gas, plastics, antifreeze, detergent | Industrial and commercial |
| Propane (LPG) | $C_3H_8$ | Cook stoves, heating, barbeques, refinery fuel gas, petrochemical feedstock | Residential, commercial, and industrial |
| Butane | $C_4H_{10}$ | Petrochemical feedstock, lighter fluid, synthetic rubber, fuel blending | Industrial and transportation |
| Isobutane | $C_4H_{10}$ | Refinery inputs, aerosols, refrigerants, petrochemical feedstock | Industrial |
| Pentane | $C_5H_{12}$ | Natural gasoline, solvents, blowing agent for polystyrene foam | Transportation, industrial, and commercial |
| Pentanes plus | $C_5H_{12}$ and heavier | Gasoline, ethanol blends, oil sands production | Transportation and industrial |

C, carbon; H, hydrogen; LPG, liquefied petroleum gas.

*Sources:* US Energy Information Administration, "What Are Natural Gas Liquids and How Are They Used?," April 20, 2012, https://www.eia.gov/todayinenergy/detail.php?id=5930; International Energy Agency, "The Future of Petrochemicals," 2018, https://www.iea.org/reports/the-future-of-petrochemicals.

Condensate output worldwide is on the rise. Historically, the Middle East and Russia were the largest producers. But in the United States, production "more than doubled" between 2008 and 2017 and is now the top global NGL producer due to the boom in oil and gas fracking.[79] Production is increasing in Australia too, due to growing interest in developing gas fields rich in condensates. Condensate demand is rising, as they can be used to make petrochemicals, and more studies are needed to ascertain the lifecycle GHG emissions of these resources.

## Light and Tight Oil: Another Fracking Feast

Fracking not only liberates gas (as discussed previously) but also produces large volumes of light oil and condensates. Oil fracking emerged in the 1860s to increase oil production by detonating dynamite and nitroglycerine downhole.[80] Next, acid was injected to melt rather than crack the source rock starting in the 1930s. A decade later, mixtures of sand and gelled gasoline were used to prop open rocks. It was projected that this technology would have been applied nearly 1 million times by the late 1980s.[81] In the 2010s, modern-day fracking along with horizontal drilling breakthroughs took aim at light tight oils in gas-rich liquid plays. Prime candidates—including North Dakota's Bakken formation, West Texas's Permian Basin, and South Texas's Eagle Ford Basin—began to undergo intense development.[82]

Table 2.4 illustrates the dramatic differences in gas and condensate contents of fracked resources.[83] Unconventional plays with high gas levels require data on gas compositions and sufficient pipeline takeaway capacity to minimize GHG emissions. Economics are at odds with methane management because oil is priced higher than gas.[84] Therefore, leaky systems that burn off unwanted gas or deliberately vent methane into the atmosphere are profitable. Unprepared, careless, or corrupt operators that emit a lot of potent methane emissions pose significant climate risks.

## Extra-Heavy Oil: Old and Dense

At the opposite end of the hydrocarbon spectrum from light tight oil lie heavy oils that contain so much carbon they are semisolid. Generally, the more carbon

Table 2.4  Relative Shares of Oil, Gas, and Condensate from Texas's Eagle Ford Basin

| Resource Type | Energy Content | | | Total Production | |
|---|---|---|---|---|---|
| Zone Type | Oil | Condensate | Gas | Liquids | Gas |
| Black Oil | 84% | 1% | 15% | 79% | 21% |
| Volatile Oil | 37% | 26% | 37% | 52% | 48% |
| Condensate | <1% | 21% | 79% | 21% | 79% |
| Gas | <1% | 5% | 95% | 9% | 92% |

Notes: Gas production includes methane as well as ethane, propane, and butane. Further, asset specific, gas speciation is required to accurately model methane leakage from gas, condensate, and light oil assets.

Source: Abbas Ghandi et al., "Energy Intensity and Greenhouse Gas Emissions from Crude Oil Production in the Eagle Ford Region," September 2015, https://www.researchgate.net/publication/303592051_Energy_Intensity_and_Greenhouse_Gas_Emissions_from_Crude_Oil_Production_in_the_Eagle_Ford_Region_Input_Data_and_Analysis_Methods.

an extra-heavy oil contains, the higher its lifecycle emissions of $CO_2$ and black carbon.[85]

These ancient oils have spent a very long time underground subjected to heat and pressure. Degradation occurs when their lighter hydrocarbon contents are eaten by bacteria, are dispersed or dissolved by water, or disperse into their surroundings.[86] What remains is tarry, extra-heavy oil and semisolid bitumen, each with its own properties. Heat, steam, and chemicals are used to either reject their extra carbon or add hydrogen.[87] Table 2.5 provides various climate-related details about different types of extra-heavy oil and bitumen.

The Western Hemisphere contains nearly 70 percent of the global supply of technically recoverable heavy oil and over 80 percent of its technically recoverable

Table 2.5 Characteristics of Different Extra-Heavy Oils and Bitumen

| Category | Extra-Heavy Oil | Raw Bitumen | Diluted Bitumen | Synthetic Crude Oil (SCO) | Partially Upgraded Bitumen |
|---|---|---|---|---|---|
| Classification | Heavy, sour crude | Bitumen | Heavy, sour crude | Light, sweet crude | Medium, sour crude or medium, sweet crude |
| API Gravity | 4°–17° | 8° | 20°–22° | 26°–33° | Variable |
| Sulfur Content | 3–4% | 5% | 3–4% | 0.1% | Variable |
| Water and Solids | 10–25% | 2% | 0.1% | 0% | Variable |
| Heavy Metals[a] | High | High | Variable | n/a | Variable |
| Acidity (TAN) | High | High | High | n/a | Variable |
| Upgraded[b] | No | No | No | Yes | Partial processing |
| Highest GHG Emissions | Production and refining | Production | Refining and shipping | Production | Production, shipping, and refining |

[a] Heavy metals include vanadium, nickel, iron, and others.

[b] Upgrading involves preprocessing to remove excess carbon and other impurities in the form of petcoke before the refining stage.

API, American Petroleum Institute; GHG, greenhouse gas; TAN, total acid number.

*Source:* Author's estimations based on *Oil Sands Magazine,* https://www.oilsandsmaga-zine.com/technical/bitumen-upgrading; Jacobs Consultancy, March 2018, https://albertainnovates.ca/wp-content/uploads/2018/07/Bitumen-Partial-Upgrading-March-2018-Whitepaper-2433-Jacobs-Consultancy-FINAL_04July.pdf; Manik Talwani, Rice University Baker Institute, 2002, https://scholarship.rice.edu/handle/1911/91524.

bitumen.[88] Venezuela's Orinoco Belt (the world's single largest petroleum deposit) has the largest store of extra-heavy oil.[89] The majority of bituminous oil is buried in Alberta, Canada. As early as the sixteenth century, what was then called black pitch was reported along the banks of the Athabasca River.[90] In the 1960s, open pit mining began to produce oil sands. Today, most production in Alberta involves in situ steam-assisted gravity drainage, similar to Venezuela's approach, which uses heat and steam to move viscous oil over long distances with downhole pumps.[91]

Foreign assistance has facilitated these highly complex techniques in Venezuela.[92] But saturated markets, low oil prices, transport bottlenecks, and political instability have led to recent production cuts, buying time for cleaner technological breakthroughs.[93] The prospects for low-GHG impact using various techniques, such as in situ gasification methods to extract hydrogen from oil sands and nanotechnology and microbial methods to transform oil sands into light oil and natural gas underground, could sequester its own massive carbon content.[94] This may be the best way to safely access such a massive resource base from the world's heaviest oils.

## Ultradeep Oils: Extremely Far Down

Deepwater exploration has been underway for forty-five years, and recent developments are increasing unconventional oil reserves and production worldwide. In 2012, Exxon drilled the world's deepest oil well on Russia's Sakhalin Island— 40,502 feet below the surface.[95] Brazil and the US Gulf of Mexico hold the world's largest recoverable reserves.[96] As long as the associated gas in these fields is not vented or flared, ultradeep oil tends to have relatively low GHG emissions given the high natural pressure that facilitates oil extraction.

The use of floating production storage and offloading (FPSO) vessels has become increasingly common, especially for the production of deepwater and ultradeep offshore oil and gas resources that are largely exported. FPSO vessels have processing equipment on deck and hydrocarbon storage below board, are moored in place, and can rotate freely like a weathervane. An FPSO vessel offloads processed oil and gas to either a tanker or pipeline.[97] The largest climate risk from FPSO vessels is methane leakage and flaring, and NASA has developed new techniques that can now monitor methane over reflective water surfaces using satellites and remote sensing devices.[98]

## Arctic Oil: Back to the North Pole

More oil than gas is currently produced from the fragile Arctic. Together, just over 1 million barrels per day of oil and NGLs are marketed by the United States,

Russia, and Norway.[99] Installing infrastructure and operating there elevate climate risks due to melting ice and permafrost that then release significant amounts of stored $CO_2$ and methane.

Although the first Arctic oil deposits were discovered in 1920 in Canada's Northwest Territories, it was not until the early 1960s that BP drilled the first oil wells in the US Arctic (in Alaska). A decade later, the Trans-Alaska Pipeline was completed, sending Arctic oil to the United States. Exploratory drilling followed in the Canadian Arctic (on Melville Island) and the Russian Arctic (in Western Siberia). The first shipment of Russian Arctic oil to Europe commenced in 2014. Back in 2000, oil was discovered in the Norwegian Arctic (in the Barents Sea). The Arctic also contains several large deepwater oil fields with some of the world's greatest drilling depths.[100]

Ongoing development in the Arctic has been erratic. In 2015, for example, Shell pulled out of the Arctic citing disappointing prospects, high costs, and mounting public opposition.[101] BP sold off its assets in 2019.[102] Conversely, in 2020, the Russian government began offering new incentives for Arctic oil and gas development.[103] And ConocoPhillips, Equinor, and others have continued to pump oil and gas from Alaska's North Slope over the past forty years while Hilcorp Energy is upping its presence. Despite recent well failure linked to thawing permafrost,[104] development is more easily expanded once infrastructure is in place in sensitive ecosystems.[105]

## Depleted Oil: Boosting Recovery

As oil and gas reservoirs decline, it takes more effort to extract their remaining resources. EOR is the third stage of the extracting process, following immediate (primary) and subsequent (secondary) extraction efforts.[106] Depending on the characteristics of a given reservoir, production costs, and market conditions, EOR can liberate hard-to-produce oil remaining in place. There are between 350 and 400 active EOR projects around the world that account for 2 million barrels of oil per day.[107] The three commercialized EOR categories include thermal recovery, gas injection, and chemical injection. EOR tends to require large energy inputs and have elevated climate risks unless GHGs are sequestered.

Thermal recovery uses steam, fire, or electric heat to make viscous oil flow. California's heavy oil is injected with steam that operators make using natural gas. Substituting concentrated solar steam can significantly lower GHG emissions.[108] Fire flooding, another thermal method, ignites the oil underground to make it flow by cracking heavy hydrocarbons, vaporizing lighter hydrocarbons, and turning entrained water into steam. Such in situ combustion may sound novel, but it is the oldest thermal recovery technique in the book and has been applied over the past century in Russia, Romania, India, Kazakhstan, Azerbaijan, Canada, and the United States.[109]

Gas injection uses $CO_2$, natural gas, nitrogen, or exhaust flue gases that expand in a reservoir and push out additional oil. Gases also dissolve in oil to make it less viscous. In the United States, gas injection dominates most EOR production. Recently, EOR using $CO_2$ has emerged because it simultaneously increases oil production and sequesters carbon. Its climate risks depend on the source of $CO_2$. Anthropogenic $CO_2$ from manmade sources such as direct air capture or industrial exhaust have net-negative GHG emissions. But $CO_2$ obtained from natural reservoirs or producing acid gas exacerbates climate problems.

Chemical injection introduces manufactured molecules, such as polymers or detergents, into a reservoir to enable oil flow. This EOR technique is more expensive, is less predictable, and has greater climate uncertainties compared to other methods. Additional techniques are under development, including microorganisms (microbial EOR) that digest and transform the oil itself and enhance production as well as nanotechnology—using metal oxides, organic particles, and inorganic particles—to increase mobility, reduce viscosity, and shift internal reservoir forces.[110] Expanding data analytics by companies like Amazon and others could help further improve cutting-edge EOR methods aimed at recovering not only depleted oil but also a growing share of the world's remaining oil in place.[111] Future EOR techniques that access only hydrogen and leave the carbon in place would be a game changer for slashing the GHG emissions of such oil and gas extraction.

## Kerogen: Unformed and Immature

Kerogen contains solid hydrocarbons woven into sedimentary rock. It takes millions of years to mature and turn into oil and gas.[112] By baking (or retorting) kerogen into select liquid hydrocarbons, engineers can accelerate this natural process.[113] Turning immature oil shale into synthetic crude oil, gasoline, or diesel takes huge amounts of energy. Therefore, GHG emissions from mining, retorting, and refining oil shale are estimated to be as much as twice those of conventional oil.[114]

Oil shales are buried around the globe, but Estonia accounts for the majority of the world's current oil shale production.[115] Elsewhere, it takes high oil prices to spur oil shale development. In the late 1970s, the US Congress created a synthetic fuels program to jumpstart production of oil shale, which was subsequently shuttered when oil prices fell in the mid-1980s. In the mid-2000s, the US government awarded six oil shale leases on public lands in Colorado and Utah. The companies involved pulled out one by one, turning instead to fracking shale oil and developing deepwater gas.[116] A sustained effort requiring ongoing government commitments and financial incentives will be needed for companies or countries to make long-term commitments to future oil shale developments around the world.

Other hurdles besides climate risks and costs exist. Oil shales contain toxic levels of sulfur, nitrogen, arsenic, iron, and nickel. Underground (in situ) operations that use earthen chambers to retort oil shale can contaminate groundwater. Above-ground (ex situ) methods have waste disposal problems, damage habitats, consume significant water in arid regions, and have restoration issues like coal mining does.

These are all important considerations because oil shale resources in place could contain more energy than the combined total of all other oil sources on Earth—roughly 30 trillion barrels of oil.[117] But the climate and other environmental risks of oil shale mean that this supply may come at too high a cost.

## Biofuels: From Plants to Fuels

Aside from the aforementioned categories, there are other subtypes of unconventional hydrocarbons that must be converted into a usable state using various methods. These include biofuels as well as CTLs and GTLs.

Biomass-based liquid fuels (biofuels) have also existed since cars were first introduced in the early twentieth century. Henry Ford himself once claimed that "the fuel of the future is going to come from fruit [because] there is fuel in every bit of fermentable vegetable matter."[118] Biofuel sources include food, food waste, nonedible plants, grass, and algae.

All biofuel production employs chemical reactions and heat to break down hydrocarbons and refine them into petroleum products. In 2019, just under 2 million barrels per day of ethanol was produced fermenting sugarcane in Brazil and starchy crops (such as corn) in the United States.[119] Meanwhile, nearly 1 million barrels per day of biodiesel is currently being made from seed oils, including palm oil in Indonesia, the United States, Brazil, and numerous EU nations.[120] In the United States, older refineries are being renovated to make renewable diesel and jet fuel from waste oils and animal fats.[121] First-generation biofuels require large energy inputs and have not been found to reduce GHG emissions compared to conventional oil.[122] Next-generation biofuels that convert cellulose in the cell walls of plants could reduce climate impacts and food security concerns, but air pollution could remain challenging because ethanol contributes to local smog formation.[123]

## GTLs and CTLs: Turning Gas and Coal into Petroleum Fuels

Liquid fuels have a higher market value than natural gas. This economic proposition has spurred the complex conversion of gas to liquids, especially in countries that do not have ample oil resources, such as Germany and South Africa. Modern GTL technology is based on the Fischer-Tropsch process,[124] which builds simple gaseous hydrocarbons into long-chain liquid fuels using a catalyst.[125] The world's largest GTL plant is located in Qatar and processes 1.6 billion cubic feet per day of gas from twenty-two offshore wells into 120,000 barrels per day of NGLs and ethane and 140,000 barrels per day of other assorted GTL products such as jet fuel and diesel.[126] Different pathways of GTLs have variable GHG emissions. But higher energy inputs can increase the climate impacts of GTLs upward of 25 percent compared to conventional oil.[127] Methane leakage is a major determinant of climate risk.

Meanwhile, coal is mostly used worldwide to generate electricity, but CTLs can also be obtained by converting this solid hydrocarbon, directly or indirectly, into diesel and gasoline. Direct liquefaction uses solvents to dissolve coal under high temperatures and pressure into a liquid that is refined into fuels. Indirect methods gasify coal and turn it into syngas, which is then converted into liquid fuels using the Fischer-Tropsch process. Countries with large coal reserves—such as China, India, the United States, and Australia—have advanced CTL production capabilities. The high economic costs and environmental impacts of CTLs have limited their applications worldwide. Efforts are underway to combine coal and biomass to convert these resources into liquids. When coupled with successful CCS, CTL could emit 5 to 10 percent fewer GHGs than conventional oil does. Without CCS, the climate penalty of CTL could be "twice" as high as that of conventional oil.[128]

Unconventional oils, like their gas counterparts, are likely to account for a growing share of future production. This general trend means that unless the outsized climate impact of such resources is closely charted, the industry is at risk of greatly increasing their climate impact throughout the oil supply chain from production to processing to shipping.

## *The Climate Footprints of Conventional and Unconventional Hydrocarbons*

The oil and gas industry currently emits a reported 5.2 gigatonnes (Gt) of $CO_2$e emissions annually by producing, transporting, and processing hydrocarbons.[129] This figure excludes emissions from end-use consumers like motorists, truckers, and airlines. It also significantly undercounts the contribution from methane emissions.[130] Unconventional oil and gas have wide-ranging GHG intensities that are previewed here and analyzed in detail in forthcoming chapters.

Collectively, if unconventional hydrocarbons (including both oil and gas) replace conventional resources by midcentury without effective mitigation measures in place, supply-side climate footprints are estimated to triple in size. Figure 2.2 plots the nearly 200 Gt $CO_2e$ that are poised to be cumulatively emitted from the development of technically recoverable unconventional oil and gas between 2030 and 2050 if current demand for these resources persists. GHG emissions of this magnitude represent a major share of the remaining carbon budget of safe levels of emissions that can be produced while still limiting global warming.[131]

## *The Ultimate Unconventional*

All this talk of the climate footprints of unconventional forms of oil and gas raises the question of whether or not there may be other, more climate-friendly

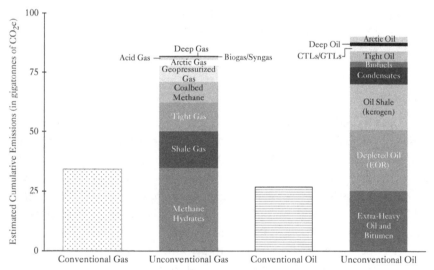

**FIGURE 2.2** Estimated Cumulative Industry GHG Emissions for Conventional versus Unconventional Oil and Gas Resources (2030–2050)

*Notes:* Industrial oil and gas sector emissions plotted do not include end-use fuel consumption. The numbers in Table 2.2 were used to calculated 10 percent of technically recoverable oil and gas through 2040 without climate mitigation measures imposed. Adjustments were made to volumes for methane hydrates (lowered due to technical barriers), EOR (lowered due to depletion rate), oil shale/kerogen (lowered due to technical barriers), and condensates (increased due to fracking outside the United States). Small sample of industrial supply-side GHG emissions only (production, refining, processing, and shipping) used to calculate average emission intensities. $CO_2e$, carbon dioxide equivalent; EOR, enhanced oil recovery; GHG, greenhouse gas.

*Sources:* Author's calculations, using the OCI+ model.

alternatives. In 1980, I worked in a chemical engineering laboratory at the University of Colorado researching hydrogen—a simple, plentiful, carbon-free molecule with high energy content that was all the rage following the second energy crisis and resulting spike in oil prices. Back then, economics were not on hydrogen's side. The precious metal catalysts, like platinum, that I used to boost the efficiency of chemical reactions were rare and costly. Hydrogen's prospects are much more favorable today, given cheap renewable electricity, more catalyst options, and major climate concerns.[132]

Hydrogen can be made from renewables, and it emits only water when burned, eliminating carbon from the energy equation. Hydrogen can also be used to produce heat, store surplus renewable power, and serve as a chemical feedstock, and can be used as a component of fuel cells (for chemical batteries) to generate electricity (for both vehicles and stationary sources).

This simple element does not exist alone, however; it is locked up in enormous quantities of water, hydrocarbons, and other organic matter.[133] Its climate risks depend on the carbon content of its source; the energy intensity of the separation processes used; and the leakage rate of $CO_2$, methane, and other GHGs. The seemingly harmless water vapor that hydrogen forms when combusted is itself a weak GHG, as discussed in chapter 3.

## Shades of Hydrogen

Hydrogen is produced via different pathways.[134] Producing what is termed *black hydrogen* entails cracking coal and oil, while another method is to split methane gas via steam methane reforming (SMR). Its large energy inputs and vented $CO_2$ generate high GHG emissions. Gray hydrogen also uses fossil fuels and vents $CO_2$, but it substitutes renewable energy to generate steam, slightly lowering its GHG emissions. Natural gas (SMR) accounts for roughly half of current hydrogen production worldwide, while the rest is made from oil (30 percent) and coal (18 percent).[135] New petrochemical pathways with uncertain climate risks could capture hydrogen shed when ethane is cracked to form ethylene.

Blue hydrogen is formed by splitting methane gas using catalysts while sequestering or utilizing the carbon.[136] Because it still involves methane, blue hydrogen presents considerable climate risks. By contrast, green hydrogen produced from the electrolysis of water powered by wind, sun, or other clean forms of electricity presents the fewest GHG emissions. Green hydrogen exemplifies a truly circular economy whereby water yields hydrogen, which is then combusted back into water.[137] Table 2.6 estimates GHG emissions from several hydrogen pathways using different feedstocks and processes.[138]

Table 2.6  Estimated Lifecycle GHG Emissions for Hydrogen Production
Pathways

| Production Pathway | Gaseous Hydrogen | Liquid Hydrogen | Commercialization Timeframe | Operational Scope |
|---|---|---|---|---|
| Grid Electrolysis | 285 | 385 | 5 years | Decentralized |
| Animal Manure | 150 | 219 | 20 years | Decentralized |
| SMR | 121 | 191 | Now | Centralized/ decentralized |
| Landfill Gas | 57 | 116 | 5 years | Decentralized |
| Biomass Gasification | 45 | 85 | 10 years | Centralized |
| Solar Electrolysis | 25 | 76 | 10–20 years | Centralized |
| Wastewater Treatment | 20 | 75 | 10–20 years | Decentralized |

*Notes:* Assumes well-to-tank GHG emissions for centralized production pathways. GHG estimates from Intergovernmental Panel on Climate Change, AR4 systematically undercount emissions because the GWP used for methane is out of date.

GHG, greenhouse gas; GWP, global warming potential; SMR, steam methane reforming.
*Source:* Marshall Miller, Arun Raju, and Partho Sorothi Roy, "The Development of Lifecycle Data for Hydrogen Fuel Production and Delivery," University of California, Davis, National Center for Sustainable Transportation and Institute of Transportation Studies, October 2017 (Used CA-GREET Tier 2, Life Cycle Analysis Model), https://escholarship.org/uc/item/3pn8s961.

## Shifting the Focus on Unconventionals

Oil and gas have dominated energy and commodities markets for well over a century, but hydrogen possesses the flexibility and protection humanity will need in the future. Hydrogen can be used as energy carriers (fuel), storage devices (batteries), and feedstocks (inputs) to produce petrochemicals, fuels, and other goods.[139] If produced using renewables, green hydrogen can cut carbon out of the process and safeguard the climate.

Oil refineries are today's largest consumers of hydrogen (generated using SMR) to crack and treat crude to meet low-sulfur fuel specifications. Hydrogen consumption in US refineries increased by 60 percent between 2008 and 2014, and demand keeps growing.[140] This may explain why oil companies, such as Shell and Equinor, are investing in a clean hydrogen future.[141] Other industrial actors—automakers, chemical manufacturers, and steelmakers, as well as governments and banks in Europe, Asia, and elsewhere—foresee a future for green hydrogen production from excess renewable energy generation.[142]

Someday, hydrogen fuel will be key to powering rockets. Green hydrogen markets could become sizable, if and when commercial space travel takes off. Hydrogen for midcourse space refueling could be accomplished with technologies to convert water into hydrogen- and oxygen-based fuels.[143]

Academic research and government research and development (R&D) are underway on breakthrough technologies that could attract oil industry capital to transition energy markets.[144] Refining oil and natural gas underground using nanoscale microbes, catalysts, and biotechnology could sequester the carbon in place and produce green hydrogen from fossil fuels.[145] These areas of R&D are central to a clean-energy paradigm shift in the world marketplace.

Shrinking the climate footprints of unconventional oil and gas supplies is a top near-term priority. Over the long term, it will be critical to prevent the bounty of unconventional hydrocarbons from wreaking havoc on the climate. As discussed in chapter 3, new tools have been developed to assess the GHG emissions of different types of oil and gas, identify where in the petroleum supply chain emissions can be reduced, and help guide the world along the path to a clean energy transition.

## 3

# A Better Way to Measure Climate Footprints

THE WIDELY DIVERGING climate footprints of various oils and gases (especially unconventional ones) make clear how vital it is to find ways to accurately measure these differences. In 2013, I hosted a symposium on unconventional oil at the Carnegie Endowment for International Peace, where I was a senior associate in the Energy and Climate Program. That morning, I stood before a packed room in the nation's capital to tee up the conversation on uncovering the unknowns about oil.[1] I arranged several panels and keynote addresses with experts from industry, government, academia, and nongovernmental organizations (NGOs). Top energy journalists served as moderators.[2]

Everyone in the room asked the same burning question: What does the transforming oil and gas landscape mean for climate change? The answer is complicated. The climate risks of hydrocarbons are not a one-size-fits-all proposition. As the preceding chapters have shown, their differences are much greater than their similarities.

We do know that long-held assumptions that a barrel of oil or a cubic foot of gas emits a set amount of carbon dioxide ($CO_2$) are not valid.[3] The varying climate footprints of oil and gas can amount to much more than the carbon contained in these raw materials. Lifecycle greenhouse gas (GHG) emissions—from generating energy and leaking pollutants in the early stages of extracting and processing oil and gas to the later stages of delivering and consuming a multitude of petroleum byproducts—could be just as divergent as the resources themselves are. The question remained, then, are the differences between the GHG emissions of different types of oil and gas large enough to matter?

To answer this question, I assembled a team of academic research partners. Together, we would use scientific methods, collect empirical data, and gather measurable evidence to quantify and compare the GHG emissions of twenty-first-century oil and gas in a systematic way.

In this chapter, I describe the development, design, data inputs, emission drivers, and uncertainties of the tool we developed to accomplish this task. The Oil Climate Index plus Gas (OCI+) is a first-of-its-kind assessment tool for estimating the lifecycle GHG emissions from different oil and gas resources.[4] Its findings decisively overturn the widely held (but mistaken) beliefs that oils (and gases) all have basically the same climate impacts and that the transportation sector (mainly motorists) is responsible for essentially all petroleum sector emissions. In reality, some barrels of oil and cubic feet of gas pose far greater climate risks than others and needless amounts of GHGs are emitted and leaked by the oil and gas industry itself (supply-side Scope 1 and 2 emissions[5]) before fuels even reach customers (Scope 3 emissions).[6] Armed with this knowledge, climate actions can focus on those oil and gas assets with large industrial climate footprints while we work to pivot the entire market to a durable net-zero emission energy transition.

## *Barreling Ahead*

Immediately following the 2013 Carnegie symposium, I approached one of the panelists—Adam Brandt, a professor of energy resources at Stanford University.[7] We agreed that the petroleum sector's total climate impacts could not be estimated by simply counting the carbon contained in the oil or gas itself. Take, for example, a barrel of heavy Midway-Sunset oil (California's most produced crude). According to the Environmental Protection Agency's (EPA's) calculations, one barrel is estimated to emit 448 kilograms of carbon dioxide equivalent ($CO_2e$) in emissions when counting all of the carbon contained in that one barrel of Midway-Sunset oil.[8] In reality, a barrel of Midway-Sunset oil emits an estimated 765 kilograms $CO_2e$ (an increase of over 70 percent) when *all* emissions from its production, refining, shipping, and wide-ranging end uses are considered.[9]

Adam's research identified vast differences in upstream oil GHG emissions (from wellhead to refinery gate),[10] which he projected using a novel GHG-estimating model.[11] The California Air Resources Board formally adopted this tool to implement the state's low-carbon fuel standard (LCFS).[12] This policy encourages the use of alternative transport fuels with lower GHG emissions by imposing on oil refiners mandatory emission cuts with tradeable GHG credits.[13]

Such "product-centric"[14] policies, however, can easily overlook GHG emissions from system leakage and petroleum coproducts, emissions that do not stem directly from consuming transport fuels.[15] So-called well-to-wheel analyses do not offer comprehensive, process-level details, and they omit portions of the oil and gas supply chain.[16] Such product-centric analyses make consequential errors that tend to underestimate the variation in total petroleum emissions.

Poor boundary choices that focus on select transport fuels miss the climate effects of coproducts like petroleum coke, heavy residual fuels, and petrochemical feedstocks whose GHG emissions and market values do not correlate well with those of gasoline and diesel.[17] Moreover, using averages rather than more detailed assessments of representative practices does not capture the full range of observed variability in emissions levels.[18]

A more comprehensive method of assessing the differences between oil and gas resources is called for. Adam agreed that a resource-centric, barrel-forward approach was needed to count *all* the emissions in the whole barrel to highlight the GHG savings potential in the entire oil and gas supply chain.

## Assembling the OCI+ Team

As we began adapting Adam's upstream production GHG model, the next step was to model the GHG emissions of oil refineries. We did not have to search long for a project partner. Joule Bergerson from the University of Calgary and her colleagues were in the early stages of developing a model that indicated that emissions of the refining process varied as much as production-stage GHG emissions.[19] Joule was eager to join the OCI+ research team.

Jonathan Koomey was our fourth and final partner.[20] Jon and I had gone to graduate school together; worked at the Lawrence Berkeley National Laboratory under the same mentor, Art Rosenfeld; and had successively taught at the Yale School of Forestry and Environmental Studies (since renamed the Yale School of the Environment). Although we pursued different energy paths—mine paved with hydrocarbons and his amid electrons and data—two decades later, we would be reacquainted. It was clear that Jon's broad-based energy and environmental expertise would greatly benefit our OCI+ team. With all the researchers in place, we set out to develop a new tool to estimate the total lifecycle emissions of the oil and gas sector.

Parsing oils by their climate impacts allows multiple stakeholders, each with their own objectives, to consider climate risks in prioritizing the development of future oils and gases and the adoption of greater policy oversight over today's oils and gases. While stakeholders' priorities vary, all actors would be better served by accurate, transparent measures of climate risk associated with different oil and gas resources.

What follows is an explanation of the various models underpinning the OCI+, how they work, the data they use, and what drives GHG emissions in each stage of the oil and gas lifecycle.

## Constructing the OCI+

The OCI+ is a metric that takes into account the total lifecycle GHG emissions of individual oils—from upstream extraction to midstream refining to downstream end uses. It offers a powerful yet user-friendly way to compare oils and gases and assess their particular climate impacts before development decisions are made and once operations are underway.

The OCI+ uses the following open-source tools to evaluate actual emissions levels associated with an individual resource's supply chain. The underlying OCI+ models currently run on an Excel-based platform, but future versions could be programmed to speed up execution and enhance functionality.[21] The index's main three components are Adam's model for measuring upstream emissions from exploration until the petroleum resources enter the refinery, a model Joule developed for gauging the midstream emissions of the refining process, and a model Jon and I constructed to assess the downstream GHG emissions of the shipping and end uses of the resulting petroleum products.

- **The Oil Production Greenhouse Gas Emissions Estimator (OPGEE)**: Led by Adam Brandt, this model estimates upstream emissions from oil and gas exploration, drilling, production, separation, processing, and transport to the refinery inlet and gas distribution system.[22]
- **The Petroleum Refinery Lifecycle Inventory Model (PRELIM)**: Spearheaded by Joule Bergerson, this tool estimates the emissions and petroleum product yields of refining crude oil.[23] It is the first open-source refinery model that estimates energy and GHG emissions associated with various crudes processed by different refinery configurations using different processing equipment.
- **The Oil Products Emissions Module (OPEM)**: Led by me and Jonathan Koomey, this module estimates the emissions from the transport and end use of all petroleum products yielded by a given oil or gas. An overriding goal of OPEM is to include (and thereby avoid) carbon leakage from petroleum coproducts.[24]

Putting these pieces together, Figure 3.1 illustrates a flowchart of the OCI+'s three underlying emission estimation models. Details on each model follow.

The OCI+ analyzes total GHG emissions using a mass-balance approach, which means that all the carbon that goes in must come out. Therefore, all coproducts are counted, and none are lost or hidden. Two functional units (the underlying metrics or bases) enable different comparisons. Lifecycle GHG levels can be measured per barrel of oil equivalent (BOE) of crude and gas produced. Alternatively, emissions can be assessed per megajoule of energy content for all

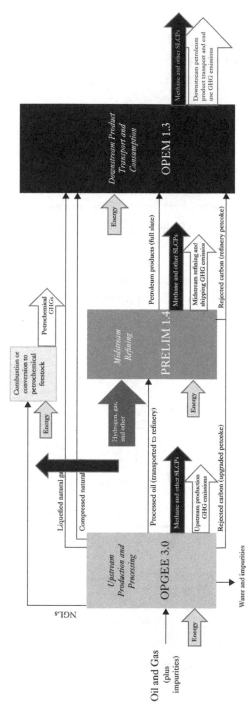

**FIGURE 3.1** Simplified Process Flowchart of the Oil Climate Index + Gas (OCI+) Model

*Notes:* OPGEE assumes that any NGLs removed upstream along with the processed gas are not sent to the refinery. Any remaining liquid hydrocarbons that cross the field boundary, however, make their way to the refinery and are modeled in PRELIM. NGLs, natural gas liquids; OPGEE, Oil Production Greenhouse Gas Emissions Estimator; PRELIM, Petroleum Refinery Lifecycle Inventory Model; SLCPs, short-lived climate pollutants.

*Source:* Author's depiction.

final petroleum products. These units can then be converted to estimate emissions per dollar value of all petroleum products sold. The adaptability of this model was essential to presenting its findings in multiple ways, particularly ones that convey the economic factors at play.

## Modeling Upstream GHG Emissions

Unearthing oil and gas deposits and preparing them for shipment to a refinery, petrochemical plant, or gas distribution system is the first step in the petroleum value chain. The processes involved differ depending on the hydrocarbons extracted. These are the upstream processes that OPGEE seeks to forecast. It is an open-source model that is free to download, modify, and use, as long as users make their assumptions known if they publish the results.[25] Figure 3.2 lays out a simplified schematic of the OPGEE model.

The upstream GHG impacts of different oils and gases vary significantly. OPGEE has been run on nearly 9,000 global crudes from ninety countries, using data from nearly 800 references, including government sources, scientific literature, and publicly available technical reports.[26] Proprietary databases can be used

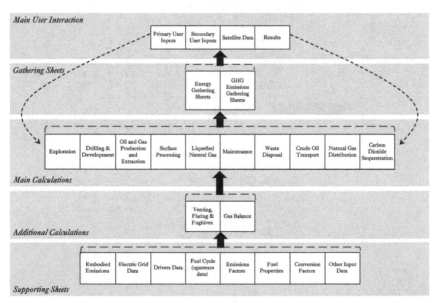

**FIGURE 3.2** Simplified Schematic of the Oil Production Greenhouse Gas Emissions Estimator (OPGEE) Model

GHG, greenhouse gas.

*Source:* Author's adaptation based on OPGEE 3.0aBETA Documentation.

to supplement these data when information is unavailable in the public domain.[27] OPGEE contains more upstream oil runs than any other modeling effort to date.

In the most recent version, OPGEE 3.0a, it is possible to estimate upstream GHG emissions from gas as well as oil production.[28] When gases are separated from liquids (oil and water), they are then treated, gathered, dehydrated, decontaminated (removing excess $CO_2$), and demethanized (removing methane). Heavier gases or natural gas liquids (NGLs) may be recirculated to lift further resources out of the ground or shipped to industrial customers. Methane (or marketable natural gas) is compressed and shipped to various end users. Throughout the various upstream processes, computations are made for gases (like methane, ethane, and other volatile organic compounds [VOCs]) that are flared, vented, and released as fugitive emissions.

## What Drives Upstream Emissions?

GHG emissions result at every upstream stage due to combustion and leakage, which are influenced by several factors. The characteristics of a given hydrocarbon resource—such as gas content, water content, carbon content, and field age—determine the extraction techniques and surface processing required. GHG-sensitive resources include those with high gas-to-oil ratios, flaring-to-oil ratios, water-to-oil ratios, and low American Petroleum Institute (API) gravities (heavy oils). Upsets and nonroutine operations—such as blowouts, well workovers, and poorly maintained flares—can significantly elevate GHG levels. Emission drivers include pumping and compressing fluids and gases, generating steam, removing water and contaminants, leaking methane, and flaring and venting associated gases.

Where oil and gas deposits are located—the geography and surrounding ecosystem (whether it be desert, Arctic tundra, jungle, forest, or offshore)—influences how disruptive extraction is to land use. When resource development changes land use, this can affect the land's biological (soil and plants) capacity for carbon storage. The more naturally stored carbon that is released, the more GHGs that are emitted. OPGEE accounts for land use–related GHG emissions from oil and gas development based on the land's carbon richness and development intensity. When appropriate, data for a particular region can be input to overwrite the model's fixed factors.

Once the oil and gas are extracted, they require various degrees of additional processing, whether on-site or at another nearby or distant facility. An oil field's location, its distance from transport hubs, and refinery selection determine how the crude is shipped; the resulting transport emissions can be estimated accordingly. A gas field's location, the availability of on-site processing equipment and types of byproducts that result, existing infrastructure in place, and purchasing arrangements determine its transport emissions that can be estimated. Sometimes pipelines, railroads, or trucks ship the oil and gas over land. In other instances, barges move oil over inland waterways, and seaborne shipments use marine vessels. The model uses standard routes as default values, which users can change for particular distances and modes. Figure 3.3 displays the relative share of highly

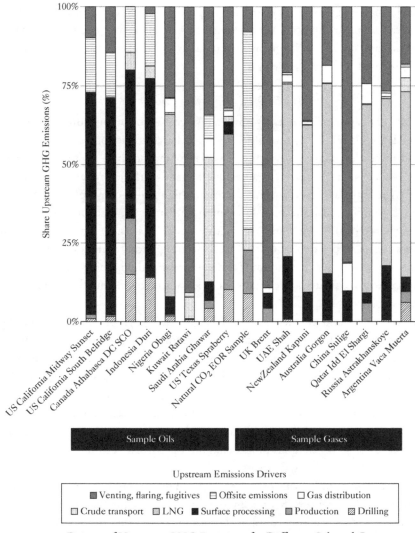

**FIGURE 3.3**  Drivers of Upstream GHG Emissions for Different Oils and Gases
EOR, enhanced oil recovery; GHG, greenhouse gas; LNG, liquefied natural gas.
*Source:* OCI+ Preview Web Tool, 2020.

variable upstream emission drivers for producing a range of different oil and gas assets.

## Modeling Midstream GHG Emissions

Once the oil and gas has been extracted and moved to a refinery, the midstream stage of the petroleum value chain begins. Refining used to be a simple process that involved heating up and boiling oil to separate out its main components. But the changing nature of oils demands corresponding changes in refinery operations.

By adjusting for various refinery configurations and processing techniques, the PRELIM model analyzes how crude quality and refinery units affect energy use and GHG emissions. PRELIM can run numbers on a single crude or a blend of oils, and when combined with OPGEE, the model sheds light on the midstream emissions in the broader lifecycle of oils and the liquids contained in gases. The model influences the OCI+ in two important ways. It estimates midstream GHG emissions, and it outputs a given refinery's petroleum product yield.[29] The type and volume of products vary depending on the input crude and the refinery's design.

Oils vary significantly in their midstream GHG emissions. PRELIM has been run on 343 crude oils processed in 478 refineries located in eighty-three countries, representing 93 percent of global crude oil refining throughput in 2015.[30] Public oil assays, analyses of crude oil composition at various distillation temperatures designed to simulate the inner workings of a refinery, are used wherever possible, and proprietary databases are used to supplement these data when public information is unavailable. This model represents more refining oil runs than any other similar modeling effort to date.

### *Matching Oils to Refineries*

Matching oil characteristics with refining infrastructure to meet end-use product demand is the ultimate goal of all refiners. Every refinery is unique in terms of the combination of equipment it uses, the blends of crudes it is optimized for, and ultimately the products it sells.

Figure 3.4 illustrates the three major refinery configurations, from simplest to most complex, assessed by PRELIM—hydroskimming, medium conversion, and deep conversion. PRELIM also contains a total of eleven processing unit combinations within these basic refinery categories. One deep-conversion refinery configuration, for example, might employ a coking unit to reject high levels of carbon in the form of petcoke, while another may use hydrotreating to add hydrogen to

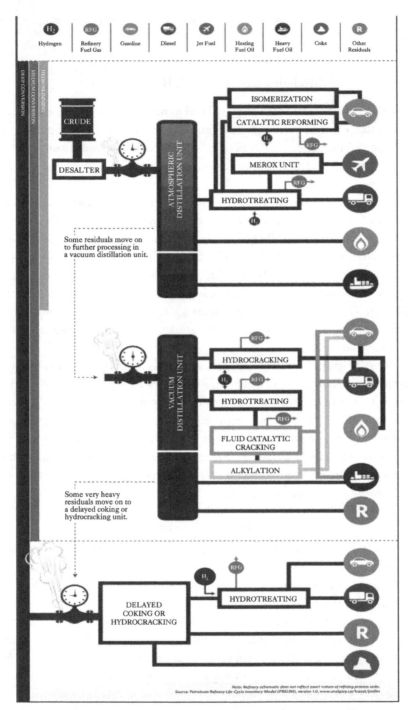

**FIGURE 3.4** Simplified Refinery Configurations in the Petroleum Refinery Lifecycle Inventory Model (PRELIM)

*Source:* Deborah Gordon and Eugene Tan, "Different Oils, Different Climate Impacts," Carnegie Endowment for International Peace, 2015, https://carnegieendowment.org/files/DifferentOils_Print.pdf

crude to remove sulfur and other contaminants and manufacture gasoline and other petroleum products.

The characteristics of individual process units incorporated into the PRELIM model were obtained from published literature and input from industry experts. Technically, each crude can be blended and processed in many different refinery configurations, but in practice crude oils are best matched to certain configurations. PRELIM selects the default refinery configuration that best suits a given crude oil based on its properties (API gravity and sulfur content). As such, light and sweet (low-sulfur) crudes are best processed in simpler refineries, and heavy and sour (high-sulfur) crudes are ideally directed to complex deep-conversion refineries. The following rules of thumb illustrate the basic principles at play:

- **Deep-conversion refinery**: heavy crude under 22 degrees API with any sulfur level
- **Medium-conversion refinery**: medium, sweet crude (22 to 32 degrees API, with less than 0.5 percent sulfur content by weight); medium, sour crude (22 to 32 degrees API with more than 0.5 percent sulfur content by weight); and light, sour crude (over 32 degrees API with more than 0.5 percent sulfur content by weight)
- **Hydroskimming refinery**: light, sweet crude over 32 degrees API and less than 0.5 percent sulfur content by weight

Many experts think that a crude oil's API gravity and sulfur content are reliable predictors of a refinery's GHG emissions. This, however, is a fallacy that has long hampered the collection of the full range of data needed to model refining emissions. While API gravity and sulfur are good indicators of the GHG emissions stemming from a *default* refinery type and end use, they alone are not sufficient for conclusively determining midstream refinery GHG emissions.

### What Drives Midstream Emissions?

PRELIM reveals that a number of factors lead to elevated GHG levels during midstream petroleum operations. Crude quality, the selected process units employed (the refinery configuration), and the energy efficiency of the process units all play important roles in determining the energy requirements and resultant emissions of an individual crude (or a crude blend). Individual oils are rarely refined in isolation, and PRELIM's GHG-level estimates can be proportionally assigned to individual oils in a mixture of crudes fed into a refinery.

While inputs of heat, steam, and electricity influence refinery emissions, the amount of hydrogen required to process each crude is also a major driver of refinery GHG emissions. How hydrogen is generated largely determines the relative share of this emissions driver. The heavier the crude, the more hydrogen is usually utilized during refining. Figure 3.5 displays the relative share of highly variable, midstream emissions drivers for refining different crude oil assets.

## Modeling Downstream GHG Emissions

Downstream GHG emissions encompass the marketing (transport) and consuming of natural gas and petroleum products.[31] Transport can involve ocean vessels, pipelines, railcars, barges, or tanker trucks. The consumption of petroleum products meets demand for billions of kinds of goods and services in every economic sector. Transport movements and end uses are difficult to precisely track. The OCI+ sets default values that users can modify.

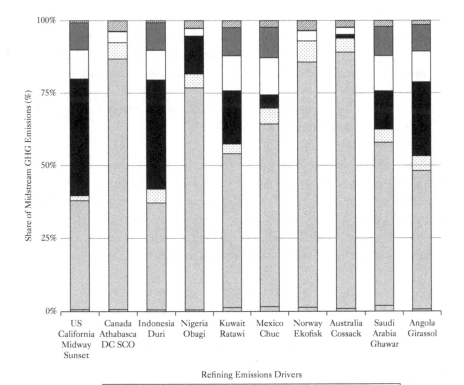

**FIGURE 3.5** Drivers of Midstream GHG Emissions for Refining Different Oils
GHG, greenhouse gas; SMR, steam methane reforming.
*Source:* OCI+ Preview Web Tool, 2020.

### Assumptions about Downstream Transport

OPGEE models the transport of crude oil to the refinery entrance and processed gas on to the distribution hub. It also accounts for petcoke transport, when it is removed via upstream upgrading. OPEM models all petroleum product transport including petcoke when it is produced in the refinery. Default transport distances are set in OPGEE and OPEM. It is assumed that all crude oil is transported to Houston refineries by pipeline (US crudes) or marine vessel (imported crudes)[32] and that gas is transported via pipeline within each continent to a major regional hub.[33] (Natural gas converted to liquefied natural gas [LNG] for intercontinental transport employs ocean vessels and liquefaction and regasification operations.) Volumes transported are provided by PRELIM for refined products and OPGEE for coproducts removed upstream, such as some amounts of petcoke and NGLs.

Multiple modes are typically used to deliver petroleum products, which users can input in OPEM. For example, exported gasoline can move from the refinery via pipeline to a marine terminal, be exported by ocean vessel, be put into a pipeline at its destination, and then be moved by a tanker truck to refueling stations. Alternatively, domestic gasoline can be put in a pipeline and shipped to another city where it is then loaded onto a tanker truck to various gasoline stations. There are different modes and distances that petroleum products can follow, but the highest GHG emission intensities tend to occur in the final miles due to the low fuel efficiency of heavy-duty trucks. Current default assumptions in OPGEE and OPEM result in estimated transport emissions that minimally amount to an estimated 1 to 2 percent of oil lifecycle GHG emissions.[34]

### Assumptions about Downstream End Use

While transport emissions are typically minor relative to those stemming from other parts of the petroleum lifecycle, GHG emissions derived from end-use consumption can dominate the lifecycle of oil and gas production. Prior calculations assessing the petroleum lifecycle have historically compared oil to alternative transport fuels and compared GHG emissions predominantly on gasoline and diesel yields. However, end-use GHG emissions are highly variable when one fully accounts for the consumption of all petroleum products, including petrochemical feedstock and bottom-of-the-barrel byproducts like petcoke, fuel oil, bunker fuel (known as bunker C), and asphalt. Figure 3.6 illustrates a simplified flowchart of the OPEM model for downstream emissions.

### What Drives Downstream Emissions?

Four variables drive GHG emissions from the transport of petroleum products: mode, distance, fuel used, and the mass of the product. Depending on the

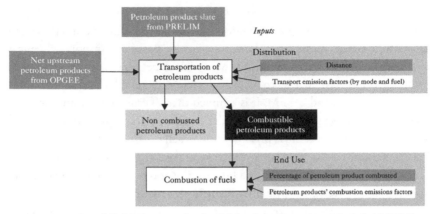

**FIGURE 3.6** Simplified Schematic for the Oil Products Emissions Module (OPEM)

*Note:* The transportation of crude oil from the field to the refinery and natural gas and upstream NGL shipping GHGs are calculated in Oil Production Greenhouse Gas Emissions Estimator (OPGEE), but they are reconciled in OPEM. PRELIM, Petroleum Refinery Lifecycle Inventory Model.

*Source:* OCI+ Preview Web Tool, Methodology, OPEM Model.

fuel that is used, the typical range of transport options for GHGs from the least emissions intensive to the most starts with ocean vessels, pipelines, trains, and barges before ending with tanker trucks. The distance traveled, delays incurred, and fully loaded return journeys also drive the transport share of GHG emissions.

However, it is difficult to ascertain precise modes, distances, fuels, and other logistical details for these figures. There is no global agency or group to collect and audit data that involves multiple private actors controlling highly dynamic logistics systems driven by changing supply and demand. This lack of centralized oversight matters because there is great variance in many of these variables, including the routes taken. For example, there are five different sea routes—ranging from 9,500 to 16,940 nautical miles—that a tanker of liquefied petroleum gas (LPG) can take from Texas to Japan alone (the largest importer of US products), each running through a different global choke point with its own geopolitical challenges.[35] This multiplicity of fluctuating variables is true of nearly every route and transport permutation, a level of complexity that compounds the challenge of tabulating accurate downstream emissions totals.

End-use GHG emissions are driven by fuel type, the uses various petroleum products are put to, specific fuel quality, vehicles' fuel efficiency and level of maintenance, and local conditions such as traffic congestion. Many, but not all, petroleum products derive their highest value by being burned. Table 3.1 charts the different emissions levels generated by combusting equivalent volumes of

**Table 3.1  End-Use Emissions Factors of Refined Petroleum Products**

| Petroleum Product | Combustion GHG Emissions (kilograms of $CO_2e$/gallon) | GHG Emissions Relative to Gasoline (+/− %) |
|---|---|---|
| Petrochemical Feedstocks[a] | 1.3 | −86% |
| LPG | 5.7 | −35% |
| NGLs | 6.3 | −29% |
| Natural Gas[b] | 7.5 | −15% |
| Gasoline | 8.8 | 0 |
| Jet fuel | 9.8 | 11% |
| Diesel | 10.2 | 16% |
| Fuel Oil | 11.0 | 25% |
| Liquid Heavy Ends (Residual Fuels) | 11.3 | 28% |
| Petroleum Coke | 14.7 | 67% |

[a] GHGs for petrochemical feedstocks consider emissions from converting ethane to ethylene.

[b] Natural gas combustion emissions are listed in liquid terms (per BOE).

BOE, barrel of oil equivalent; GHG, greenhouse gas; LPG, liquefied petroleum gas; NGLs, natural gas liquids.

*Source*: Environmental Protection Agency, *Emission Factors for Greenhouse Gas Inventories*, 2018, https://www.epa.gov/sites/production/files/2018-03/documents/emission-factors_mar_2018_0.pdf

different petroleum products, as well as the differences generated by different fuels relative to gasoline.

## *Calculating Carbon Dioxide–Equivalent GHG Emissions*

Having calculated more accurate estimations of the emissions profiles of the full spectrum of petroleum products, the next step is to ascertain their respective impacts on the climate. To sum up the effects of different GHGs with varying climate forcing properties,[36] the Intergovernmental Panel on Climate Change (IPCC) uses global warming potentials (GWPs) over set timeframes. Table 3.2 identifies current GWP multipliers for aggregating long-lived ($CO_2$) and short-lived climate pollutants (SLCPs) over periods of 20 and 100 years.

The OCI+ and each of its three underlying models estimate $CO_2e$ levels of GHG emissions; it currently tabulates estimates for $CO_2$, methane, and nitrous

Table 3.2 Global Warming Potentials of GHGs from Oil and Gas Systems

| GHGs Emitted | Oil and Gas Sources | Estimated Lifetime | 100-Year $GWP_{100}$ | 20-Year $GWP_{20}$[a] |
|---|---|---|---|---|
| Carbon Dioxide $(CO_2)$[b] | All oil and gas combustion | n/a (centuries) | 1 | 1 |
| Methane $(CH4)$[b] | Natural gas systemwide leakage; oil and gas combustion | 12 years | 34 | 86 |
| Volatile Organic Compounds (VOCs) | Condensates, gas, light oil, petrochemicals | 5 hours to 60 days | 5 | 14 |
| Ethane $(C_2)$ | Same as VOCs | 58 days | 13 | n/a |
| Propane $(C_3)$ | Same as VOCs | 13 days | 12 | n/a |
| Butane $(C_4)$ | Same as VOCs | 1 week | 8 | n/a |
| Nitrous Oxide $(N_2O)$[b] | All oil and gas combustion | 121 days | 298 | 268 |
| Nitrogen Oxides $(NO_x)$ | All oil and gas combustion | 22 days | −11 | 19 |
| Ozone $(O_3)$ | Formed by VOCs + $NO_x$ and intermediary reactions | 22 days | n/a | 65 |
| Carbon Monoxide (CO) | Incomplete oil and gas combustion | 45 days | 5 | 19 |
| Black Carbon (BC) | Extra-heavy oil, fuel oil, diesel, and petcoke | 100 years | 460– 900 | 1,600– 3,200 |
| $CH2F2$ (R32) | Industry refrigerants | 5 years | 677 | 2,430 |
| $CHF2CF3$ (R125) | Industry refrigerants | 28 years | 3,170 | 6,090 |
| Hydrogen $(H_2)$ | Future transition fuel | 2.5 years | 6 | n/a |
| Water Vapor $(H_2O)$ | All fossil fuel combustion | Several days | <1[c] | <1[c] |

[a] The GWPs cited consider climate-carbon feedbacks. Without these feedback values, the respective 100- and 20-year methane GWPs are 28 and 84 and the $N_2O$ GWP is 264. Note that GWP for methane in particular have been increased over time, starting at 21 in 1995.

[b] These GHGs and their 100- and 20-year GWPs are currently used in the OCI+ webtool.

[c] The $GWP_{100}$ for near-surface emitted water vapor is in the range [−0.001, +0.0005] and the $GWP_{20}$ is [−0.004, +0.002].

Notes: The GWP values are rounded to the next whole number.

GHG, greenhouse gas; GWP, global warming potential; OCI+, Oil Climate Index plus Gas.

Sources: Gunnar Nyhre and Drew Shindell, IPCC AR5, https://www.ipcc.ch/site/assets/uploads/2018/02/WG1AR5_Chapter08_FINAL.pdf; "Anthropogenic and Natural Radiative

Forcing," Tables in Appendix 8A, 2013, https://www.ipcc.ch/site/assets/uploads/2018/02/ WG1AR5_Chapter08_FINAL.pdf; Climate and Clean Air Coalition; Olvind Hodnebrog et al., "Lifetimes, Direct and Indirect Radiative Forcing, and GWPs of Ethane, Propane, and Butane," 2018, https://rmets.onlinelibrary.wiley.com/doi/full/10.1002/asl.804; Leide Timma et al., "Temporal Aspects in Emissions Accounting," Energies, 2020, https://www.mdpi.com/ 1996-1073/13/4/800/pdf; Aiswarya Rogothaman and William Anderson, September 2017, https://www.researchgate.net/publication/319911732_Air_Quality_Impacts_of_Petroleum_ Refining_and_Petrochemical_Industries; Richard Derwent et al., "Global Environmental Impacts of the Hydrogen Economy," 2006, https://www.geos.ed.ac.uk/~dstevens/ Presentations/Papers/derwent_ijhr06.pdf; Steven Sherwood et al., "The Global-Warming Potential of Near-Surface Water Vapour," 2018, https://iopscience.iop.org/article/10.1088/ 1748-9326/aae018/pdf

oxide. The model provides sliders to assess the lifecycle emissions of GHGs using the GWPs of these two timeframes. Future versions of the OCI+ could incorporate others like black carbon, carbon monoxide, nitrogen oxides, and VOCs— four additional GHGs that are prevalent in the oil and gas sector.

The IPCC has increased these multipliers over time as more data becomes available.[37] Researchers now think that the current GWP approach of uniform multipliers masks the true behavior of much more potent SLCPs. As such, alternative approaches are under development.[38] Increasing methane's GWP will elevate OCI+ estimates for gas and light oil with high leakage potential. Given the high GWP for black carbon, adding this pollutant to future versions of the OCI+ is expected to raise the GHG emissions of heavy oils. In sum, while it was once thought that $CO_2$ dominated the incidence of global warming, the more scientists learn about the climate-forcing nature of short-lived GHGs, the greater the role they are thought to play in climate change. As such, given the key role methane, VOCs, black carbon, and other non-$CO_2$ emissions play in the oil and gas sector, the current OCI+ likely undercounts GHG emissions. The OCI+ estimates will be updated over time as new GWP assumptions come to the fore.

## Model Data and Uncertainty

Whether global oil and gas production and consumption return to record levels, wane, or fluctuate in the future, the increasingly complex and unconventional nature of these hydrocarbons necessitates a better understanding of the oil and gas supply chain, especially as it transforms in the future. To make its assessments, the OCI+ models require trustworthy, open-source, standardized data. The more data that is gathered and input into the OCI+, the less uncertain its estimates are. And as new and updated information is gathered, the OCI+ can be rerun to reassess results.

To date, the oil industry has generally not been forthcoming with detailed information about their operations and future plans for their expansion. Of course, data can also be purchased from firms like Wood Mackenzie and IHSCera when modeling all oil and gas operations worldwide, but these input data are pricey and cannot be made public. General public data can be obtained from news sources, industry announcements, and academic journals. And more specific oil and gas data on companies and specific operations is expanding as investors, regulators, and the public demand greater transparency and remote sensing data is relayed from satellites and other sources. Remote sensing is also working to improve government inventories where open-source data is housed and the OCI+ can access these government records.

The OCI+ also uses remote sensing data inputs. Tower-based measuring stations, drive-by detection, and flyover techniques using satellites, aircraft, and drones are collecting data on GHGs. Governments, companies, and private citizens are involved, and some of these data are open source.[39] For example, the National Aeronautics and Space Administration's (NASA's) Carbon Monitoring System (CMS) shares data with stakeholders from its satellites, flyovers, and other missions.[40] CMS is a multi-million-dollar initiative established following a 2011 pilot program to use NASA satellites to support national and international policy, regulatory, and management activities.[41] In addition to measuring methane, $CO_2$, and other GHGs, satellites also monitor gas flares, as discussed later.[42] Satellite data can also provide intel about installed equipment holding out the possibility of using machine learning to identify and count various oil and gas emission sources like compressors, storage tanks, or heat exchangers from space.

If they are open source, other data streams that the OCI+ could use—like asset-grade and blockchain data—are coming online too. These data are obtained directly from their respective sources, and this information is verifiable and immutable, so any alterations can be detected. While not all of these asset-grade data are publicly available, some new firms are planning to share their data with civil society actors.[43] These collections of data could someday offer near-instantaneous recordkeeping, greater transparency, and increased security.[44] Once they are made public, such advances in data capture can spur new thinking on calibrated regulatory approaches, streamlined management practices, and asset investment strategies.[45]

## OPGEE Data

OPGEE utilizes up to sixty data inputs, from simple entries like the country where an oil or gas field is located to hard-to-obtain information like a given field's productivity index (expressed in daily production per unit pressure).[46] Key variables include steam-to-oil and water-to-oil ratios, flaring rates, venting rates, system pressures, crude gravity, gas composition, gas-to-oil ratio, and production rates. Users do not need to identify every input value for a given inquiry because OPGEE functions with limited data. The model has comprehensive, smart defaults that fill in missing data. Reasonable estimates based on empirical analyses, technical references, and published journals are assigned to fill in for missing data.

The largest source of uncertainty in OPGEE is the lack of public information on global oil fields and their contents. Many operators in regions around the world are not required to formally publish such data. Such omissions mean that data are not consistently and routinely reported year over year, introducing uncertainty with respect to updating emissions over time. Imprecise data reporting introduces additional uncertainty. Errors in applying the model can lead to further uncertainty. In its open-source terms of use, OPGEE requires that all users that publicly release results publicly share their input data so that use of the model can be validated and any user errors can be identified and corrected.

## PRELIM Data

Refineries use heat to separate oil and gas into various hydrocarbon components. To model this process, PRELIM requires crude oil assay data that is arranged into nine standardized temperature cuts to mirror real-world refining operations.[47] Assays are typically published only when a crude variety is widely marketed, a condition that applies to only a small subset of global oils that trade globally in large volumes. The majority of crude assays are not made public, especially when oils are used domestically and not exported. Assays are also required for wet gas and condensate assets that contain some oil.

Data availability and quality are ongoing issues because PRELIM is sensitive to an oil's composition. Uncertainty ensues when assays are outdated, unreliable, or not standardized. For example, PRELIM transforms assays that do not contain certain temperature cuts, and the model uses a proxy assay when one is missing altogether. As new assays are publicly reported, PRELIM can be rerun to update the results. A global library of open-source oil assays would make it possible to run PRELIM on every oil resource (current and prospective) worldwide.

## OPEM Data

OPEM data inputs entail a detailed product yield (in volume or mass) per barrel of processed oil and gas. Product volumes are provided by OPGEE and PRELIM (as discussed earlier), and emission factors for shipping and end use are obtained from US government sources.[48] Default assumptions are entered for distances traveled to end users.[49]

The main OPEM uncertainties involve uncertainties about product output derived from PRELIM, combustion emission factors in locations beyond the United States, and transport specifications for marketing petroleum products.

All models as well as measurement systems always contain a degree of uncertainty. This does not necessarily diminish their value in decision-making. Instead, in an effort to reduce uncertainty, models are refined and new versions are developed. This is the case with OPGEE, PRELIM, and OPEM, which continue to be updated to better reflect the engineering systems they model using newly gathered data. Opportunities to fine-tune assumptions and reduce uncertainty also arise in response to publishing OCI+ estimates. For example, operators reach out to provide better data, scientists conduct comparative analyses with their satellite measurements, governments collect additional data, and civil society actors focus their attention on higher emitting assets. In other words, despite their uncertainty, OCI+ estimates promote ongoing efforts to zero in on GHG emissions.

## *The Role of Remote Sensing*

The data limitations in the OCI+ improve as more open-source data is gathered and published. Since operators are unlikely to be the guaranteed source of these data, especially if policymakers do not require data transparency, other sources are needed. Here, government scientists can fill the void with a growing stream of remotely sensed data from a variety of instruments that can observe industry practices without their involvement.

In November 2019, I was invited to present the OCI+ to a large group of NASA scientists at their workshop on CMS applications.[50] For three days, scientists interacted with researchers, like me, who are using their data on GHGs. Their reaction to the OCI+ confirmed for me that the model has added value in the panoply of GHG-estimating tools. Top-down remote sensing, bottom-up reporting, regional inventories and assessments, and lifecycle GHG models like the OCI+ are part of a multipronged mitigation approach. Taken together, these methods mutually reinforce accuracy and knowledge on climate risks from the oil and gas sector.[51]

Between satellites, planes, drones, and ground monitors, GHG remote sensing is on the rise worldwide. Table 3.3 offers a sampling of past, present, and future remote-sensing missions to measure and attribute methane and other GHG emissions.

Table 3.3 Future, Present, and Past GHG-Measuring Satellites and Aircraft (as of 2021)

| Institution Satellite Name | Country of Origin | Type (# of) Instruments | Spatial Resolution (Pixel Area) | GHGs Detected | Revisit Rate | Launch Dates |
|---|---|---|---|---|---|---|
| *Public-Private-NGO Partnership* | | | | | | |
| Carbon Mapper | US | Satellite ecosystem (2+) | 30 m x 30 m | Methane, $CO_2$ | Daily to weekly | 2023 |
| *Non-Governmental (NGO)* | | | | | | |
| GeoCARB | US | Satellite | 5 to 10 km | Methane, CO, $CO_2$ | Daily | 2022 |
| MethaneSat | US | Satellite | 400 m x 100 m | Methane, others? | 7 days or less | 2022 |
| *Commercial* | | | | | | |
| Bluefield | US | Satellites, aircraft | 20 m x 20 m | Methane | As ordered | 2019, ongoing |
| GHGSat | Canada | Satellites (>3), aircraft | 50 m x 50 m, 25 m x 25 m | Methane, $CO_2$ | As ordered | 2015, 2019, 2020, ongoing |
| DigitalGlobe | US | Satellite | 3.7, 30 m? | Methane, petcoke | As ordered | 2014, ongoing |
| *Government* | | | | | | |
| MERLIN | Germany / France | Satellite | 120 m | Methane | 28 days | 2025 |
| CO2M (Sentinel 7) | EU | Satellite | 2 km x 2 km | CO2, NO2, Methane | Global coverage in 5 days | 2025 |
| GOSAT-GW | Japan | Satellite | 1-3 km x 10 km | $CO_2$, methane, ozone, $NO_2$ | 3 days | 2023 |
| Feng Yun 3G | China | Satellite | <3 km | CO2, Methane, NO2, CO | n/a | 2023 |
| TanSat-2 | China | Satellites (constellation, 6) | < 2 km x 2 km | CO2, CH4, CO | 3-5 days | 2022 |
| MetOp | EU | Satellite | 75 km x 75 km | Methane, $CO_2$, $NO_2$ | Daily, 2 days | 2021 |
| Microcarb | EU | Satellite | 2 km x 2 km | $CO_2$ | 21 days | 2021 |

(*Continued*)

Table 3.3 Continued

| Institution Satellite Name | Country of Origin | Type (# of) Instruments | Spatial Resolution (Pixel Area) | GHGs Detected | Revisit Rate | Launch Dates |
|---|---|---|---|---|---|---|
| EnMAP | Germany | Satellite | 30 m x 30 m | $CO_2$, methane, ozone | 27 days | 2021 |
| TROPOMI | Netherlands | Satellite | 7 km x 7 km, 5.5 km x 3.5 km | Methane, ozone, $NO_2$, ammonia | Daily | 2018 |
| GOSAT-2 | Japan | Satellite | 9.7 km | $CO_2$, methane, ozone, $NO_2$ | 3 days | 2018 |
| Gao-Fen 5 | China | Satellite | 10.3 km | $CO_2$, methane | Daily | 2018 |
| Feng Yun 3D/F | China | Satellite | 10 km | $CO_2$, Methane, $NO_2$, CO | Monthly | 2017, 2023 |
| TanSat | China | Satellite | 1 km x 2 km, 2 km x 2 km | $CO_2$ | 16 days | 2016 |
| OCO-3/OCO-2 | US | Satellites (2) | 2.25 km x 1.29 km | $CO_2$ | 16 days | 2014, 2019 |
| AVIRIS-NG | US | Aircraft | 4 m – 20 m | Methane | As ordered | 2012, ongoing |
| VIIRS | US | Satellite | 375 m, 750 m, resampled to 500 m and 1 km | Gas flares, fires, lights | 4 days | 2011 |
| GOSAT | Japan | Satellites (3) | 10.5 km | $CO_2$, methane, ozone, $NO_2$ | 16 days | 2009 |
| TES-Aura | US | Satellite | 0.53 km x 0.53 km | $CO_2$, methane, ozone, $NO_2$, $N_2O$ | 16 days | 2004 |
| SciSat-1 | Canada | Satellite | >500 m | $CO_2$, methane, $N_2O$ | Annual | 2003 |
| AIRS-Aqua | US | Satellite | 13.5 km | $CO_2$, methane, ozone | 16 days | 2002 |
| EnviSat | EU | Satellite | 30 km x 60 km | Methane, $CO_2$ | 35 days | 2002 |
| Hyperion | US | Satellite | 30 m | Methane | 16 days | 2000 |

*Source*: Carbon Mapper, Personal Communication, June 19, 2021.

*Notes*: m=meter; km=kilometer.

A remote-sensing network, in theory, can pick up any emission signal anywhere on the planet at any time. In practice

e, these sensors are best suited to detecting regional hot spots, quantifying emissions rates, and guiding ground-based follow-up studies.[52] These satellite-based sensors are limited by the instruments onboard, their positioning, cloud cover, signal strength over natural background emissions in the atmosphere, array of sources clustered together, and an adequate inventory of existing emission sources. Current techniques rely on complex atmospheric inversion models to decipher and estimate sources of transport-prone, rapidly dispersed emissions amid background GHG concentrations (the natural and manmade emissions that are present).[53] Future detection methods are under development, including some designed to use methane isotopes,[54] ethane,[55] and other tracer gases.[56]

Private firms are acquiring and analyzing raw government satellite data, and this can introduce errors because raw data must first be carefully cleaned up to remove data gaps, remove outlying data, eliminate mistaken observations over the earth's reflective surface, and use extreme care in complex atmospheric computations. For example, in May 2020, the European Space Agency released a global map with elevated methane readings compared to the year prior, and a private data firm set about analyzing and selling the data.[57] When I asked a renowned expert about the validity of the private firm's data, he said that companies advertise methane satellite data, but buyers have to take care because firms' estimates can be off by a factor of three or more.[58] In response, data providers are improving their analytics and scientists are automating methods that speed up satellite data analysis with a user-friendly inversion tool for cleaned-up satellite data.

The OCI+ currently uses satellite data and other remote-sensing data as model inputs, including from a satellite for flaring data shared by NASA and the National Oceanic and Atmospheric Administration (NOAA) and another satellite called the Tropospheric Monitoring Instrument (TROPOMI) that maps methane hot spots. The NASA-NOAA satellite known as the Visible Infrared Imaging Radiometer Suite (VIIRS) was originally launched to observe light sources on Earth like city lights and forest fires. In 2015, scientists discovered that VIIRS also detects the radiant emissions from gas flares used by petroleum systems.[59] The TROPOMI satellite was developed jointly by the Dutch and the European Space Agency. In addition to using satellite data as inputs, the OCI+ is being used to improve remote sensing by improving GHG inventories of existing oil and gas emissions sources that scientists require in their inversion models. And the OCI+ is also being used to affirm satellite readings during oil and gas events like blowouts.[60] Thus, not only is remote sensing being used to swap out new remote-sensing data for model inputs that were previously filled in using defaults, but also the OCI+ is being used to calculate and verify the remote-sensing data.

## Spotting Flares from Space

Gas flaring has a unique heat signature that can be detected via the VIIRS satellite.[61] Nearly 7,500 individual flare sites were detected worldwide in 2012, burning an estimated 143 billion cubic meters of gas (nearly 4 percent of global gas production).[62] Despite efforts to curtail this wasteful practice (discussed in chapter 7), in both 2018 and 2019, global gas flaring was up 3 percent a year.[63] Flares in the United States, Russia, Iraq, and Iran account for the greatest volumes of natural gas being burned worldwide.[64] Increased fracking (in the United States) and political unrest in regions around the world may be the impetus of increased flaring when normal operations and shipping arrangements are upended. Flaring can indicate a state in crisis or an economic decision to prioritize more profitable oil over less profitable gas production.

The data obtained by VIIRS provides the flaring-to-oil ratio input that OPGEE uses. Overlaying VIIRS data on a map of OCI+-modeled GHG emissions creates a visualization of the climate risks generated by flaring.[65] NASA now conducts an ongoing monitoring, reporting, and verification (MRV) system of global flaring sites through CMS using VIIRS to remotely sense time-series data.[66] Work is underway to arrange for these data to offer site-specific tracking of flares, including estimates for the volume of gas burned and the ability to distinguish routine versus nonroutine flaring.[67]

## Mapping Methane Hot Spots

The OCI+ also benefits from data generated by space-based sensors that can be used to measure methane hot spots. A few years ago, I received a call from a gas company that had used an early version of the OCI+ to show how much lower their GHG emissions were than their competitors'. (While I was pleased that they found the model useful, I voiced my concern because they omitted methane in the OCI+'s application to rank their own and others' operations—a big measurement oversight.). In our meeting, I was told that the company's competitors were covertly releasing methane. After BP's Deepwater Horizon oil spill in 2010, regulations were adopted limiting the amount of gas that companies could flare and vent from US offshore platforms.[68]

But detecting violations of this law proved challenging. A Coast Guard boat equipped with Light Detection and Ranging (LiDAR) equipment, remote sensing that uses infrared lasers to detect methane, would make a single daily pass to record visible flaring and invisible methane from each platform.[69] But some operators wanted to discard more gas than permitted. To avoid a citation, as soon as the coast was clear, they turned off the flare's pilot ignition and expelled invisible gas into the air, effectively avoiding detection. Breaking this law saved companies real money. Not only could they produce more oil but also they avoided paying royalties on the gas they emitted.

In part to remedy this problem, instead of relying on one-off LiDAR passes, regulators now use satellites to map methane hot spots, while the OCI+ is mapping these hot spots alongside global oil and gas assets.[70] The TROPOMI satellite began reporting in 2018.[71] For example, TROPOMI reported leakage from a gas well blowout in Ohio in early 2018, and together with a tracer transport simulation, scientists quantified the emission rate and total methane release from the accident.[72] But TROPOMI's coverage is course. Like a wide-angle camera lens, it cannot zoom in and pinpoint equipment leaks. Other satellites in this emerging remote sensing ecosystem, like Carbon Mapper, have "portrait" lens capabilities with higher resolution that can measure individual leaks.[73]

Historically, problems existed quantifying methane located over water. The ocean absorbs sunlight, preventing satellites from detecting the light reflecting off the earth's surface. Scientists have developed new "glint mode" observations— satellites equipped with sensors that point to bright spots over oceans where solar radiation is directly reflected off the earth's surface.[74] This means that one-third of current oil and gas production, most refineries, and all shipping conducted offshore and near coastlines are now detectable.

## Tracking GHGs from Above

In addition to satellites, low-flying, high-resolution, access-limited instruments can be used to verify oil and gas emissions. This localized approach can help reconcile top-down with bottom-up GHG estimations.[75] Using aircraft to fly over target sites and drones to collect data can police super emitters in targeted regions.[76] However, time and resources limit researchers' abilities to locally examine widespread areas. Regional conflicts, political tensions, and other reasons may also prevent flyovers in regions that contribute most heavily to GHG emissions. Still, pairing flyover data with satellite data, emissions inventories, and OCI+-type modeling offers a fairly reliable way to see the big picture.

Top-down sensors, bottom-up reported data, and GHG-estimating models can work in tandem to assess oil and gas emissions. It will take a network of different measurement, reporting, and computational tools, each with their different capabilities, to quantify and attribute GHG emissions to different sources. None of these methods are foolproof and each has associated uncertainties. Taken together, however, various emission detection and assessment methods can be mutually reinforcing. The OCI+ model not only employs reported and remote sensing data in its calculations but also can be used to fill data gaps, provide intel in areas where data transparency is lacking, and run scenarios (where data does not yet exist) to identify future sensor and monitoring opportunities that may arise.

## Jointly Modeling GHG Emissions and Other Air Pollutants

The OCI+ was originally developed to account for the lifecycle emissions of GHGs. But many GHGs simultaneously pollute the air and pose serious health and social justice risks, especially for low-income individuals. These copollutants increase mortality rates, raise health care costs, damage crops, lower agricultural yields, erode infrastructure, and impair visibility. Likewise, many air pollutants contribute to climate change. These two manmade atmospheric concerns—air pollution and climate change—are inextricably linked and exacerbate one another.[77]

Reducing air pollution can also protect the climate, just as reducing GHG emissions can improve air quality. For example, particulate matter (PM) from the partial burning of diesel fuel in engines is a known carcinogen that sickens and kills people.[78] PM (in the form of black carbon[79]) is also a powerful GHG that circulates in the atmosphere, ending up in the North and South Poles, landing on ice and snow, darkening surfaces, and causing less sunlight to be reflected back into space, thus contributing to global warming. As darkened snow and ice melt, such warming increases. Wetter, warmer weather in turn spurs plant growth and casts shadows that further darken the earth's surface, leading to additional warming. Reducing PM levels, then, lessens morbidity and mortality from air pollution as well as GHG emissions.

The oil-fueled petroleum sector reportedly contributes the most to net global warming effects over a twenty-year timeframe, due to high emissions levels of $CO_2$, methane, ozone-forming air pollutants (like VOCs and nitrogen oxides), and black carbon (PM).[80] Compared to other economic sectors, like power generation, oil and gas have lower emissions of pollutants that cool the atmosphere in the short term, including sulfates, aerosols, and organic carbon. In other words, oil and gas have the largest effects on both air quality and climate change in the short term.

Comanaging climate change and air quality is a high priority, especially in Asia where regional pollution is driving the policy agenda more so than climate change. The 2008 Olympics in Beijing spotlighted the city's intense air pollution, and the government adopted Los Angeles's 1984 strategy to simply halt driving, leading to less smog and blue skies during the games. But the pollution returned thereafter.[81] Today, China's cities suffer some of the world's worst air pollution, which compete closely with India's metropolitan centers.[82] Elsewhere, the Middle East (particularly Saudi Arabia and Kuwait), with their major oil and gas operations, have major air pollution problems. Developing a unified front to jointly reduce air pollutant and GHG levels could be more effective than focusing on climate change alone.

Several air pollutants are already included in one or more of the OCI+'s underlying models. Once a pollutant is computed in all of the three models, the OCI+ can model out corresponding estimates (see Table 3.4).

## Massive Mismatch in Climate Risks

The OCI+ highlights three central facts. First, massive tonnes of $CO_2$e emissions will be emitted from consuming oil and gas in the decades ahead, much more than the atmosphere can safely accommodate, as shown in Figure 2.2. Second, different barrels of oil and cubic feet of gas have widely varying GHG emissions, and it is the industry's responsibility to reduce the GHG emissions intensity from the various hydrocarbons they produce, process, refine, and ship. Third, petroleum's varying climate impacts are not currently recognized or priced into the market value of competing crudes, natural gas, or their products. As such, we need to both consume less oil and gas overall and reduce the emissions intensity of each barrel and cubic foot we continue to use.

Analysis of thousands of oil and gas resources modeled to date reveals that their emission differences are far greater than currently acknowledged. Large emissions ranges exist whether values are calculated per BOE, per megajoule of products, or per dollar value of products, and these emissions ranges are expected to grow as new, unconventional oils are identified.[83]

Large variations in GHG emissions exist upstream, where the oil with the highest emissions intensity has approximately fifteen times the emissions of the lowest-intensity oil.[84] These differences are also evident in upstream gas production and processing emissions, which vary by a significant order of magnitude.[85] Midstream refining emissions exhibit large variations too, as the oil with the highest emissions intensity produces approximately *fourteen times* the emissions of the lowest-intensity oil.[86] The differences stemming from the downstream shipping of GHGs are estimated at a factor of two for oil and compressed gas and significantly higher for LNG.[87]

These supply-side climate impacts are the responsibility of the oil and gas industry.[88] In the *World Energy Outlook 2018*, the International Energy Agency (IEA) used the OCI+ and its underlying models to estimate supply-side emissions intensities for all global oil and gas resources. Figure 3.7 illustrates the broad range of climate risks posed by the oil and gas industry.

## Table 3.4  Air Pollutants in OCI+ Models (as of 2020)

| Air Pollutant Emitted | Oil and Gas Sources | OPGEE | PRELIM | OPEM |
|---|---|---|---|---|
| Carbon Dioxide | Oil and gas combustion and system leakage | Y | Y | Y |
| Methane ($CH_4$) | Gas leakage and oil and gas combustion | Y | Y | Y |
| All VOCs[a] | Condensates, gas, light oil, and petrochemicals | Y | Y | N |
| Nitrous Oxide ($N_2O$) | All oil and gas combustion | Y | Y | Y |
| Nitrogen Oxides ($NO_x$) | All oil and gas combustion | N | Y | N |
| Ozone ($O_3$)[c] | Photochemical reaction of VOCs and $NO_X$, with methane as intermediary | N | Y | N |
| Carbon Monoxide (CO) | Incomplete oil and gas combustion | Y | Y | N |
| PM[b] | Extra-heavy oil, fuel oil, diesel, and petcoke | N | Y | N |
| Sulfur Dioxide ($SO_2$) | High-sulfur oil and gas | Y | Y | N |
| Hazardous Air Pollutants (HAPs)[d] | Impurities in oil and gas | N | Y | N |
| Ozone Depletion (CFC-11) | Refrigerants in oil and gas processing and insulation in oil and gas systems | N | Y | N |

[a] VOCs include ethane, propane, butane, and a long list of other hydrocarbons contained in oil and gas.

[b] Includes $PM_{2.5}$ and $PM_{10}$.

[c] Photochemical ozone-forming potential is considered since ozone is not directly emitted and is formed via photochemical reactions of VOCs and $NO_x$.

[d] Includes several heavy metals, aromatics, and other chemical compounds that are known and suspected carcinogens and ecotoxins that pollute air and water (also known as air toxins).

OCI+, Oil Climate Index plus Gas; OPGEE, Oil Production Greenhouse Gas Emissions Estimator; OPEM, Oil Products Emissions Module; PRELIM, Petroleum Refinery Lifecycle Inventory Model.

*Sources*: Author's inputs based on documentation from OPGEE, PRELIM, and OPEM models. For more information, see Congressional Research Service, "Methane and Other Air Pollution Issues in Natural Gas Systems," September 17, 2020, https://fas.org/sgp/crs/misc/R42986.pdf

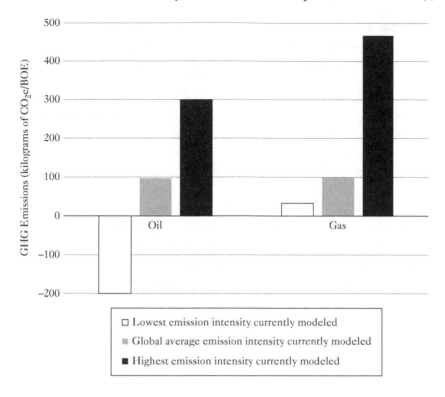

**FIGURE 3.7** Estimated Ranges of Currently Modeled Emissions Intensities of Global Oil and Gas Supplies

*Note:* Estimates assume 100-year GWP. Values increase substantially using 20-year GWPs. See OCI+ Web Tool for details.

*Sources:* International Energy Agency, *World Energy Outlook 2018,* Table 11.1, Figure 11.6 (for global average), https://www.iea.org/reports/world-energy-outlook-2018/oil-and-gas-innovation; OCI+ Preview BETA Web Tool, (for lowest- and highest-emission-intensity oil and gas). BOE, barrel of oil equivalent; $CO_2e$, carbon dioxide equivalent; GHG, greenhouse gas.

Although it is outside the direct control of the oil and gas industry, the OCI+ estimates a significant variation in the end-use (Scope 3) GHG emissions of various oils, gases, and related products. The oil with the highest emissions intensity from end-use consumption has approximately 65 percent more emissions than the lowest-intensity oil does, and for gas the corresponding variation is over 100 percent.[89]

## From a Theoretical Model to Real-World Impact

The OCI+ traces the overall GHG footprints for different oils and gases, which are large and could get bigger over time as unconventional oil and gas development grows. Several companies have engaged in OCI+ development and have helped improve the underlying models, including Chevron and Exxon. Other companies have put the OCI+ into action. For example, Saudi Aramco, Norway's Equinor, and Texas-based Southwestern have used the OCI+ models for strategic planning. While it is unclear to what extent individual companies have developed their own models, global energy consultancies—such as Rystad, Baker Hughes, and Wood Mackenzie—are using the OCI+ to advise their clients in the oil and gas industry.

OCI+ applications by industry, government, and civil society actors affirm the need to better assess and differentiate oil and gas GHG emissions. The most climate-intensive oils and gases, covered in the next chapter, require special attention from operators and investors, government policymakers, and civil society alike. Assessing the climate footprints of oil and gas resources is essential to finding a way to balance the enormous economic value that petroleum delivers with the equally massive threats these hydrocarbons pose to the global climate.

*4*

# Curbing the Climate Footprints
# of Oil and Gas

GIVEN HOW DRASTICALLY the emissions and potency of different oils and gases can vary and how prominent a market role less clean, unconventional varieties are likely to play, there is a clear need for creative solutions to combat climate change. And because hydrocarbons are not going away anytime soon, supply-side solutions by the oil and gas industry need to be part of the equation.

In the summer of 2012, I joined a group of energy experts from think tanks on a research trip to Fort McMurray in Alberta, Canada, to tour several oil sands operations. As we flew into the region, jet-black seas of solid petroleum coke (petcoke) unfolded beneath us.[1] The look of scorched Earth reinforced my climate concerns about this solid, residual byproduct of oil. Throughout our visit, my colleagues asked interesting questions on energy security. In contrast to their questions on political economy, I took a different—technical—approach aimed at fully accounting for all of the greenhouse gas (GHG) emissions in a barrel of bitumen.

Earlier that year, I had begun to investigate fuel-grade petcoke because I did not think that standard calculations of the lifecycle GHG emissions of extra-heavy oil did justice to the impact of this little-known, highly polluting, bottom-of-the-barrel petroleum byproduct.[2] In permit documents, it was claimed that the controversial new oil sands pipeline—Keystone XL—would have as much as 20 percent higher GHG emissions than conventional oil does.[3] But this figure seemed low. It only considered a portion of the emissions, whether inadvertently or on purpose. The emissions of the petcoke byproduct the oil sands produce were omitted, even though petcoke's combustion alone emits more GHG emissions than coal does, elevating oil sands' lifecycle emissions.[4]

At the time, little Albertan petcoke was finding its way to market, given the long distance to mostly Asian end users and degraded nature of this fuel source. Nevertheless, oil sands operators were required by Alberta law to preserve their carpeted petcoke for future consumption, as something akin to a strategic petroleum reserve.[5] In other words, none of this petcoke is considered *permanently* sequestered, even if it is used for reclamation.

The petcoke problem only got worse when fracking took off in North Dakota, after which light tight oil was mixed in to dilute Canadian bitumen (or dilbit). Instead of upgrading bitumen in Alberta and storing petcoke, dilbit slurries were shipped to US refineries, shifting the petcoke problem to Illinois, Indiana, and Texas, where it could be readily exported overseas to places with less stringent environmental protection measures than the United States. Consequently, US petcoke exports, which fluctuate from month to month, have increased 60 percent between January 2012 and 2020.[6] While petcoke exports declined along with overall oil demands during the pandemic, they shot back up 30 percent between August 2020 and April 2021.[7]

Petcoke is one example of how easy it is for oil and gas sector emissions to hide in plain sight and go unattended. The power of lifecycle accounting is that it can assess each stage in the oil and gas value chain—production, processing, refining, shipping, and end uses. In this chapter, the Oil Climate Index + Gas (OCI+) is used to identify where the oil and gas sector's GHG emissions are highest and how to reduce them. While there is no silver bullet to eliminate this large climate footprint, numerous opportunities exist to shrink the industry's intensifying climate risks. This is a hopeful story. There are even more solutions available than we originally imagined.

## *Parsing Real-World Solutions*

Most people do not have a clear sense of where GHG emissions come from. Consumers, especially motorists, have long been told that they are the main culprits of climate change. But full responsibility does not rest on the public's shoulders. Figure 4.1 shows that, depending on the resource selected and the operations employed, up to half of total lifecycle GHG emissions in some oil and gas categories may be emitted by petroleum producers and refiners themselves. Supply-side oil and gas GHG emissions arise when producers, refiners, and shippers consume fossil fuels in their supply chains, leak GHGs systemwide, operate fields and equipment beyond their useful lifetimes, transfer properties or trade assets to negligent operators, and market dirty byproducts like petcoke and other heavy residual fuels.

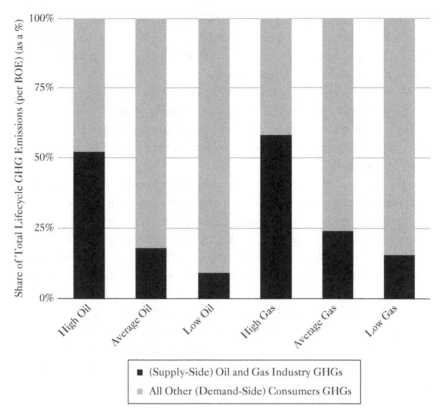

**FIGURE 4.1** Industry versus Consumer GHG Emissions for Select Oil and Gas Assets

*Notes:* High GHGs assume 20-year GWPs and low GHGs assume 100-year GWPs. The supply-side GHG numbers for average oil and gas use are sourced from the 2018 International Energy Agency report and the end-use GHGs are sourced from the US Treasury 2017. High-GHG oil and gas include petcoke and LNG shipping in supply-side estimates. Even lower supply-side GHGs using sequestered $CO_2$ EOR (with negative production GHGs) are not plotted. BOE, barrel of oil equivalent; GHG, greenhouse gas; LNG, liquefied natural gas.

*Sources:* Author's estimates, Oil Climate Index + Gas Preview Web Tool, 2020, https://dxgor-don.github.io/OCIPlus/; International Energy Agency, *World Energy Outlook 2018*, https://www.iea.org/reports/world-energy-outlook-2018/oil-and-gas-innovation; John Horowitz et al., "Methodology for Analyzing a Carbon Tax," US Treasury, Office of Tax Analysis, Working Paper 115, January 2017, https://www.treasury.gov/resource-center/tax-policy/tax-analysis/Documents/WP-115.pdf

## *The Origins of Demand-Side Strategies*

Wearing a tan sweater, then-president Jimmy Carter gave his first fireside chat two weeks into his administration in February 1977.[8] He stressed that, in response to the nation's real energy problem and what he termed its "permanent" oil and gas shortage, Americans needed to make sacrifices to voluntarily conserve petroleum. If citizens just turned down their thermostats and took personal steps to reduce their oil and gas demands, Carter claimed that US energy savings would exceed total energy imports from foreign countries.

These were highly uncertain times for the oil and gas sector amid (temporarily) dwindling domestic supplies and unstable imports. Carter's energy address—bookended by two supply shocks to global oil markets—convinced my generation to prioritize demand-side petroleum strategies like driving less, biking more, and purchasing more efficient cars.

Like others, I continued to pin my hopes on demand-side oil and gas solutions when climate concerns rose to the fore in the late 1980s.[9] For example, in 1989, while working at the Lawrence Berkeley National Laboratory, I developed a novel demand-side policy that was the brainchild of my mentor, Dr. Arthur Rosenfeld, the "godfather of energy efficiency,"[10] and his colleague Amory Lovins. Our idea was to offer a special kind of rebate (feebate) to consumers who bought fuel-efficient cars paid for by fees on those who bought inefficient models. This revenue-neutral, self-financing, incentive-based approach was an alternative to raising the gasoline tax, an unpopular disincentive that most politicians shunned.[11] The US Environmental Protection Agency (EPA) funded my research, and after a stint at the California Senate Office of Research, I helped craft the bill and find bipartisan cosponsors and sought to get the legislation passed in California. After our DRIVE+ feebates bill passed in the California legislature—but was vetoed by then-governor George Deukmejian—the effort spawned the adoption of similar bills in other states and countries.[12]

Many experts continue to focus on demand-side strategies for achieving the United Nations' (UN) Sustainable Development Goals (SDGs). The UN, for example, calls for consumers to waste less; use energy efficiently; bike, walk, or use public transportation; recycle; plant trees; and avoid using plastic bags.[13] The Intergovernmental Panel on Climate Change (IPCC) calls for exponentially lower oil and gas demand (87 percent and 74 percent, respectively) by 2050 to avert the imminent consequences of continued atmospheric warming.[14]

Demand-side strategies might have succeeded alone in reducing GHGs in a world where the supply of oil and gas was actually in terminal decline. But up until the 2020 pandemic, oil and gas demand was on the rise globally. Having hit a low in the second quarter of 2020, oil consumption is projected to increase over 20 percent, hitting an all-time high by the end of 2022.[15] Global gas consumption is projected to rise as well. If such demand growth resumes, it will take complementary demand-side and supply-side strategies to reduce oil and gas sector emissions. In other words, it is not enough to battle back petroleum demands;

abundant oil and gas availability calls for serious supply-side strategies to combat climate change.

## Zeroing in on Supply-Side Strategies

The oil and gas industry has played its hand shrewdly, allowing countless governments and nongovernmental organizations (NGOs) to reiterate Carter's core message that simply using less will solve the world's energy (and climate) problems. The problem is that, over the past forty-plus years, global demand for oil and gas has continued to rise, not fall.

This calls for a change in approach. We can no longer wait for motorists to alter their travel behavior and vehicle purchase habits as long as the oil and gas industry sits on the sidelines. Drivers compose just part of the market. Countless consumers worldwide demand a multitude of petroleum products beyond gasoline, including those who fly (jet fuel); order packages for delivery (diesel); heat their homes (heating oil); run factories (fuel oil); generate electricity (natural gas); use cook stoves (liquefied petroleum gas [LPG]); do house renovations (roofs and insulation); pave roads (asphalt); and purchase everyday products like medicine, paint, and fertilizer (petrochemicals).

The oil and gas industry lies at the center of petroleum's economic appeal and is crucial to reducing petroleum's climate footprint. The International Energy Agency (IEA) recently underscored this way of thinking when it wrote, "Minimizing emissions from core oil and gas operations should be a first-order priority for all, whatever the transition pathway."[16] Furthermore, the IEA has identified cost-effective measures to reduce the emissions intensity of delivered oil and gas operations overall by 45 percent by 2030.[17] Further, the IEA assumes three important facts. First, oil and gas are not going away as global populations grow and the worldwide economy expands. Second, oil and gas are integral to existing energy and economic systems. And third, the solution necessarily involves reducing oil- and gas-related GHG emissions in line with international targets.

While there are openings for GHG reductions throughout the oil and gas sector, different industry actors are better positioned to pursue some strategies than others. To be successful, industry actors need to match up with the climate solutions that best suit their strengths. In that spirit, the OCI+ can help identify ways forward for shrinking the petroleum industry's climate footprint.

Table 4.1 reviews numerous industry strategies (with corresponding OCI+ estimates) that responsible parties can employ to reduce the intensity of GHG emissions—including ones for upstream producers; midstream refiners and processing facilities; and downstream shippers, traders, and retailers. Specific

Table 4.1 Supply-Side GHG Mitigation Strategies by Industry Subsector

| Oil and Gas Supply-Side Strategy | Upstream Companies | Midstream Companies | Downstream Companies | Est. Reduction in GHG Intensity (kilograms of $CO_2e$/BOE) |
|---|---|---|---|---|
| *Leading Upstream Strategies* | | | | |
| Eliminate routine gas flaring and venting in production* | X | | | −210 |
| Use renewable electricity in upstream inputs* | X | | | −70 |
| Use only manmade (not natural) $CO_2$ in EOR projects | X | | | −290 |
| Pump and reuse water most efficiently | X | | | −50 |
| *Leading Midstream and Downstream Strategies* | | | | |
| Employ green hydrogen in refining | | X | | −40 |
| Lock up carbon in noncombustible end uses | | X | X | −180 |
| Sequester heavy residuals for reclamation and reuse | | X | X | −90 |
| Reduce GHG emissions from LNG conversion and shipping | | X | X | −120 |
| *Leading Lifecycle Oil and Gas Strategies* | | | | |
| Decommission legacy assets with highest GHG emissions | X | X | X | −460 |
| Minimize fossil fuel inputs in all oil and gas operations* | X | X | X | > −140[a] |
| Employ leak-free equipment in all oil and gas operations* | X | X | X | −130 |
| Operate permanent, leak-free CCS systems | X | X | X | [b] |
| Avoid operating in sensitive ecosystems | X | X | X | −170 |

[a] Modeled substituting solar steam for natural gas, but the GHG reduction potential is larger when all fossil fuel inputs are minimized systemwide.

[b] Wide range in GHG reductions depending on CCS specifications.

*Notes:* All values rounded. Strategies not additive because they apply only to individual BOEs of the particular resource modeled, except systemwide actions noted with *. Assumes $GWP_{20}$ applied (methane = 86), except for decommissioning, which uses $GWP_{100}$.

BOE, barrel of oil equivalent; CCS, carbon capture and storage; $CO_2$, carbon dioxide; $CO_2e$, carbon dioxide equivalent; EOR, enhanced oil recovery; GHG, greenhouse gas.

*Source:* Author's estimations using OCI+ Preview Web Tool, https://dsgordon.github.io/OCIPlus/

examples are detailed in subsequent sections regarding leading ways to mitigate GHG emissions each step of the way.

## Mitigating Upstream GHG Emissions

There are several promising strategies for cutting GHG emissions from upstream oil and gas production. Four representative examples include eliminating routine flaring and venting, preferentially using renewable alternatives to generate onsite electricity, using only manmade carbon dioxide ($CO_2$) in enhanced oil recovery (EOR) projects, and reusing and pumping produced water more efficiently.

### *Eliminate Routine Flaring and Venting*

Flaring—the burning of unwanted gas—is an extremely carbon-intensive industry practice. This technique is employed upstream (and elsewhere in the supply chain) to provide backup safety protection if there is too much gas to safely handle and pressure builds up in the system, from wellheads, refineries, and elsewhere. Venting—the intentional release of gas—is a highly methane-intensive industry practice. Numerous devices are designed upstream and elsewhere systemwide through the supply chain to release gas from pipelines, storage tanks, valves, and other equipment, including pressure relief valves, tank hatches, and pressure-sensing (pneumatic) controllers.

Because $CO_2$ is a less potent GHG than methane, flaring poses less immediate climate risk than venting gas does. Nevertheless, flares are prone to overuse and are often poorly maintained, which can increase methane emissions and black carbon formation. While flaring and venting cannot be entirely eliminated, according to OCI+ calculations, strictly limiting routine flaring and venting in the production process can reduce lifecycle emission intensity by upward of an estimated 40 to 50 percent.[18] This is the case offshore of Norway, where operators routinely reinject their gas rather than flaring or venting it.[19]

However, most oil and gas operators or governments do not routinely report the volumes and composition of gas that they flare and vent. Notably, a given country's flaring and venting practices can shift markedly over time. Between 2014 and 2018, for example, the volumes of gas flared in Libya and Iran increased 62 and 42 percent, respectively, while Kazakhstan reduced its flaring volume by 49 percent, as reported by the Visible Infrared Imaging Radiometer Suite (VIIRS) satellite.[20] While volumes of vented gas are not verifiably reported at present, future detection and quantification could be made possible by the presence of more methane satellites, as discussed in chapter 3.

Companies alter their flaring and venting activities due to a variety of factors, including infrastructure, operational, economic, geopolitical, or regulatory situations.[21] For example, when infrastructure is lacking and there is not sufficient pipeline takeaway capacity, operators may resort to flaring and venting their gas. Or in an attempt to maximize profits, old or malfunctioning equipment like inefficient flares or faulty pressure relief valves remain in place. Likewise, flaring and venting can run amok when local conditions are unstable and careful gas management is difficult due to ongoing disruptions.

Some of these situations, like when local conflicts upset operations, are out of companies' direct control. Yet cleaner alternatives to flaring and venting do exist. For example, Chevron replaced old, inefficient flares located off the shores of Angola with modern flares; began shipping gas onshore for treatment; and then sent it on to a gas-to-liquid (GTL) plant for marketing.[22] Even in places where gas collection equipment is not available, flaring is not the only recourse. In such cases, mobile storage devices can be deployed to gather, move, and empty gas contents into permanent gathering systems. North Dakota operators, for instance, have used mobile storage when no pipelines were available.[23] New efforts are underway to store excess gas underground until it can be used onsite or someone wants to buy it.[24]

Another solution involves using the captured natural gas to power microturbines. Small, stationary, onsite power-generating sources can supply onsite electricity, provide local power, or supply regional electric utility grids.[25] Microgrid technologies can handle gases with wide-ranging compositions, such as wet gas that is rich in ethane or sour gas from high-sulfur fields. That said, microturbines that use these off-spec gases need emission controls to further reduce local air and climate pollutants.

In practice, eliminating routine flaring and venting can make companies money over time, but there is often resistance to spending manpower and laying out capital up front. The onus is often on policymakers to require cleaner solutions. The success of the VIIRS satellite to identify and quantify emissions from flares around the globe highlights the need to monitor, report, and verify whether operators are following best practices or not. This is one of the motivations for deploying a host of new methane and other GHG satellites to identify and promote policymaking.

### *Use Renewable Electricity Upstream*

Drilling, extraction, and surface processing equipment require a lot of energy for heat, steam, and power. Although fossil fuels like diesel, natural gas, and even petcoke have long supplied upstream energy, renewable alternatives are increasingly available. This consideration is even more relevant because many operations are remotely located and renewable energy could be a more reliable option than unreliable power grids.

Norwegian oil and gas platforms in the North Sea, for example, are retooling to be powered by electricity generated onshore from hydropower. This means that natural gas can be reinjected instead of used to generate power, which reduces its lifecycle emission intensity by an estimated 15 percent, according to OCI+ calculations.[26] Renewables and low-GHG electricity could also be employed for other uses beyond upstream activities throughout the oil and gas supply chain. Other examples are discussed in more detail later.

### *Use Only Manmade $CO_2$ for EOR*

EOR operations using $CO_2$, discussed in chapter 2, are a common way to increase oil mobility and facilitate extraction.[27] In 2017, there were nearly 100 EOR projects employing $CO_2$ worldwide, which produced an estimated 1.6 million barrels per day of oil.[28]

This practice has been around for decades. Fifty years ago, the first such commercial EOR project in West Texas recovered $CO_2$ from the exhaust of regional plants that were processing natural gas.[29] While the pilot program was technically successful and environmentally beneficial, by the late 1970s, it was replaced by pipelines that gathered $CO_2$ from natural underground sources in Colorado and New Mexico. Today, using natural $CO_2$ for EOR is the norm, and natural sources of $CO_2$ are more commonly used than industrial sources. Existing infrastructure in place to use natural $CO_2$ makes it cheap and easy to maintain the status quo. Dozens of companies are engaged worldwide in $CO_2$ EOR, but only a handful of them aspire to be climate leaders and may wholly convert their practice to use manmade $CO_2$, as discussed later. This is regrettable because the source of $CO_2$ matters considerably for climate change. After all, the OCI+ estimates that using manmade $CO_2$ can reduce lifecycle GHG intensity by an estimated 50 percent.[30]

Collecting and burying manmade $CO_2$ from waste flue gases or direct air capture can result in net-negative GHG emissions, while dislodging naturally stored $CO_2$ buried in subsurface reservoirs increases GHG levels.[31] The former can have major climate benefits. Take the United Arab Emirates, for example, where the Abu Dhabi National Oil Company (ADNOC) currently captures nearly 1 million metric tonnes of $CO_2$ annually from its Emirates Steel facility, compresses and dehydrates it, and sends it by pipeline for EOR operations at the Murban Bab and Rumaitha oilfields.[32] ADNOC's practice actually decreases $CO_2$ levels in the atmosphere, while ExxonMobil is essentially removing naturally stored $CO_2$ and then redepositing it, which does not decrease GHG levels.

Contrast ADNOC's use of captured manmade carbon, for example, with ExxonMobil's LaBarge Gas Field in Wyoming that contains 65 percent $CO_2$ and 21 percent methane (some of the lowest hydrocarbon content of any natural gas produced worldwide).[33] Since 1986, ExxonMobil has produced LaBarge's highly acidic gas, stripped off its $CO_2$, reinjected it for EOR, and pumped any excess $CO_2$ underground. The EPA has implemented a monitoring, reporting, and verifying (MRV) plan to try to prevent $CO_2$ emissions from leaking and being vented.[34] (Admittedly, it is unlikely that ExxonMobil would produce unprofitable LaBarge gas at all if its profitable EOR operations were not located nearby.)

Even though EOR requires significant fossil fuel inputs, if done efficiently using industrial (manmade) sources of $CO_2$, EOR not only reduces GHG levels but also could spur market uptake of large-scale carbon capture and storage (CCS) by permanently storing $CO_2$ deep underground in geological formations.[35]

### Reuse and Pump Produced Water More Efficiently

Water exists naturally in oil and gas reservoirs at varying volumes and with different characteristics, depending on geologic conditions. Moreover, some hydrocarbon extraction techniques involve injecting water, which further affects the amount and quality of produced water. Once it is brought to the surface, water must be separated from oil and gas for hydrocarbons to be further processed. Pumping all of this water during oil and gas extraction is also ripe for ecofriendly reforms. Fracking in West Texas's Permian Basin not only produces oil and gas but also generates an OCI+-estimated 9 million barrels of water per day, enough to fill nearly 320,000 Olympic swimming pools a year.[36] While different basins' water contents vary widely, there are many other waterlogged oil and gas fields worldwide that contain upward of 100 barrels of water for every barrel of oil.[37] In fact, the US petroleum industry produces more water than oil, condensates, or gas.[38] Entrained water is undesirable in part because it weighs down each barrel of oil and it requires extra energy to lift, separate, and reinject (if regulations allow).[39] Pumping water very efficiently can reduce total lifecycle emission intensity by an estimated 8 percent, according to the OCI+.[40]

If produced water cannot be reinjected into the reservoir, contaminants like dissolved salt, grease, and radioactive elements need to be removed before it can be recycled or reused as potable water.[41] However, when such water is reinjected, pumping large volumes of produced water back underground appears to be linked to earthquakes in some places that had never previously experienced them.[42] This is thought to be the case because reinjecting wastewater increases the pressure exerted on rocks in the reservoir. If faults exist, this can exacerbate underground instability, leading to earthquakes.[43]

The prevalence with which watery oil and gas are produced is on the rise, so avoiding assets with high water content may not always be practical. As the industry ventures into more complex formations and proliferates unconventional techniques like fracking, and even as conventional wells age, greater volumes and varying-quality water will be produced.[44] Moreover, efforts are underway to reclaim valuable impurities in produced water to make it profitable. For example, produced water in some shale formations contains lithium, which some producers seek to market for manufacturing rechargeable batteries.[45] Such breakthroughs could further increase future volumes of produced water and drive up GHG emissions. This trend underscores a pressing need to first install high-efficiency pumps and employ renewable electricity instead of diesel fuel to run treatment equipment.

## Mitigating Midstream and Downstream GHG Emissions

There are also many novel ways to cut GHG emissions from gas processing, oil refining, and shipping. Five of the most promising methods include employing green hydrogen, paying a premium for low-GHG crudes, locking up carbon in noncombustible end uses, sequestering heavy residuals for reclamation and reuse, and reducing GHG emissions from liquefied natural gas (LNG) conversion and shipping.

### *Employ Green Hydrogen in Refining*

Refinery operations have not changed much in the past century and are long overdue for a revamp.[46] But refineries are the workhorse of the industry, churning out millions of barrels of petroleum product each day for constant consumption. They cannot afford downtime to renovate. And if they do, competitors will pick up the slack, yielding profits to other companies. Refineries also operate on low-cost margins (with high volumes), which means that capital expenditures to existing facilities are actively avoided.

Although revamps are not free or easy, refineries can cut their GHG emissions by over half by installing new technology to cogenerate electricity using waste heat, capture carbon from flue gases, and replace steam methane reforming (SMR) with renewable hydrogen.[47] Emissions savings can add up quickly given that the industry refines roughly 100 million barrels of oil globally each day.[48]

Low-carbon hydrogen production carries a lot of potential for mitigating high GHG emissions since refineries consume a lot of hydrogen. Not only can green hydrogen reduce lifecycle emission intensity by an estimated 8 percent in nearly every refinery but also this shift would provide a bridge to a low-carbon hydrogen economy.[49] As discussed in chapter 2, hydrogen is currently produced by SMR.[50] This outdated, GHG-intensive process has been propped up by abundant supplies of natural gas.[51] In refineries that continue to use SMR to generate hydrogen and employ coking to crack oil molecules apart, CCS technologies could further reduce the lifecycle GHG emissions of oil and gas, as discussed in more detail later.

*Lock Up Carbon in Noncombustible End Uses*

Finding more ecofriendly ways to repackage various byproducts of petroleum refining is another promising avenue for lowering GHG emissions. This task calls to mind a memorable line from the 1967 film *The Graduate*: "There's a great future in plastics."[52] Fast forward half a century, and today petrochemicals are essential ingredients in all high-tech products and most consumables in everyday life, including plastics, batteries, insulation, and more. Fed by fracking, a practice that has unleashed a bounty of light oils and wet gas, petrochemical businesses are expanding worldwide and finding new uses for condensates and natural gas liquids (NGLs). For over a century, refining has sought to take conventional oil and break it down (using heat, steam, and pressure) into various petroleum fuels. The proliferation of unconventional oil and gas holds out the promise of reconfiguring processes to turn hydrocarbons into durable stuff rather than fuels.

Some ways of packaging such petroleum byproducts carry less of a climate footprint than others. When refined into LPG and gasoline, NGLs are burned as fuels with outsized GHG emissions. But NGLs, natural gas, and even petcoke can also be converted into various noncombustible, durable products, and when they are, their estimated emission intensity can be reduced by some 40 percent, according to the OCI+.[53] Take petcoke, for example, which can conceivably be processed into activated carbon to purify water rather than burned as a high-GHG, degraded fuel for electric power.

When it comes to plastics, there is a growing array of types, each with their own lifecycle GHG lifetimes. The more disposable the plastic and the fewer its use, the greater the likelihood of its disposal and disintegration into microplastic particles that pollute the air, land, and water. Although petrochemicals tend to lock in GHGs, some do so more permanently than others. Building insulation, clothing, and paints, for example, tend to have longer lives. Disposable plastic straws, bags, and water bottles, on the other hand, have larger GHG footprints and raise other environmental concerns because they are only used once and produce additional emissions when they are later dumped or burned as garbage.[54] The climate footprints of plastics are even larger when extra methane is leaked along their supply chains. On the other hand, petrochemicals in paint or insulation have long lifetimes, and if their associated methane emissions are minimized, so are their climate effects.

*Sequester Heavy Residuals for Reclamation and Reuse*

Reducing the climate effects of especially harmful byproducts of heavy oils like petcoke is a high priority. Residual byproducts (like petcoke) always seem to accompany oil and gas, and some of these residuals are of higher quality (with lesser climate impacts) than others. Since petroleum flows in such large volumes, residuals build up quickly, making it difficult to safely dispose of undesirable buildups. Markets handle these castoffs by setting prices low enough that someone, somewhere will buy them. But not everything *should* simply be burned, especially substances that impose undue environmental harm like the excess carbon being wrung from the world's heaviest oils.

As for petcoke, although it is too polluting to be burned in wealthy Western countries, it is often exported to less affluent nations and blended with coal to generate dirty power.[55] Combusting petcoke emits $CO_2$, black carbon, particulate matter (PM), and sulfur oxides, which pollute local air, break down into toxic organic aerosols, and raise average global temperatures. Petcoke also contains toxic heavy metals like vanadium and nickel. Keeping a lid on all these harmful effects by permanently sequestering all of the petcoke a barrel of Canadian oil sands produces would, based on OCI+ projections, reduce this resource's GHG emissions by a level nearly equivalent to a barrel of conventional oil.[56]

Alberta's high-sulfur bitumen makes an especially damaging form of fuel-grade petcoke.[57] Other heavy oil suppliers and refiners in Venezuela, California, and Texas also market fuel-grade petcoke.[58] Global reserves of extra-heavy oil and bitumen are so large that the best long-term solution would be to either develop in situ production techniques that keep the carbon in the ground and produce only gas or renovate refineries to add green hydrogen rather than coke heavy oil and eliminate petcoke altogether.

### Reduce GHG Emissions from LNG Conversion and Shipping

Trimming the emissions generated by LNG would be another step in the right direction. While oil travels around the globe, gas has historically remained closer to home for local and regional end uses.[59] In 2000, roughly 80 percent of natural gas was consumed locally where it was produced, and the rest traveled to select destinations via regional pipelines.[60] But gas volumes are increasing in regions that cannot use it all, are stuck with unreliable regional pipeline infrastructure, or can obtain higher prices elsewhere. These market forces create incentives to ship gas farther, even across oceans.

But the climate footprint of LNG is quite substantial. In 2018, 430 billion cubic meters (44 percent of total international gas trade) was liquefied in the form of LNG and moved by ocean tankers.[61] This trend has major implications for climate change. Pipelines are still required to move gas to coastal areas where it is then liquefied and kept very cold while in transit—an energy- and climate-intensive operation. Moreover, regasifying LNG upon arrival and moving it to its ultimate destination via pipelines (or trucks) further elevate risks of methane leakage.

Whether LNG is shipped over short distances (Algeria to Spain) or 100 times farther (Norway to China), the GHG emissions associated with liquefaction dominate its climate footprint.[62] By comparison, compressed pipeline gas has an estimated lifecycle emission intensity that is 20 percent lower than that of LNG shipped by ocean vessel.[63] While climate safety may dictate that gas remain a regional fuel, the prospects of more compressed pipeline gas are dim because pipelines are fixed and limit whom gas can be marketed to. Dirtier LNG, on the other hand, offers flexible, long-distance trade that markets can arbitrage with the goal of raising gas prices.

## Mitigating GHG Levels throughout the Supply Chain

There are ways to cut supply-side GHGs that apply equally as well to all industry operations regardless of whether companies produce, process, refine, or transport oil and gas. Four promising approaches that should be pursued include decommissioning high-GHG-producing legacy assets, minimizing fossil fuel inputs in oil and gas operations, employing leak-free equipment, and avoiding operations in sensitive ecosystems.

### Decommission Legacy Assets with the Highest GHG Emissions

Once-plentiful petroleum resources in long-used fields, plays, and formations can deplete over time, and the consequences can be considerable. Just because oil and gas keep trickling out does not mean that the energy benefits of such resources always outweigh their climate costs, especially as time goes on. Emissions tend to increase as resource deposits are depleted, as the characteristics of a field change, and as equipment ages.[64]

Aside from the host of other variables that affect petroleum's climate footprint, there is another: the age of (and remaining deposits left in) the field it is taken from. Take the case of Brent—the internationally recognized oil basket that benchmarks the pricing for around 70 percent of all global crudes.[65] This namesake UK oil field in the North Sea has seen better days. It no longer produces the light, sweet oil that made it famous. Today, Brent produces four times more gas and NGLs than oil,[66] and its gas quality has degraded so much that producing a barrel of Brent gas emits excessive volumes of $CO_2$ and methane. Decommissioning the Brent site and replacing it with a new, low-emitting gas asset would reduce lifecycle climate intensity by over 50 percent, according to the OCI+'s findings.[67]

Other aging oil and gas fields worldwide can have similar climate tolls as their deposits deplete. Brent is a prime example of why emissions must be tracked on an ongoing basis. Accurate accounting requires temporal data collection and periodic reassessments to update findings on climate intensities and GHG inventories. Updated records on GHG emissions should be factored into deciding when the lifecycle of a given site's oil and gas assets should end and should inform when fields are shut down and attending facilities are decommissioned—processes for which adequate funds should be set aside at the outset.[68]

### Minimize Fossil Fuel Inputs in All Oil and Gas Operations

As previously mentioned, petroleum companies shoulder high energy costs and generate substantial GHG emissions from fossil fuels even before their end products go to market. Such vertical integration, whereby a company owns and supports its own supply chain, is a testament to the highly profitable genius of John D. Rockefeller. But this lucrative model imposes heavy negative externalities, a burden that is poised to grow even more as unconventional hydrocarbons are unearthed.

In response, the oil and gas industry can substitute alternative sources of renewable energy in place of its hydrocarbon inputs. For example, concentrated solar arrays are being installed in petroleum patches to generate steam to extract heavy oil in places like California and the Middle East.[69] Utilizing solar energy instead of natural gas saves money and cuts the emissions intensity of production.[70]

The industry can leverage renewable energy even more in additional production and refining operations.[71] Not only would doing so cut lifecycle emission intensity by over 20 percent, based on OCI+ projections, but also such investments could help cross-train the industry's workforce to install and maintain equipment for solar, wind, and other renewable energy sources.[72] Such technology reassignments also would infuse large sums of capital from the oil industry into renewables within their own operations, facilitating the transformation of twentieth-century petroleum companies into twenty-first-century energy companies.

### Employ Leak-Free Equipment in All Oil and Gas Operations

Although oil and gas infrastructure is ideally designed to be composed of entirely closed systems, it can still unintentionally spring leaks. Aging, widespread, remote, complex, and costly infrastructure is not always economical to fix because these GHG emissions are not priced into the company's bottom line. In other words, the valves, seals, pumps, controls, and other equipment needed to plug up old infrastructure can cost more to fix and replace than the companies would recoup in product sales. The advent of fracking has exacerbated this problem because these operations are smaller and less capital intensive, and have shorter lifetimes because production is expected to decline more quickly than in large-scale oil and gas projects.

Preventing fugitive emissions now is a high priority in process design. Employing safer, leak-free equipment—such as that used in refineries where leaks can cause explosions and fires—could prevent the release of gases in both upstream and downstream oil and gas operations, saving an estimated 14 percent in lifecycle emissions intensity, according to the OCI+.[73] These efforts will be increasingly important if the fracking boom experienced in the United States and Canada picks up in other countries like Mexico, the United Kingdom, Poland, Russia, China, India, and Australia.

Beyond climate concerns, fracking operations are also being contested on other grounds, including water quality, air quality, truck traffic, soil contamination, agricultural damage, earthquakes, cancer morbidity, and other public health problems.[74] Fracking bans are on the rise.[75] And greater gas production from fracking also risks renewable energy uptake in electricity production, where gas competes head to head with solar and wind energy.[76]

Still, fracking is drawing needed attention to the oil and gas industry's sizable methane emissions. Some scientists believe that the recent spike in global methane emissions is coming from US oil and gas fracking.[77] Others estimate that, over the past decade, North America has contributed more than one-third of the total increase in global methane emissions.[78] Fortunately, compared to many other supply-side oil and gas climate strategies, reducing leaks (especially fugitive emissions) is feasible, affordable (sometimes even profitable), and only getting easier.[79]

### Operate Permanent, Leak-Free CCS Systems

CCS entails removing man-made $CO_2$ from the atmosphere or industrial waste streams and permanently storing it underground.[80] (Additional efforts are underway to find other uses (beyond underground storage) for the $CO_2$ that is removed.[81] Repurposing one industry's captured carbon and supplying it to others in efficient, ecofriendly ways is the goal of the circular economy.) CCS is part of a larger effort known as carbon dioxide removal (CDR), which includes biological carbon removal, direct air capture, and other technologies.[82] CDR technologies, if applied successfully to achieve net-zero emissions (permanent reductions in atmospheric $CO_2$ levels), are thought to be a major backstop that will ultimately be needed to prevent catastrophic rises in global temperatures.[83]

The resulting GHG savings can vary widely, depending on the resource in question and the operations involved. For example, comparing identical fields, one where manmade carbon is captured and the other where it is not, is estimated to cut total lifecycle GHG intensity in half, based on OCI+ data.[84] By contrast, if Saudi Arabia's Ghawar Oil Field were to offset its own production

GHG emissions, CCS would reduce its lifecycle emissions intensity only by an OCI+-estimated 4 percent.[85]

The oil and gas industry already employs many CCS-related techniques from pump systems and pipelines to subsurface analysis and EOR using $CO_2$. Companies are conducting CCS pilot programs—using various techniques like reinjecting gas in offshore production, CCS-equipped refining, and CCS by cement and fertilizer plants. Although CCS has gotten a lot of recent attention, these practices date back decades and have been bolstered by government research and development and investments.[86]

Oil companies are not only funding CCS demonstrations but also working on other climate (geo)engineering techniques. The petroleum industry thinks these technologies can offset oil and gas GHG emissions and extend the lifetime of its operations. If a global price on carbon materializes, the industry could also profit handsomely from CCS. In other words, CDR and other geoengineering efforts[87] could become industrial policy that is largely the purview of the oil and gas sector.

### Avoid Operating in Sensitive Ecosystems

As they say in real estate, three things matter most: location, location, location.[88] Likewise, the particular location of oil and gas development matters because certain natural ecosystems are at higher risk of damaging the climate. Biomes like the Arctic and rainforests carry high climate risks when dirt paths give way to wider paved roads, followed by electrical lines and traffic lights. Streams and rivers are dredged to accommodate barges. Coastlines are altered and dredged to build shipping terminals. And natural habitats are impacted when mangroves, wetlands, and swamps are lost.

Some oil and gas deposits are buried in ecosystems where the terrain—rainforests, boreal peat bogs, and permafrost-laden tundras—sequester GHG emissions. Removing petroleum resources amid rising global temperatures can disrupt sensitive ecosystems and release climate-forcing gases. Take Russia's Yamal Peninsula, for example, which has a landscape pockmarked with craters that explosively open, venting $CO_2$ and methane.[89] Likewise, petroleum operations in the rainforests of the Amazon, Sumatra, and other similar places threaten sensitive ecosystems that boast more biodiversity than anywhere else on the planet.[90] Unlike the harsh environment in the Arctic where oil and gas development is difficult and the required infrastructure tends to be isolated, seasonal, and small scale, rainforests are wet and warm year-round, which allows developers to deforest broad areas of land.[91]

Typically, oil and gas drilling is estimated to have minimal climate impacts over the equipment's lifetime.[92] However, initial development in the Arctic or rainforests can carry large climate intensity penalties that are some twenty times greater than those in less sensitive ecosystems.[93] Avoiding sensitive ecosystems can reduce upstream emissions in these fragile locations by an estimated 0.3 tonnes for each barrel of oil equivalent (BOE) of oil and gas produced.[94] When methane and $CO_2$ releases are factored in, as permafrost melts and soil is disrupted, oil and gas operations in sensitive ecosystems could trigger feedback loops with large climate impacts.

### Additional Supply-Side Strategies

Aside from the mitigation measures discussed previously, there are numerous other strategies to further reduce supply-side oil and gas GHG emissions. Table 4.2 presents a sampling of additional measures and identifies which oil and gas actors these changes would involve. It will take the full complement of supply-side strategies to meet the Paris Agreement's goals to mitigate the effects of climate change.

## Cumulative Mitigation Potential of Supply-Side GHGs

Reductions in GHG emissions throughout the petroleum supply chain can contribute meaningfully to climate change mitigation goals. Studies cite potential emissions savings as great as 160 gigatonnes (Gt) carbon dioxide equivalent ($CO_2$e) from oil production,[95] plus 50 Gt of $CO_2$e from oil refining by 2050.[96] Factoring in gas (based on its relative consumption levels compared to those of oil) adds an estimated emissions savings of 100 Gt of $CO_2$e by 2050. Altogether, this would entail a total of possible emissions reductions from supply-side oil and gas operations amounting to over 300 Gt by 2050.

Compare these numbers to the total levels of global GHG emissions. By 2100, cumulative emissions from fossil fuel combustion and industrial processes worldwide (which have been accumulating since the preindustrial period in the mid-1800s) are projected to reach 2,700 Gt of $CO_2$e (see Figure 4.2). The IPCC, the UN-affiliated body tasked with formulating recommendations on climate mitigation measures, has called for emissions reductions of nearly 50 percent (from 2017 levels) by 2030 and carbon neutrality by 2050 to limit the earth's average temperature rise to 1.5 degrees Celsius (°C).[97] Reaching this goal reportedly would require a reduction of approximately 1,900 Gt of $CO_2$ in all energy-related emissions by the end of the twenty-first century.[98]

**Table 4.2  Other Supply-Side GHG Mitigation Strategies by Affected Actor**

| Oil and Gas Supply-Side Strategy | Upstream Companies | Midstream Companies | Downstream Companies | Government | Civil Society |
|---|---|---|---|---|---|
| Increase fracking efficiencies | X | | | | |
| Increase efficiencies of ultradeep drilling | X | | | | |
| Electrify processing equipment | X | X | | | |
| Utilize excess gas in microturbines | X | X | | | |
| Produce low-GHG synfuels from CCS | | X | X | | |
| Minimize petrochemical super pollutants | | X | X | | |
| Use lifecycle GHG shadow price to plan projects | X | X | X | | |
| Compensate employees for low-GHG performance | X | X | X | | |
| Prohibit venting of methane and $CO_2$ | X | X | | X | |
| Use underground in situ petroleum conversion | X | X | | X | |
| Generate steam with solar energy | X | X | | X | |
| Install cogeneration capacity | X | X | | X | |
| Collect and report standardized assays | X | X | | X | |
| Collect and report speciated gas compositions | X | X | | X | |
| Set carbon price (low-GHG oil/gas premium) | | X | | X | |
| Require climate-operating best practices | X | X | X | X | |
| Couple vehicle electrification with refining | | X | X | X | |
| Report GHGs on operating and equity basis | X | X | X | X | X |
| Update GHG reporting and use current GWPs | X | X | X | X | X |
| Transparently disclose/digitize data collection | X | X | X | X | X |
| Disclose climate risks caused by companies | X | X | X | | X |

CCS, carbon capture and storage; $CO_2$, carbon dioxide; GHG, greenhouse gas; GWPs, global warming potentials.

*Sources:* Author's estimations. For additional strategies, see Figure 2 in Deborah Gordon and Stephen D. Ziman, "Petroleum Companies Need a Credible Climate Plan," Carnegie Endowment for International Peace, November 2018, https://carnegieendowment.org/files/Gordon_Petro_Companies_Need_Climate_Plan_Nov2018.pdf

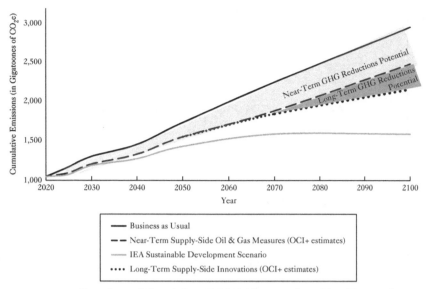

**FIGURE 4.2** Cumulative Oil and Gas GHG Reductions from Near-Term and Long-Term Supply-Side Measures

$CO_2e$, carbon dioxide equivalent; GHG, greenhouse gas; IEA, International Energy Agency; OCI+, Oil Climate Index + Gas.

*Sources:* Author's estimates using the OCI+. See Table 4.3 for assumptions.

How much can the oil and gas sector itself contribute to this goal? Notably, experts' ability to quantify the scale of emissions from supply-side oil and gas operations worldwide continues to evolve. Greater industry transparency can help experts and other stakeholders improve their knowledge and grasp solutions more readily. There is still work to be done on this front. Nevertheless, ample opportunities do exist for reducing the petroleum sector's climate footprints. The OCI+ team as well as many other researchers, including experts at the IEA, are working to quantify the potential for reductions in the oil and gas sector's supply-side GHG emissions.[99] It is estimated that the oil and gas industry could reduce its annual GHG emissions by some 4 Gt by 2025 and by nearly 9 Gt by 2040.[100] Table 4.3 provides more details.

As such, the potential mitigation potential of 9 Gt $CO_2e$ in the oil and gas sector by 2040 corresponds closely with the IEA's sustainable development scenario, which over the same timeframe calls for estimated cuts of 11 Gt of $CO_2$ from all oil and gas emissions sources.[101] This mitigation potential represents a 35 percent reduction in projected 2040 oil and gas sector emissions.[102]

Figure 4.2 gives a sense of how readily supply-side oil and gas GHG reductions may be able to help close the gap between business as usual and the IEA's Sustainable Development Scenario (SDS) over the near term and the long term.

Table 4.3  Oil and Gas Sector GHG Mitigation Strategies and Estimated Emissions Savings (2025 and 2040)

| Mitigation Strategies[a] | GHG Baseline (Gt) | Mitigation Potential (%) | GHGs Saved[d] (Gt) | Comments |
|---|---|---|---|---|
| Minimize methane[b] (2025) | 4.4 | −45% | (1.9) | Masnadi, *Science* 2018 |
| Minimize methane[b] (2040) | 6.8 | −75% | (5.1) | IEA −50% at no net cost |
| Eliminate $CO_2$ venting[c] | 0.2 | −50% | (0.1) | Author estimates, OCI+ |
| Renewables/electrification[c] | 2.6 | −40% | (1.0) | Author estimates, OCI+ |
| Reduce LNG emissions[c] | 0.4 | −75% | (0.3) | Assume 2× by 2040 |
| Industrywide CCS (2025) | 2.9 | −25% | (0.7) | Overlaps with electrify ops |
| EOR using $CO_2$ (2020) | | | (0.2) | IEA WEO 2018 estimate |
| CDR, CCUS, and EOR using $CO_2$ | | | (2.4) | IEA estimate |
| No petcoke combustion (2025) | 0.2 | −100% | (0.2) | *Oil & Gas Journal* estimate |
| Renewable hydrogen (2040) | 0.4 | −100% | (0.4) | Relates to petcoke ban |
| **2025 GHG Mitigation Potential** | | | (3.7) | Sum 2025 strategies |
| **2040 GHG Mitigation Potential** | | | (8.6) | Sum 2040 strategies |

[a] Certain mitigation strategies overlap, so to avoid double counting, they are not considered additive.

[b] $CO_{2e}$ from methane corrected for twenty-year GWP (=86).

[c] These actions assume 2025 and 2040 timeframes.

[d] Figures may not add up to 100% due to rounding errors.

CCS, carbon capture and storage; CCUS, carbon capture utilized but not stored; CDR, carbon dioxide removal; $CO_2$, carbon dioxide; EOR, enhanced oil recovery; GHG, greenhouse gas; GWP, global warming potential; IEA, International Energy Agency; LNG, liquefied natural gas; OCI+, Oil Climate Index + Gas; WEO, *World Energy Outlook*.

*Sources:* International Energy Agency, *World Energy Outlook 2018*, Chapter 11, https://www.iea.org/reports/world-energy-outlook-2018/oil-and-gas-innovation; OCI+ Preview, 2020, https://dxgordon.github.io/OCIPlus/; *Oil & Gas Journal*, 2018 Worldwide Refining Survey, https://www.ogj.com/ogj-survey-downloads; Masnadi, *Science*, 2018, https://science.sciencemag.org/content/361/6405/851.summary; Jing, *Nature Climate Change*, 2020, https://www.nature.com/articles/s41558-020-0775-3.

Through 2100, measures undertaken by oil and gas suppliers are estimated to result in considerable cumulative emissions reductions, as represented by the shaded areas in the figure between the top black line (business as usual) and the bottom dotted line (SDS).[103] Efforts to accomplish the remainder of cumulative emissions reductions (depicted as the white space above the SDS curve) will be dependent on CDR technologies and demand-side strategies, especially coordinated efforts to support a long-term, comprehensive, clean energy transition.[104]

## Influencing Oil and Gas Supply-Side GHGs

The deployment of supply-side oil and gas measures to reduce emissions can produce rapid, durable GHG savings over the next few decades. Such measures are a high priority. They can buy time for longer-term supply-side mitigation measures to be developed, for demand for hydrocarbon-intensive products to shift, and for a worldwide energy transition to ramp up.[105]

Tools like the OCI+ illuminate how stakeholders can successfully reduce the climate risks that oil and gas pose. The long list of tasks to be accomplished includes helping investors make realistic asset valuations and helping industry devise sound infrastructure plans, guiding policymakers to set standards and price GHGs accurately, and giving civil society actors the information they need to advocate for and offer incentives for industry to make wise energy choices. Specific supply-side mitigation efforts entail assessing corporate GHG emission reports that currently rely on self-reported data and methods that companies choose without sufficient third-party verification.[106] And the OCI+ can highlight when companies move to transfer their dirtiest assets to other, less responsible operators to wipe them off their books.[107] Supply-side mitigation efforts also involve projecting climate risks from future oil and gas investments so that infrastructure can be evaluated in terms of its climate fitness and so it can be stress-tested before capital commitments are made. The OCI+'s ability to help validate and project emissions levels will be instrumental as humans and the petroleum products they use push the earth closer to its climate warming limits.

Supply-side oil and gas GHG mitigation strategies present opportunities to target select industry assets controlled by thousands of companies rather than focusing mainly on the demands of billions of consumers worldwide. Oil and gas industry actors (a diverse group) are the topic of the next chapter. Success on one given front—one asset operated by one company in one country—may not be easily replicated globally. As such, it will be critical for experts and decision makers to use the OCI+ to assess and prioritize supply-side oil and gas mitigation measures while simultaneously advancing plans on demand-side measures that reduce oil and gas consumption. Evidence-based strategies offer a prudent way to shrink this key sector's massive climate footprint.

# PART II

## *The Players*

# 5

# Industry: Turning the Titanic

PRESSURE ON THE oil and gas industry to act more decisively to combat the effects of climate change is long overdue. In the summer of 1988, sickening clouds of smog blanketed much of the United States. I was living in Washington, DC, and was working on Capitol Hill for Representative David Skaggs, who served on the House Committee on Science, Space and Technology.[1]

My most vivid memory from that summer on the Hill was attending a congressional hearing on global warming with testimony by Dr. James Hansen (then director of the National Aeronautics and Space Administration [NASA] Goddard Institute for Space Studies). I had first learned about the greenhouse effect while studying chemical engineering in college when Hansen published his seminal 1981 paper in *Science* forecasting that manmade global warming would "emerge from the noise level of natural climate variability by the end of the [twentieth] century."[2]

The prospect of climate change set off alarm bells in my ears that never stopped ringing. In his testimony, Hansen even tied climate change to that summer's heat wave.[3] At the time, however, poor local air quality was a more pressing national concern. (I even remember suffering an acute case of bronchitis from swimming outdoors.)

Instead of acting on climate change, experts called for a reduction in gasoline volatility (its evaporative characteristics) to lower emissions from cars and trucks and reduce ground-level ozone—a move that would also inadvertently shave off some volume of climate-forcing gases. While the regulators in charge previously had been focused squarely on automakers to clean up the air, they now turned their sights on petroleum refiners. On the heels of Californian state regulators, the US Environmental Protection Agency (EPA) set out to reformulate gasoline.[4]

Oil companies en masse resisted, stating it was not technically possible or affordable to reformulate cleaner gasoline—except the Atlantic Richfield

Company (ARCO), the one company that was already refining cleaner gasoline *before* doing so was required.[5] Not only was cleaning up gasoline doable, but also this reformulation only cost 2 cents a gallon more to make.[6] Once ARCO committed, the government adopted new regulations, and every oil company followed suit.

Which of today's industry leaders will step up and be tomorrow's low-greenhouse-gas (GHG) energy companies? And how do we stop leading companies from simply transferring their dirtiest assets to another cadre of industry laggards?[7]

Climate change is a long-term societal risk calculated by businesses that seek short-term petroleum commodity profits. Talk is cheap, and companies can claim they are making headway without offering proof. As such, it is difficult to separate a company's ecofriendly rhetoric from action, confirm constant progress amid moving company targets, and gain access to data to verify corporate actions. Without leadership to turn this titanic-sized industry in a more environmentally responsible direction, the world will stay on course for a severe environmental reckoning.

This chapter introduces the large and varied ensemble of actors that make up the oil and gas industry. Durable climate progress rests on a better understanding of the industry's structure and the positioning of its different actors. The membership of this ensemble extends far beyond large international companies that are household names to include lesser-known national oil and gas companies and countless small and independent producers, refiners, and shippers worldwide. This cast is also joined by ancillary actors—from investors and consultants to traders and assurance agents—who round out the industry. These oil and gas actors contribute different GHG emissions that they self-report in ways that are inconsistent and that therefore do not allow comparisons. Notably, no single company is currently positioned to meet the climate change targets of the 2015 Paris Agreement.

Concrete examples showcase how some companies are better positioned than others to be climate leaders. While it will not be easy to reform this mature industry, tangible progress is possible. A series of corporate benchmarks is provided for companies that intend to improve their climate leadership position. The chapter concludes with the most pressing question for climate progress: will it ultimately be more productive to partner with or defeat the mighty oil and gas industry?

## *The Structure of the Petroleum Industry*

The oil industry reaches far beyond ExxonMobil, the most renowned petroleum company, heir to the Standard Oil empire.[8] Unlike Exxon, most companies are not household names nor do they operate a corner gas station. Still, all of these firms play vital roles in one or more ways.

Figure 5.1 depicts the three main functions and ancillary activities that make up the oil and gas industry. Upstream operators explore for and produce oil and gas. Midstream companies move and store oil and gas and may do some processing as well. Downstream entities refine oil and also market petroleum products.[9] Several other entities do not fall neatly into these supply chain categories. Traders store and ship crude between upstream producers and downstream refiners and also ship refined petroleum products to end users. Gas is moved regionally by various pipeline operators and a growing number of corporate entities are expanding worldwide operations to liquefy, ship, and regasify gas. Petrochemical companies acquire feedstock from upstream producers to manufacture petroleum-based commodities. Oil and gas service companies work alongside upstream, midstream, and downstream operators, installing and maintaining equipment. Individual companies throughout the supply chain engage consultancies and join industry associations. Atop the entire petroleum value chain sit investors, shareholders, and governments who infuse the industry with money and receive dividends, collect royalties, or redistribute rents.

Not all companies partake in all aspects of the oil and gas business. Some are vertically integrated across the value chain, while others focus on a specific unit operation. Companies are typically identified by their ownership and reach. Supermajors like ExxonMobil and Shell are international oil companies (IOCs) that are publicly traded and globally engaged. State-owned enterprises (SOEs) like PetroChina and Gazprom are national oil companies (NOCs) that

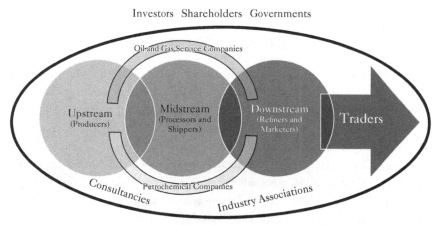

FIGURE 5.1 Actors in the Oil and Gas Supply Chain
*Source:* Author's depiction.

are directly affiliated with a specific country.[10] Independents like Marathon Oil
and EOG Resources are privately held companies that operate on a more limited
geographic basis. The newest arrangement is international national oil companies
(INOCs) like Saudi Aramco and Qatar Petroleum, which are majority SOEs that
also acquire outside investments and release public stock offerings.[11]

## Assets in the Trillions

This expansive array of companies—large and small, public and private, domestic
and international, and vertically integrated and operationally differentiated—all
have at least one business interest in common: hydrocarbons. Together, companies engaged in the oil and gas industry had an asset valuation of over $10 tril-
lion (in 2017).[12] For comparison, after rising nearly 20 percent between 2017 and
2018, the net income of the 150 largest global oil and gas companies plummeted
nearly 60 percent in just two years amid the global pandemic in 2020 and then
rose by 73 percent in 2021.[13] Although their fortunes dip from time to time, these
companies continue to wield tremendous clout worldwide, especially when it
comes to climate change.

Table 5.1 summarizes the asset values of different types of companies based on
their roles in the supply chain as of 2017.[14] NOCs make up the lion's share of the
industry's asset value today at 45 percent.[15] Supermajors are about half as wealthy
as the NOCs, estimated at 20 percent of total assets. The remaining industry
actors—independent producers, independent refiners, chemical companies,
shippers, oil service companies, and traders—each make up under 10 percent of
industry assets. Despite their smaller share of the overall pie, the industry could
not function without their distinct roles in the oil and gas value chain.

## Deep Pockets, Tight Margins

The markets for oil and gas have historically been more valuable than those for all
other raw materials combined,[16] making up a significant share of the global economy.[17] The industry's network of infrastructure is extensive, is constantly under
development, and contains trillions of dollars in assets.[18] Successful development
fills investors' pockets and national coffers.

Despite its outsized reach and massive pocketbooks, the oil and gas sector
operates on tight margins, especially when it comes to refining. Small changes
in demand (driven exogenously by demographics, macroeconomics, and policy
decisions) can cause price volatility that has drastic short-term impacts. As such,

Table 5.1 Where Companies Fit into the Oil and Gas Supply Chain (2017)

| Company[a] | Country | Estimated Assets (in billions of USD) | Upstream | Midstream | Downstream | Other |
|---|---|---|---|---|---|---|
| *Supermajors* | | | | | | |
| Royal Dutch Shell | Netherlands | $407 | x | x | x | Chemicals |
| ExxonMobil | US | $349 | x | x | x | Chemicals |
| BP | UK | $277 | x | x | x | Chemicals |
| Chevron | US | $254 | x | x | x | Chemicals |
| Total | France | $243 | x | x | x | Chemicals |
| Eni | Italy | $138 | x | x | x | Chemicals |
| Equinor | Norway | $111 | x | x | x | Chemicals |
| *Other IOCs[b]* | | | | | | |
| Occidental | US | $84 | x | x | | Chemicals |
| ConocoPhillips | US | $73 | x | | | |
| Repsol | Spain | $72 | x | x | x | Chemicals |
| Suncor | Canada | $71 | x | x | x | |
| *NOCs[c]* | | | | | | |
| PetroChina | China | $631[d] | x | x | x | Chemicals |
| Gazprom | Russia | $316 | x | x | x | Chemicals |
| Petrobras | Brazil | $251 | x | x | x | Chemicals |
| Rosneft | Russia | $212 | x | x | x | Chemicals |
| PdVSA | Venezuela | $190 | x | x | x | |
| Pemex | Mexico | $108 | x | x | x | Chemicals |

*(Continued)*

Table 5.1 Continued

| Company[a] | Country | Estimated Assets (in billions of USD) | Upstream | Midstream | Downstream | Other |
|---|---|---|---|---|---|---|
| Lukoil | Russia | $90 | x | x | x | Chemicals |
| Pertamina | Indonesia | $56 | x | x | x | Chemicals |
| NIOC | Iran | n/a | x | x | x | Chemicals |
| *INOCs* | | | | | | |
| Saudi Aramco | Saudi Arabia | $359[c] | x | x | x | Chemicals |
| KPC | Kuwait | $135[c] | x | x | x | Chemicals |
| Qatar Petroleum | Qatar | n/a | x | x | x | Chemicals |
| *Midstream Companies* | | | | | | |
| Enbridge | Canada | $121 | | x | | Oil pipelines |
| TransCanada | Canada | $78 | | x | | |
| KinderMorgan | US | $74 | | x | | |
| Maersk | Denmark | $49 | | x | | |
| PJSC Transneft | Russia | $32 | | x | | |
| *Other Refiners* | | | | | | |
| Sinopec | China | $226 | | | x | Chemicals |
| Vitol[f] | Switzerland | n/a | x | x | x | Traders |
| Gunvor | Switzerland | $63 | x | x | x | Traders |

| Company | Country | Value ($B) | | | |
|---|---|---|---|---|---|
| Phillips 66 | US | $54 | x | | |
| Valero | US | $50 | x | | |
| Marathon | US | $14 | x | | |
| *Service Companies* | | | | | |
| Siemens | Germany | $163 | | | Consult |
| Schlumberger | US | $71 | | | Consult |
| UOP/Honeywell | US | $58 | | | Consult |
| Baker Hughes | US | $52 | | | Consult |
| Halliburton | US | $26 | | | Consult |
| Transocean | Switzerland | $26 | x | x | |

[a] While collectively they amount to a nominal share of the oil and gas industry's asset valuation, there are thousands of small independent oil and gas companies (family operations and "wildcatters") in the United States and elsewhere.

[b] There are at least 100 midsized companies that, in the aggregate, are valued at over $1 trillion. They are too large to be "wildcatter" independents but not large and integrated enough to be supermajors.

[c] This is a subset of the estimated seventy-one NOCs operating in sixty-one countries.

[d] PetroChina includes CNPC in its financial entries.

[e] 2018 data.

[f] In 2019, Vitol sold a stake of its company to ADNOC.

INOCs, international national oil companies; IOCs, international oil companies; NOCs, national oil companies.

*Sources:* Energy Intelligence, "Big Oil's Emissions," *EI New Energy*, May 9, 2019; *Oil & Gas Journal*, OGJ150, September 3, 2018, https://digital.ogj.com/ogjournal/20180903/MobilePagedReplica.action?pm=2&folio=Cover#pg1; *Oil & Gas Journal*, "O&GJ Refining Survey 2018"; OGJ100, September 3, 2018, https://digital.ogj.com/ogjournal/20180903/MobilePagedReplica.action?pm=2&folio=Cover#pg1; NRGI National Oil Company Database, https://resourcegovernance.org/analysis-tools/publications/national-oil-company-database; Platts Top250 Rankings, https://www.spglobal.com/platts/top250

oil and gas companies and their affiliated businesses are laser-focused on short-term profits.[19]

Corporate sustainability, which calls for a more long-term mindset that companies can no longer publicly deny, takes a back seat to the immediate bottom line. Yet climate change poses an existential threat to oil companies' profit-making model, striking at the heart of the petroleum industry's very being. With damages escalating dramatically over time, the intergenerational costs of global warming are at odds with the long-term returns required to cover the cost of major upfront capital investments as well as the need to meet short-term demand for petroleum. The diverse range of corporate actors in the industry differ markedly along financial and climate dimensions, as discussed later.

## Different Actors, Different Emissions

NOCs tend to emit more GHGs for similar levels of asset value than IOCs. Independent producers create fewer emissions than independent refiners. And chemical companies and oil and gas traders and shippers operate somewhere in the middle. By far, those that provide the highest asset value for the least climate impact in this sector are oil and gas consultancies and field service providers. Although they do not account for their clients' GHG emissions, these actors play an outsized role in decisions that can have tangible climate-impacting results. Figure 5.2 plots this relationship between the levels of GHG emissions and asset values for dozens of oil and gas companies, disaggregated by industry actor. This information offers insights on which parties are most responsible for high GHG emissions levels and where to focus attention on mitigation countermeasures to the greatest effect.

## Responsible Parties

Assigning GHG emissions to the responsible party depends on the particular assets involved. For example, some producers ship their own oil and gas, while other producers may sell their oil and gas to traders or source shipping out to independent midstream companies. Table 5.2 details estimated ranges of total lifecycle emissions assigned to responsible parties. For example, oil and gas producers may account for as little as 4 to 7 percent of the lifecycle GHG intensity if the asset they extract readily flows and they use low-emitting operations.[20] Conversely, emissions can increase dramatically to as high as 31 to 63 percent of GHG emissions when they recover difficult, unconventional assets and do not take precautions to operate with extreme care.[21] The same holds for refiners.

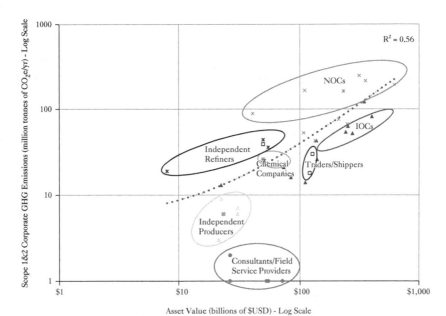

**FIGURE 5.2** GHG Emissions versus Asset Values for Various Oil and Gas Companies

GHG, greenhouse gas; IOCs, international oil companies; NOCs, national oil companies.

*Sources:* Author's calculations; Energy Intelligence, "Big Oil's Emissions," *EI New Energy*, May 9, 2019, https://static1.squarespace.com/static/5b85ce643917ee06a7c16d3f/t/5f4808de-95ab552bfc3658a2/1598556382887/EI+8.2020.pdf; CDP, Corporate GHG Table, https://www.cdp.net/en/investor/ghg-emissions-dataset; and individual company websites.

### Table 5.2  Total Lifecyle GHG Intensity by Industry Actors

| Responsible Industry Actor | Crude Min. | Oil Max. | Natural Min. | Gas Max. | Sample Oil and Gas Resources Averages and/or Ranges[d] |
|---|---|---|---|---|---|
| Oil and Gas Producers[a] | 4% | 31% | 7% | 63% | 19% |
| Oil Refiners[b] | 1% | 32% | <1% | 4% | 6% |
| Midstream Operators[c] | <1% | 9% | <1% | 13% | <1–5% |

[a] Gas producers include gas-processing GHGs.

[b] Petroleum coke emission shares added to refiner responsibility.

[c] LNG shippers represent max values.

[d] Sample includes twenty oil and gas resources in OCI+ Beta Preview.

*Note:* Shares are calculated for different resources and are not additive; assumes 100-year GWPs (20-year GWPs differ, but generally represent similar min. and max. GHG ranges).

GHG, greenhouse gas; GWPs, global warming potentials; LNG, liquefied natural gas.

*Source:* Author's estimates using OCI+ Preview Web Tool, 2020.

While midstream shippers contribute a smaller share of emissions intensity overall, there is a significant spread in GHG intensity between how they handle their lowest- and highest-emitting assets. Liquefied natural gas (LNG), for example, can significantly boost shippers' GHG footprint to upward of 13 percent of total lifecycle emissions.[22]

## Self-Reporting Corporate GHGs

Public GHG emissions reporting is increasingly becoming the norm for all corporations. Although self-reported GHG levels are technically certified, as discussed later, estimating emissions is complicated and error prone. As such, the figures that companies put forth offer a starting point for comparisons.

Many (but not all) oil and gas companies self-report their GHG emissions on an ongoing basis, usually as part of their annual, sustainability, or climate reports. Emissions reports are guided by various suggested GHG accounting methods that have been revised over the past twenty years to reflect changing oil and gas practices.[23] Still, companies have different objectives for reporting their GHG levels that can include miscellaneous emissions sources at the asset, facility, division, or corporate level.

Figure 5.3 details self-reported Scope 1 and Scope 2 emissions and GHG intensities for select oil and gas companies. These are the supply-side GHGs that are directly and indirectly emitted by the oil and gas company itself in its production, processing, refining, and shipping operations. Supply-side GHG intensity (signified by gray bars) vary by a factor of five, from high (Sontratach) to low (Equinor), with a sample mean of 52 kilograms of carbon dioxide equivalent ($CO_2$e) per barrel of oil equivalent (BOE). Absolute Scope 3 GHGs result from the consumption of the petroleum products sold by each oil and gas company. These end-use emissions are based on product sales (see the black "x" markers), which are not currently widely reported by oil and gas companies. Just over half of the twenty-six companies in this sample reported their Scope 3 GHG levels, and their annual average emissions of 555 million tonnes of $CO_2$e range by a factor of fourteen from high (the Saudi Arabian Oil Company or Saudi Aramco) to low (Repsol).

### Inventorying Company GHGs

GHG levels are typically inventoried using bottom-up analysis. This entails counting individual sources of emissions (equipment) or totaling onsite consumption of individual fuel types and applying relevant emission factors. A corporate inventory is then prepared by aggregating emissions from all reporting units.

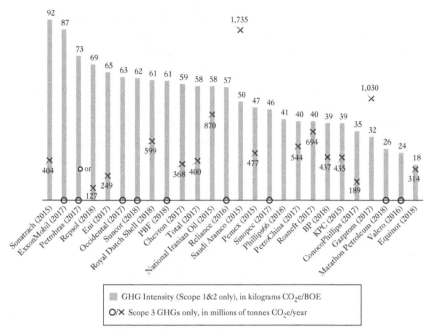

FIGURE 5.3 Self-Reported Company Scope 1 and 2 GHG Emissions and GHG Intensity (Various Years)

*Note:* Companies' Scope 3 GHGs marked with "O" indicate that these emissions were not disclosed as recently as 2019. BOE, barrel of oil equivalent; CO$_2$e, carbon dioxide equivalent; GHG, greenhouse gas.

*Sources:* Energy Intelligence, "Big Oil's Emissions," *EI New Energy*, May 9, 2019; CDP, Corporate GHG Table; CDP, "Carbon Majors Report," 2017, https://b8f65cb373b1b7b15feb-c70d8ead6ced55ob4d987d7co3fcddid.ssl.cf3.rackcdn.com/cms/reports/documents/ooo/oo2/327/original/Carbon-Majors-Report-2017.pdf; and author's calculations using data from individual company websites.

While reporting units group activities and assets to the parent company, the petroleum industry is notable for its complex ownership structures and operating agreements. Facilities are not necessarily owned and operated in their entirety by a single parent company, a feature that presents a wrinkle in GHG accounting.

As such, companies can choose themselves whether to use an equity-share or operational-control basis to document their GHG emissions. An equity-share approach accounts for the proportion of GHGs in each business unit that a company has an ownership percentage (equity stake) in.[24] The reporting of GHG levels based on operational-control accounting holds the company liable for 100 percent of the emissions for each facility it controls (either operationally or financially). The units that a company owns an interest in but does not fully control are entirely omitted from their GHG emissions reporting. There are also sizable GHG levels from third party–owned and third party–operated businesses that provide products or services exclusively to an oil company. Examples include hydrogen or electricity generation plants, chartered shipping vessels, well drilling and testing activities, and remediation work done on behalf of a single company.

Generic GHG inventory guidelines are subject to interpretation and may not convey actual emissions responsibilities because GHG levels depend not only on company structure but also on the nature of oil and gas operations and the characteristics of the hydrocarbons handled. Consequently, side-by-side climate comparisons of petroleum companies remain limited.[25]

## Gaming Emissions Reporting

The two different corporate models for tracking emissions can give rise to accounting chicanery, as a case involving BP shows. (By all means, however, BP is not the only company caught up in the gaming of emissions reporting.) In 2020, the corporation unveiled a new ambition to become a net-zero GHG company by 2050.[26] The company then announced plans to become an integrated energy company, and by 2030, it plans to cut its oil and gas production by 40 percent and its refining by 30 percent.[27] BP (along with other EU oil and gas supermajors like Shell and Total) are setting out to become low-carbon enterprises that dilute their fossil fuel profits with renewable energy revenues.[28] If this is the case, this decision is likely motivated by the fact that the European Union has limited oil and gas supplies and squelching demand is doable.[29]

But just because a company ceases its production of oil and gas does not mean that it is no longer in the petroleum business. BP CEO Bernard Looney's announcement did not reveal, for example, that BP is a joint-venture partner in the world's largest refinery (Jamnagar) in India.[30] At 1.2 million barrels per day, this single facility operated by Reliance Industries markets over 1 percent of the entire global petroleum supply.[31] BP is the Indian firm's major equity partner with plans for 5,500 new gasoline stations fueling a growing fleet of hundreds of millions of cars in India over the next five years.[32]

These kinds of business arrangements can let major companies game the system on emissions reporting. BP does report its Scope 3 emissions for the petroleum products it sells (see Figure 5.3). But, since BP does not operate the Jamnagar refinery, these massive emissions—estimated at over 200 million tonnes of $CO_2e$ per year (a level nearly equivalent to half of all BP's Scope 3 GHGs)—are not included in the company's GHG inventory.[33] Reliance Industries, however, does not report its Scope 3 GHG emissions. As BP doubles down on its GHG pledge, emissions for which it still generates profits are shuffled to an NOC with less transparent reporting standards.

All oil and gas companies should be required to report their GHG levels on both an equity and an operational basis. This is the only way to affirm whether corporations have actually reduced GHG emissions or merely shifted them to other less responsible operators. Profits made through equity shares impose real GHG impacts that must be disclosed to investors and civil society actors. On the other hand, reporting GHG levels on an operational basis provides direct oversight for investors, which comes with the added benefit of greater data transparency and the ability to change practices when responsible, high-quality management companies are at the helm and mindful of this operational reporting.

## Allocating Corporate GHGs

Once inventoried and reported, corporate GHG levels still tend to be so highly aggregated that it can be hard to establish climate priorities and develop targeted strategies. In other words, knowing where emissions reside is critical for assessing climate risks, pricing carbon, establishing research and development (R&D) priorities, and a host of other decisions. The Oil Climate Index + Gas (OCI+) is an effective tool for estimating emissions, by operation and sources, as well as those of an individual well or an entire facility. This comes in handy since not every company is sufficiently transparent when it comes to disaggregating their emissions.

For example, compare Chevron to Shell. The two companies had similar levels of direct operational (Scope 1) GHG emissions in 2019 (Chevron reported 65 million tonnes of $CO_2$e versus Shell at 81 million tonnes of $CO_2$e).[34] Table 5.3 offers insights into how Chevron and Shell each attributed their relatively consistent emissions in the prior year largely to upstream operations, but in highly varying proportions and combustion sources. Moreover, on an equity basis, Chevron and Shell had wide ranging direct emissions in 2019 of 58 and 117 million tonnes, respectively.[35]

## Stating Climate Priorities

Some subsectors of the oil and gas industry are inclined toward climate leadership, while others remain on the sidelines and avoid making climate disclosures and pledges.[36] Figure 5.4 offers a snapshot of the estimated volumes of oil and gas that are currently capitalized by companies with and without stated climate priorities.

Upstream oil and gas producers are currently split almost evenly between companies that state that climate change is a corporate priority and those that

Table 5.3  Representative Corporate GHG Allocations: Shell versus
Chevron (2018)

| By Operation | Shell Share | Chevron Share[b] | By Source | Shell Share[c] | Chevron Share |
|---|---|---|---|---|---|
| Upstream | 60% | 68% | Combustion | 15% | 72% |
| Refining | 21% | 24% | Flaring | 19% | 19% |
| Power and Integrated Gas | 19% | 4% | Process emissions | 48% | 5% |
| Transport | n/a | 3% | Venting | [d] | 3% |
| Other | n/a | 1% | Fugitive sources | 18% | 1% |

[a] Chevron shares estimated based on charts published by Chevron.
[b] Shell reports its emission sources for methane and not total $CO_2e$.
[c] Shell combines process and venting in reporting its methane emissions by source.
[d] Shell does not break out transport.

*Note:* The figure's $CO_2e$ emissions are based on direct Scope 1 GHG emissions on an operated basis.

$CO_2e$, carbon dioxide equivalent; GHG, greenhouse gas;

*Sources:* Shell, "GHG Breakdown," https://www.shell.com/sustainability/sustainability-reporting-and-performance-data/performance-data/greenhouse-gas-emissions.html; Shell, "Methane Emissions," https://reports.shell.com/sustainability-report/2018/sustainable-energy-future/managing-greenhouse-gas-emissions/methane-emissions.html; Chevron, "What We're Doing," https://www.chevron.com/corporate-responsibility/climate-change/greenhouse-gas-management.

do not. Generally, the largest IOC producers are more likely than the NOCs to acknowledge climate change and make corporate commitments, whereas the range of small independent producers runs the gamut between those that prioritize climate change and those that do not. The same general rule regarding prioritizing climate change between IOCs, NOCs, and independents applies to refiners. Meanwhile, the petrochemical companies and midstream shippers who handle the greater volumes prioritize climate change over those that process more modest volumes. But the reverse is true for oil and gas traders and service companies: the majority of them have not historically made the global climate a priority.

## *Separating Industry Leaders from Laggards*

Most industry actors are reluctant followers rather than able leaders. Once rules do change, however, companies adopt a compliance mindset, whereby they adhere to the minimum requirements necessary, the letter of the law. But to earn

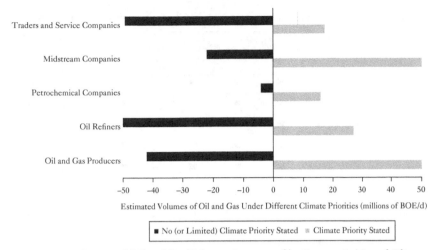

FIGURE 5.4 Estimated Oil and Gas Volumes Aggregated by Companies' Stated Climate Priorities (2018)

BOE, barrel of oil equivalent.

*Sources:* Author's calculations using individual company annual financial reports, along with other databases, including *Oil & Gas Journal*, OGJ150, September 3, 2018; Energy Intelligence, 2019; OGJ100, September 3, 2018; NRGI National Oil Company Database, December 2019; Adam Muspratt, "The Top 10 Oil & Gas Companies in the World: 2019," May 1, 2019, https://www. oilandgasiq.com/strategy-management-and-information/articles/oil-and-gas-companies; *Oil & Gas IQ*, May 1, 2019, https://www.oilandgasiq.com/strategy-management-and-information/ articles/oil-gas-industry-an-introduction; *Oil & Gas Journal*, "O&GJ Refining Survey 2018"; Alexander Tullo, "Top 50 Chemical Companies of 2018," *C&EN* 97, no. 30 (July 29, 2019), https://cen.acs.org/business/finance/CENs-Global-Top-50-chemical/97/i30; Statista.com

their social license to operate, oil and gas companies need to enable climate action and ultimately adhere to the spirit and not merely the letter of the law. In other words, in all they do, companies need to be intent on mitigating their climate footprints rather than loosely abiding by rules while exploiting any loopholes.

Separating the leaders from the laggards can reveal where change is happening and offer climate lessons for the industry as a whole. Table 5.4 ranks the clean energy strategies of supermajors with those of major NOCs. Such snapshots are useful, but showing real leadership requires ongoing tracking and reassessing over time. The OCI+ can be used to evaluate corporate claims. Like ARCO's clean gasoline breakthrough, climate action requires durable commitments by all industry actors.

European IOCs appear to be adapting to climate risks more rapidly than their competitors. They are the companies to closely watch.[37] Regardless, at present the stated actions of every oil and gas company are not enough to meet the climate targets of the Paris Agreement, and the manner in which

**TABLE 5.4** Ranking Supermajors' and NOCs' Clean Energy Transition Strategies

| IOCs | Company Rank[a] | Reduce Methane[b] | Source Renewable Power | Generate Solar/Wind | Deploy Centralized CCUS | Sell Low-GHG Gas[c] | Reduce CO$_2$[b] | Use CO$_2$ EOR | Sell[d] Other Clean Power | Sell[e] Electric Vehicles |
|---|---|---|---|---|---|---|---|---|---|---|
| Shell | 1 | ■ | ■ | ■ | ■ | ▨ | ■ | ▨ | ■ | ■ |
| Total | 1 | ■ | ■ | ■ | ▨ | ▨ | ■ | ▨ | ■ | ■ |
| BP | 2 | ■ | ■ | ■ | ▨ | ■ | ■ | ▨ | ▨ | ■ |
| Repsol | 3 | ■ | ■ | ■ | ▨ | ▨ | ■ | ▨ | ■ | ▨ |
| Equinor | 3 | ■ | ■ | ■ | ■ | ▨ | ■ | ▨ | □ | ▨ |
| Eni | 3 | ■ | ■ | ■ | ▨ | ▨ | ▨ | ▨ | ■ | ▨ |
| Petrobras | 3 | ▨ | ■ | ▨ | ■ | ■ | ▨ | ■ | ■ | □ |
| Chevron | 4 | ■ | ■ | ▨ | ■ | ▨ | ▨ | ▨ | □ | ▨ |
| Exxon Mobil | 5 | ■ | ■ | □ | ■ | ▨ | ▨ | ▨ | □ | □ |
| CNPC | 5 | ▨ | ▨ | ■ | ▨ | ■ | □ | ■ | □ | □ |
| Strategy Ranking[f] | 1 | 2 | 2 | 3 | 4 | 4 | 5 | 6 | 7 | |

[a] Ranked high (1) to low (6) and companies with same ranking are loosely tied.

[b] Methane and CO$_2$ emission reductions are based on the strength of intensity targets and evidence of their implementation.

[c] Low-carbon gases include low-carbon, green hydrogen.

[d] Power generation investments reflect commercial sales (and not companies' own use).

[e] Electric services include battery storage and electric vehicle charging.

[f] Strategy ranking was assessed by totaling all company strategic priorities.

Key: black square = highest rank: observed strategic investments and/or capital/operational expenditures in commercial activities; gray square = middle rank: company has announced strategy and/or minor investments, venture capital, and/or R&D spending; white square = lowest rank: limited evidence of investment activity.

CCUS, carbon capture utilized but not stored; CO$_2$, carbon dioxide; EOR, enhanced oil recovery; IOCs, international oil companies; NOCs, national oil companies.

*Source:* Adapted from International Energy Agency assessment of investment and strategic responses to energy transition by select companies, based on 2015–2019 activities. International Energy Agency, "The Oil and Gas Industry in Energy Transitions," 2020, https://www.iea.org/reports/the-oil-and-gas-industry-in-energy-transitions, and updated based on 2019–2020 company announcements.

companies report their GHG emissions does not allow for clear comparisons. All companies—even those in the lead—need to do more to meet growing energy demand in a more competitive environment while reducing their operating emissions and cutting GHG levels in their value chain.[38] Several examples follow describing actions by companies that are endeavoring to lead and others that are trailing behind.

## No Place for Shell Games

Shell is viewed as an industry leader on the climate, especially for scenario planning. In its most far-reaching scenario to date (Sky, 2019), Shell claims that it aspires to reinvent world energy supplies with carbon-free renewables, biofuels, carbon capture and storage (CCS), and green hydrogen.[39] In reality, Shell is still planning to invest billions in the future in oil and gas production—deepwater and shale resources—dwarfing these low-GHG supplies.[40] And while the company currently sells about three times more oil and gas than it produces, Shell is now hewing its global refining portfolio down to about ten large refineries compared to the roughly fifty-four it held a decade ago.[41] In 2020, for instance, Shell sold one of its dirtiest refineries in Martinez, California, to little-known PBF Energy for over $1 billion.[42] And in 2021, Shell announced that it is planning to exit its Aera partnership with ExxonMobil that operates Midway Sunset, one of California's dirtiest oil fields.[43]

While these moves reduce Shell's own net carbon footprint, they do not cut GHG emissions entering the atmosphere. To be a real climate leader that delivers the "healthy planet" that Shell has pledged, the company cannot merely sell its refineries, production facilities, and other assets.[44] Instead, the company must zero out GHG emissions in its own operations, starting with methane, and decommission rather than offload its dirtiest assets.[45] The need to avoid playing shell games with resources and properties applies to every oil and gas company.

## Making Refining Greener

Refining is the lynchpin for the clean energy transition. Because the world will rely on petroleum products for a long time to come, companies need to re-engineer refining to be a low-carbon endeavor—a major undertaking. This wholesale shift will involve more than incremental fixes like substituting renewable diesel for petroleum-based diesel (as discussed in chapter 2). To date, Repsol is the only company that has shared its climate plan for refining. By incorporating green hydrogen into its refining operations, Repsol is directly tying this noncarbon fuel and renewable energy source into its own business.

In other words, Shell may be shedding refineries, but Repsol is tackling them head on. This makes the Spanish oil company stand out among its rivals. Moreover, Repsol was the first supermajor to pledge net-zero emissions by 2050, with specific intermediary targets between now and then.[46] Shell and BP did follow suit on this climate target some months later, but they did not immediately specify tangible operating plans like Repsol did to meet their climate goals.[47]

## Pursuing Alternatives

In the summer of 2013, I received an invitation from Total to speak about the energy transition at an international conference that the company was organizing. This was the sixth year in a row that the attendees of Total University assembled to search for paradigm shifts and tipping points.[48] Then-CEO Christophe de Margerie seemed immensely excited by the prospects of finding a breakthrough solution—nanotechnologies, biocatalysts, and CCS—which he appeared to be more intent on than oil and gas.

Today, Total is a leader in the quest to reshape the future of energy.[49] In 2020, the company announced a 50 percent increase in its capital expenditures on renewables[50] and plans to halt refining in the Paris region and focus on making renewable fuels and plastics instead of petroleum products.[51] On par with its EU rivals, Total is assuming a new role as a national electricity retailer.

## Reconciling GHG Emissions

Weeks before the 2015 United Nations (UN) Climate Change Conference in Paris, the CEO of Norway's Equinor (then Statoil) issued a challenge to his 22,000 employees: "I want Statoil [now Equinor] to be a company recognized for shaping the future of energy."[52] Shortly after, Norway's oil association declared that, as the nation with the *cleanest* oil and gas in the world, it had earned the *right* to expand its operations.[53] This claim was based on estimates running three Equinor-produced oils through the OCI+. One of the oils (Ekofisk[54]) ranked among the ten lowest-emitting resources we evaluated at the time.[55]

But the industry group's bold assertion raised national concerns because Norway's oil was not estimated as the world's lowest-emitting oil. Moreover, Equinor reported that its GHG levels were even lower than the OCI+ had estimated.[56] To address these issues, in December 2016, Equinor invited Adam Brandt and me to Stavanger to reconcile its GHG levels. The technical exercise highlighted opportunities to tailor the OCI+ to unique operating conditions (such as running a platform entirely on renewable power and interconnecting gas handling between different platforms) to arrive at closely matching GHG emissions estimates. This further underscores the value of corporate data transparency.

It turns out that Equinor is not the average oil company. They are a small player in the crude market but a major global gas exporter.[57] Their production is more painstaking than most: platforms operate as integrated units, gas is routinely reinjected, renewable (hydro) electricity can be supplied from onshore, offshore wind energy development is on the rise, and flaring is closely monitored. Looking ahead, Equinor could play a leadership role in reconciling all of its global assets with the OCI+ model to further reduce the company's GHG emissions.

## Going Public

In December 2019, Saudi Aramco became the most valuable publicly traded company in history, then valued at nearly $2 trillion in its initial public offering (IPO).[58] Supplying 10 percent of the globe's oil, Saudi Aramco is the most profitable company in the world and the country's main revenue source.[59] If Saudi Aramco's IPO is aimed at diversifying the Saudi economy *away* from oil instead of leveraging oil to grow its economy, this could put the company in a global leadership position.

A recent ad campaign seems to support this move. Full-page, color ads festooned the pages of media outlets worldwide in 2019 and 2020 to promote Saudi Aramco's "real sustainability," "real innovation," and "real progress."[60] According to the OCI+, the average oil produced in Saudi Arabia is some of the world's lowest-emitting varieties, and with such affordable and large volumes, it is also the marginal barrel produced, which means that it pumps the extra barrel when the market signals it wants more.[61]

The relative ease and speed at which it can turn on or off its oil spigot compared to any other nation puts Saudi Aramco in a powerful position to influence which oil and gas assets to strand first as the world warms.[62] While other IOCs and NOCs need to work to reduce their GHG emissions, Saudi resources flow readily and take fewer energy and hydrocarbon inputs to refine. Challenges lie ahead for Saudi Aramco to further cut the GHG emissions from its oil, challenges that will require the use of CCS and other novel techniques. Instead of sparring with Russia over current oil prices, Saudi Aramco would be better served investing in climate mitigation measures that set it apart from laggards like Rosneft, Lukoil, and Gazprom, discussed later.[63]

## Rebranding Beyond Petroleum

In 2000, after a merger with Amoco, BP adopted the moniker "Beyond Petroleum" and reintroduced itself as a socially conscious company. At first this move paid off, earning BP awards, accolades in the press, rising brand awareness, and even growth in sales. Then-CEO Lord John Browne cleaned up gasoline (by buying

ARCO), invested heavily in hydrogen, and recognized global warming. Today, Lord Browne thinks the real difference at the time was changing the company's mindset.[64]

Unfortunately, this change was not permanent, and durable corporate action has not prevailed over clever ad talk.[65] As of the spring of 2020, BP's climate ambitions lag behind those of its EU competitors.[66] For example, BP's climate target covers fewer than half of its downstream activities, despite the fact that BP operates some of the largest and highest-GHG-emitting refineries in the world.[67] BlackRock and other investors are calling on BP to do more.[68] In response, BP is one of the first IOCs to consider voluntarily differentiating gas based its methane intensity, improved company practices, and stringent emissions monitoring.[69]

## Making Empty Promises

In 2020, Chevron distributed its first brochure stating its support for the Paris Agreement and promoting climate action.[70] What preceded this move, however, is a lesson in how *not* to do business better. For example, in Chevron's major Gorgon project in Western Australia, the company agreed to install carbon sequestration and storage (CCS) equipment[71] because the gas there is naturally high in carbon dioxide ($CO_2$) (some twenty-five times more concentrated than the $CO_2$ levels of Pennsylvania's Marcellus gas).[72] This venture was anticipated to be one of the industry's largest GHG mitigation projects that could reduce Gorgon's emissions intensity from production by an estimated 60 percent.[73]

But when Gorgon commenced production in March 2016, a series of technical problems prevented the simultaneous start-up of its CCS equipment until August 2019.[74] As a result, over three and a half years the company was unable to sequester a requisite one-half of the 30 million tonnes of GHGs emitted at their Gorgon facility.[75] And, in January 2021, Chevron's CCS was reportedly not working properly again.[76] Chevron used millions in taxpayer funds to build its CCS equipment, while the equipment remained shuttered even as the firm profited from gas extraction and sales. Whether this empty promise was intentional or not, Chevron skirted its permit conditions, which stated Gorgon would inject 40 percent of its emissions a year, or up to 4 million tonnes of $CO_2$.[77] With oil and gas markets in turmoil, time will tell whether Chevron will restructure its business models around climate change or maintain the status quo.[78]

## Playing Catch Up

For decades, the largest US oil company has lagged behind on climate action. This may have been a political calculation: ExxonMobil has long been considered one of the "most powerful businesses ever produced by American capitalism," and the US government has not consistently made climate policy a priority.[79]

The ground is now shifting beneath ExxonMobil. In August 2020, ExxonMobil was removed from the Dow Jones Industrial Average, the group of blue-chip stocks that it first joined in 1928.[80] Then, in 2021, an activist investor forced new directors on the board with the express goal of reducing ExxonMobil's carbon footprint.[81]

Thankfully, ExxonMobil no longer boldly denies that climate change is real and the company is beginning to acknowledge it can play a leading role in stopping the earth from warming. Yet many argue that ExxonMobil has squandered its social license to operate over years of miscalculations as the company plowed ahead, keeping its balance sheets healthy and share prices up rather than addressing its climate risks.[82] Despite beating back a 2019 lawsuit that it intentionally misled investors about climate risks, ExxonMobil may be unable to revive public trust.[83] As ExxonMobil (along with other IOCs) exits its long cycle of expensive projects, continues to confront thin margins, and prioritizes flexible capital, all eyes are on the largest IOC to see if it can reassert itself as a low-carbon energy provider.

## Too Big to Fail

China's oil and gas resources are considered strategic national assets that are controlled by its SOEs, including China National Petroleum Corporation (CNPC), China National Offshore Oil Corporation, China Petroleum and Chemical Corporation Petrochemical (SINOPEC), and Yanchang Petroleum Group.[84] While NOCs have traditionally been viewed as the custodians of their country's natural resources, China is not an oil- and gas-rich nation. But it is the world's second-largest consumer of petroleum products,[85] and China imports more oil and gas than any other nation in the world. Rather than grow increasingly dependent on petroleum product imports like gasoline and natural gas liquids (NGLs), China has been on a recent refinery- and petrochemical-building frenzy, which means the country only needs to secure crude oil from other nations and not rely as heavily on others to supply them a vast array of finished petroleum products.[86] To the extent that China can manage to consume fewer barrels of products than its refineries manufacture, it can sell these to other countries.

In 2019 alone, China's oil and gas state-owned enterprises (SEOs) increased petroleum product exports by 20 percent.[87] Between meeting domestic demand and supplying regional exports, China's SEOs are too big to fail. The economic pressures borne by China's oil and gas companies that are increasingly focused on profits from refinery expansion—a business with notoriously tight margins—could make them climate laggards. The less expensive and, therefore, more degraded the crude oils Chinese refiners procure, the more GHG intensive their operations will become and the more high-carbon byproducts like petcoke that they will manufacture and likely dump on poorer nations.[88]

## Selective GHG Accounting

As with petcoke, companies can hide GHG emissions in plain sight by selectively accounting for emissions. In 2016, Southwestern (an independent US energy company) undertook an analysis of the climate impacts of its energy assets.[89] Company representatives reached out to me because they were using the OCI+ to conduct their assessment. They shared their 131-page final report (which cannot be reproduced or distributed in any form without express permission). The company was intent on understanding the budgeted resources and carbon constraints in its operations. They used a lifecycle approach to assess global oil and gas climate impacts.

Upon further inspection, however, I noticed a glaring omission. The analysis only considered $CO_2$. Methane was entirely omitted. For a company with mostly gas and NGL assets, this oversight raised red flags. Given the high potency of the company's main product—methane, which has a twenty-year warming potential that is at least eighty-six times greater than $CO_2$—their analysis did not uncover the company's primary climate risks.[90] When I told their representatives that such selective GHG accounting was extremely problematic, it was not entirely clear to me whether this was an oversight or an intentional omission. Either way, companies cannot be permitted to hide any of their GHG emissions.

## Closing a Can of Worms

As increasingly different kinds of unconventional hydrocarbons and breakthroughs to access them enter the market, companies have specialized, focusing on certain resources and particular techniques. Occidental Petroleum (Oxy for short) is an American oil production company that operates internationally. Its forte is in $CO_2$ management, and the firm is the global leader in enhanced oil recovery (EOR) projects using $CO_2$.[91] While Oxy publicly extols the need for a lower-carbon future, its reputation to date rests on removing naturally stored $CO_2$ to enhance oil recovery. But there are two radically different faces of EOR using $CO_2$. Not all EOR projects using $CO_2$ have the same climate risks. Projects that exploit naturally occurring $CO_2$ tend to have much greater climate impacts than those that leverage manmade $CO_2$ that has already been gathered or captured.[92] The challenge facing Oxy now is whether it can successfully convert its operations to use only manmade $CO_2$. Oxy and other companies that engage in EOR using $CO_2$ need to double down on CCS and collect and use only manmade $CO_2$ for EOR to be real climate leaders.

## Industry Actors in the Wings

Oil and gas companies, for all their wealth and power, cannot supply petroleum to the market alone. They are supported by investors, advisers, traders, certifiers, and industry associations. Transparency is often lacking about who these ancillary actors are and how they influence corporate climate action. Still, it is important to understand the influence wielded by these supporting actors of the oil and gas supply chain, even those who may never come into contact with hydrocarbons themselves.

### Underwriting Oil and Gas

Investors come in all shapes and sizes. Institutional investors—commercial banks, insurance companies, pensions, hedge funds, endowments, mutual funds, family offices that handle ultra-high-net-worth investors, and sovereign wealth funds—pool their funds to purchase assets. Individual retail investors purchase stock shares of publicly and privately traded companies, whereas investment partnerships form holding companies that acquire other businesses.

Countless individuals and institutions have an investment stake in oil and gas, unless they have painstakingly divested from petroleum. Since all global enterprises rely on oil and gas, every investment has some degree of petroleum climate risk reflected in its bottom line.

#### Sampling Investors

While financial vehicles exist for anyone to invest in the oil and gas supply chain, institutional investors are the major force behind the petroleum sector. Table 5.5 summarizes the relative monetary value of the stakes owned by different investors and their stated climate commitment and fossil fuel divestment status. Institutions in North America and Europe control the vast majority of assets. While some investors have stated climate change is a priority and are engaged in climate initiatives, none have divested entirely from oil and gas.[93] Notably, investment managers rank risks from regulatory compliance second only to cybersecurity as their greatest near-term business challenge.[94]

Table 5.5 Sample of Investors' Estimated Assets, Climate Commitments, and Fossil Fuel Divestments

| Investor Type | Corporate Examples | 2018 Assets Under Management (trillions of USD) | Signatory on Climate Action 100+[a] | Divested from Oil and Gas? |
|---|---|---|---|---|
| Investment | | $67 | | |
| Managers | Fidelity | $6.9 | Yes | No |
| | BlackRock | $6.8 | Yes | No |
| | Vanguard | $5.1 | No | No |
| | State Street | $2.5 | No | No |
| | Allianz/PIMCO | $2.3 | Yes | Partially[b] |
| Pensions | | $36 | | |
| | Japan Pension Fund | $1.4 | Yes | No |
| | CalSTRS | $0.2 | Yes | Partially[b] |
| | New York City | $0.2 | Yes | Yes |
| Insurance | | $25 | | |
| Companies | AXA | $0.9 | Yes | Partially[b] |
| | Zurich Insurance | $0.4 | No | Partially[b] |
| | MetLife | $0.1 | No | No |
| Commercial | | $35 | | |
| Banks | Bank of China[d] | $7.0 | No | No |
| | JP Morgan Chase | $2.2 | Yes | No |
| | BNP Paribas | $2.2 | Yes | Partially[b] |
| Hedge Funds | | $3 | | |
| | Bridgewater | $0.1 | No | No |
| | Man Group | $0.06 | Yes | |
| Endowment | | $2 | | |
| Funds | Harvard University | $0.04 | Yes | [c] |
| | University of Texas | $0.03 | No | No |

[a] Climate Action 100+, "Investors," https://climateaction100.wordpress.com/investors/.

[b] Divestment in progress from coal, oil sands, and/or Arctic drilling, but not petroleum broadly.
[c] Planning in progress.
[d] Includes Industrial and Commercial Bank of China and Bank of China (Statista.com).

*Sources:* Wall Street Prep, "Largest Institutional Investors," https://www.wallstreetprep.com/knowledge/largest-institutional-investors/; Investopedia, "Top 5 Asset Management Firms in 2019," https://www.investopedia.com/articles/professionals/080615/5-biggest-financial-advisory-firms-us.asp; Sam Bourgi, "Who Are Institutional Investors?," *ETF Investing*, December 25, 2018, https://etfdb.com/etf-education/who-are-institutional-investors/; Charles McGrath, "Global Foundation Assets Reach $1.5 Trillion," *Pensions and Investments*, May 8, 2018, https://www.pionline.com/assets/docs/CO119162327.PDF; "Largest Banks as of December 2018, by Assets," Statista.com, https://www.statista.com/statistics/269845/largest-banks-in-the-world-by-total-assets/.

## Investor-Backed Climate Initiatives

In 2017, two major investment firms, Blackrock and Vanguard, led a shareholder charge against ExxonMobil to demand transparency and disclosure on the company's climate mitigation measures to meet the benchmarks of the Paris Agreement.[95] This move was meant to force ExxonMobil to join other oil supermajors—BP, Total, and Shell—in releasing its detailed plan to attain the 2 degrees Celsius goal under the Paris Agreement. While their move excited climate actors, in 2018 and 2019, BlackRock and Vanguard took a back seat among investment firms, voting in support of climate-related proposals only 10 percent of the time at various companies' annual shareholder meetings.[96] Just because investment firms state that climate change is a top priority, even if they are major investment shareholders, they are reluctant to strongarm corporate management to release a climate plan. Such investor initiatives, while useful, have not proven to be a silver bullet for markedly changing corporate decision-making.

A different approach is instead for investors to join forces and advocate for corporations writ large to tangibly show how they plan to curb GHG emissions rather than pressuring individual firms. As of 2019, only 38 percent of oil and gas companies performed and disclosed a climate scenario analysis.[97] The Task Force on Climate-Related Financial Disclosures (TCFD) was formed in 2015 to fill this void in the oil and gas sector and other parts of the economy.[98] The TCFD encourages firms to align their climate disclosures with investors' needs and incorporate climate preparedness into their balance sheets. The focus is on financial risks *to* firms—an estimated $43 trillion in assets that, by 2100, are at risk from climate change.[99] The TCFD urges financial decision makers to take climate change into account, but it does not force companies to act like corporate shareholder resolutions, government regulations, or other direct measures do.

Additionally, investors must wrestle with the climate risk imposed on society *by* oil and gas (and other carbon-intensive) companies. Climate Action 100+, a five-year investor-led initiative launched in 2017, complements the TCFD by assessing and tracking corporate progress on climate governance and policy matters. Over 450 investors that collectively manage upward of $40 trillion in assets have joined this effort.[100] As of 2019, 85 percent of oil and gas companies have made climate change a board-level priority, while only 62 percent have set long-term quantitative GHG targets. Companies will need to up their game to shrink their climate footprints.[101]

Divestment, removing funds from select companies, is another investor approach. As of 2018, institutional investors pledged to divest over $6 trillion from fossil fuel companies.[102] The insurance industry has led this effort, and banks are also taking part. For example, the European Investment Bank announced it will stop financing *new* oil, gas, and coal projects after 2021—although its port-folio contains plenty of legacy fossil fuel investments.[103] Each of these investor approaches—voting for shareholder proxy measures, assessing climate risk in bal-ance sheets, tracking progress on corporate climate action, and divesting insti-tutional investments—can help spur the investment community to action on climate change.[104]

### Advising Investors

When the OCI+ was first released in 2015, its findings were picked up and pub-lished (unbeknownst to me) in a widely read investment newsletter.[105] GMO, a privately owned independent firm, manages billions in assets for its clients seek-ing long-term market returns.[106] The firm's cofounder, Jeremy Grantham, believes climate change is the "race of our lives and subscribes to responsible investing that considers each company's environmental, social, and corporate governance (ESG)."[107] In his GMO newsletter, Grantham underscored the importance of using the OCI+ to assess lifecycle GHG emissions in each barrel of oil. To do otherwise, he wrote, leaves emissions from "colossal operations which look like they chew energy relentlessly" uncounted.[108] Grantham was spot on: the econ-omy is going to need as much oil as it can safely use during the low-carbon energy transition. More investors need to realize, as Grantham does, that unnecessarily choosing GHG-intensive resources that emit more climate-forcing gases for the same barrel consumed is just flat out bad for the economy, is needlessly risky for the environment, and benefits only select industry actors.

## Getting Advice

Oil and gas companies both generate and consume data on a massive scale. Industry advisers compile data, design strategies, perform services, and do analy-sis.[109] More recently, tech firms—such as Microsoft and Amazon—are supply-ing oil and gas companies with artificial intelligence, automation, and machine learning tools to accelerate fossil fuel exploration and extraction.[110] Industry advisers often control more data than most governments and structure big data better than many oil and gas companies; as such, they are playing an outsized role in petroleum markets, climate action, and the energy transition. The surfeit of disorganized, backlogged data collected by oil and gas companies over decades, once digitized and made readily accessible, can be used to improve operations and grow profits. These data can also be used to update GHG inventories and automate the OCI+ to conduct ongoing analyses over time.

But not all industry advisers are willing to share data and conduct joint research. This is unfortunate because the value added by academia (discussed in chapter 7) contributes to knowledge in different ways than the services provided by paid industry advisers. While many oil and gas advisers acknowledge climate change, their "laser focus" is on maximizing petroleum's economic value and maintaining the supply chain.[111] Their sharp focus on the bottom line encourages corporate consolidation during downturns, which can postpone maintenance, disproportionately cut environmental experts on the oil and gas sector's payroll, and reassign climate responsibilities to corporate nonexperts.[112]

## Trading Oil and Gas

While IOCs and NOCs operate their own global operations for crude trading and petroleum product marketing, there are companies (mostly privately held ones) that specialize in brokering oil, gas, and petroleum products at a global scale. These petroleum traders handle the majority of petroleum volumes traded on the global marketplace, and they finance producers in return for barrels sold.[113] While traders have not historically controlled significant volumes of oil and gas production, they are increasingly operating refineries and terminals. Their strategy is to maximize profits through arbitrage: taking advantage of differing prices for the same energy asset by simultaneously buying and selling energy commodities.

Global oil and gas commodity traders are some of the most mysterious corporations in the world. Traders are not household names—Vitol, Trafigura, Glencore, Mercuria, Gunvor, and Koch.[114] They are risk-taking operations that continually juggle billions of dollars in assets.[115] Addressing climate change is not their stated priority, although a couple acknowledge the importance of the issue.

## Certifying Agents

Look at the fine print in a given company's GHG inventory report, and you will find an assurance agent named.[116] Third parties—like Lloyd's of London, EY, and others—offer limited assurance on GHG inventories and ESG reports. Most assurance agents are careful to disclose that they are not involved in a company's GHG calculations and that they have no responsibility for the reported emissions data. Verification rests on knowledge that procedures were followed.

GHG accounting procedures date back to the 1996 Kyoto Protocol. They are viewed as a work in progress.[117] Intergovernmental Panel on Climate Change (IPCC) guidelines are established for countries, not companies. GHG protocols are not tailored to the corporate intricacies of oil and gas operations, with opaque, convoluted, internal exchanges between producers, refiners, shippers, and traders. For example, just because a gas station bears a given company's logo does not assign responsibility for all GHGs in the supply chain on the holding company.

Apportioning emissions to the responsible petroleum actor each step of the way is a complex undertaking. The goal is to fully account for all GHGs that are emitted into the atmosphere and not miss anything. Equally critical, it is important not to double-count emissions. But assurance agents *do not* verify GHGs emissions data at the facility level, the critical source of information in the oil and gas sector. Verification also contains fine print. Mentions of *uncertainty* typically appear alongside an acknowledgment that there are different measurement techniques that result in "materially different" tabulated GHG results.[118]

There are cases where verification is required for regulatory disclosure or emissions trading. For example, in California,[119] third-party emissions verification is required for sources over 25,000 tonnes a year.[120] For example, verification applied to only 60 out of 173 oil- and gas-related companies in California in 2018.[121] But in most cases, third-party verification remains voluntary. While it may reassure investors that corporate emissions claims have been independently reviewed, tools like the OCI+ can aid in double checking GHG accounting results.

## Banding Together

Oil and gas companies hold paid memberships in industry associations to protect their proprietary interests while forging joint positions on public policy issues, pooling their funds, and amassing political clout.[122] This "all for one, one for all" *Three Musketeers* approach is especially common when confronting environmental and other societal costs because the economic justification of the industry's preferred policies is more compelling when the benefits are amassed across the industry.

As of 2020, however, all IOCs and many other oil and gas actors are members of the American Petroleum Institute (API), the industry's most powerful collective voice. With some 600 members,[123] the API continues to promote petroleum consumption despite voicing climate concerns as early as in 1965.[124] While the API now acknowledges that climate change is a "serious issue," its core recommendation is to conduct "more research."[125] If the petroleum industry is going to be held accountable for mitigating climate change, the API will surely be central in developing concrete strategies to do so.

### Keeping Good Company

Beyond joining narrowly focused oil and gas interest groups, there are a handful of climate-concerned petroleum industry associations. For example, the

Oil and Gas Climate Initiative (OGCI) was formed in 2014 to accelerate the industry's response to global warming. Its twelve member companies (those listed in Table 5.2 plus Saudi Aramco and Oxy) support the Paris Agreement and are investing in low-carbon technologies and R&D to back up their climate commitments. The OGCI concentrates on four ways to mitigate climate change: improving energy efficiency and balancing GHG emissions through natural mechanisms and new engineering techniques like CCS.[126] The OGCI also aggregates corporate funds to invest in climate solutions for the oil and gas sector and has collected over $1 billion to implement and scale low-carbon oil and gas solutions.[127] Perhaps most important, the OGCI aims to build consensus among its members on climate action to accelerate a low-carbon energy transition.[128]

### Parting Bad Company

Industry coalitions can hinder an individual company's social license to operate, especially if the coalition opts for less climate action than the company itself. While some firms appear to be more out in front on climate change than their industry associations, only 8 percent of those companies surveyed ensure consistency between their climate positions and those taken by the associations they join.[129] Inconsistencies in policy positions between an individual company and an association they hold membership in may not even come to light. But, when it comes to climate change, inconsistent positions can create reputational problems.

Rather than compromise their positions, individual companies have started to drop out of industry groups because of their differences on climate change. In 2015, BP and Shell left the American Legislative Exchange Council (ALEC) specifically over its denial of climate science.[130] ExxonMobil, Dow Chemical, Occidental, and ConocoPhillips have also dropped out of ALEC. Others have indicated that they will follow suit. In 2019, Shell pulled out of another influential group called the American Fuel and Petrochemical Manufacturers (AFPM).[131] And in 2020, BP announced that it is withdrawing from the AFPM and two other trade groups, the Western Energy Alliance and the Western States Petroleum Association, over climate policy.

But parting coalitions cuts both ways. For example, in 2010, BP and ConocoPhillips left the US Climate Action Partnership because the coalition was lobbying Congress to pass GHG cap-and-trade legislation. At the time, these companies reasserted their support of their other trade groups—the API and National Association of Manufacturers—to shape climate and energy policy. Given this tug and pull, companies need to be mindful about changing public attitudes on issues like climate change when joining industry groups. Once a company decides to be a climate leader, it is worthwhile to either try to shift the coalition's stance or exit altogether if the group's positions detract from a company's efforts.

## Establishing Industry Benchmarks

The oil and gas industry has endured and prospered for over a century, sometimes weathering turbulent market swings. Those in charge adhere to internally derived corporate visions and strategies to guide them through these disruptive boom-and-bust cycles. But their job is likely to get visibly harder as tomorrow's oils and gases increasingly compete with one another for market share in a warming world. New benchmarks are needed to gauge how the oil and gas industry is faring.

Assessing which assets carry lower or higher climate risks is possible using the OCI+. And increased data disclosure and benchmarks by industry players will improve OCI+ estimates over time by helping fill information gaps. Companies must begin by taking steps to publicly release and routinely update credible climate plans, bolster internal climate governance, offer external support for the Paris Agreement, and join forces to back a clean energy transition.

### Publicly Release and Routinely Update Credible Corporate Climate Plans

Unlike nations that submit climate plans through their nationally determined contributions (NDCs), oil and gas companies do not routinely submit climate plans. Given their large GHG footprints (which can be bigger than those of individual nations), the single most critical benchmark is that companies have climate plans that are credible.[132] Such plans would enable governments to incorporate realistic projections of reductions in oil and gas emissions into the NDCs they have submitted to the UN. At a minimum, firms need to use open-source lifecycle GHG assessments, accounting transparency, and auditable data for all climate pollutants.

#### Use Open-Source Lifecycle Assessments to Evaluate GHG Levels

Because industry players expend so much energy on making and marketing oil and gas products, hydrocarbon resources with the greatest energy inputs result in elevated GHG emissions compared to those of otherwise equivalent assets. Few ordinary people are aware of this reality, which may even elude oil and gas workers. To make comparisons between different petroleum products easier and more accurate, the OCI+ applies lifecycle assessment (LCA) tools to count all of the

GHG emissions across the entire lifecycle of a barrel of oil (or cubic foot of gas). Companies that adopt this approach, using open-source tools like the OCI+, are better positioned to fully account for, better manage, and honestly inform the public about their climate risks.

## Increase Climate Transparency to Assess Progress on GHG Accounting

The dearth of data makes it difficult to assess climate progress. Companies need to gather and use valid, updated data to calculate their GHG emissions on both an operational and equity basis. They also should specify their total absolute GHG levels and emission intensities (per unit of oil and gas volume) by business unit. Companies need to take extra care not to play shell games, whereby they claim credit for emissions reductions that they merely shift to another party. Companies also need to provide sufficient information for assessing their energy and nonenergy projects, including current operations, start-ups, assets in the pipeline, and planned retirements of equipment and assets that they take offline and decommission.

### *Use Actual, Measured, Auditable Data for All Climate Pollutants*

In addition to $CO_2$, a company must report on how much methane, black carbon, and any other discernable GHG they produce from unintentional leakage (such as fugitive emissions), accidental releases (such as upsets), and regular operations (such as venting and inefficient equipment). Auditable data needs to be sufficiently detailed and disaggregated, so third-party agents can actually verify the amounts of GHGs emitted and not just that reporting procedures were followed.

## Bolster Internal Corporate Climate Governance

Climate change needs to be part of oil and gas companies' risk management frameworks for assessing financial performance and process safety. Employees need to be rewarded for their roles in adhering to climate targets. And companies should only associate with and be members of groups that make climate change as high a priority as they do.

### *Adopt a Process Safety Approach to Reduce Every Project's GHG Emissions*

A core mission of oil and gas companies known as process safety entails minimizing hazards impacting safety and business losses. While preventing such losses technically extends to environmental damage, companies must expressly develop plant systems and procedures to prevent unwanted releases of oil and gas that harm the climate. Doing so requires companies to acknowledge that climate risks are on par with safety risks, such as toxic releases, local fires, and explosions. The more knowledgeable and better prepared an operator is, the better it can manage and mitigate its GHG emissions. Moreover, during industry downturns, companies must retain their seasoned environmental specialists because it is difficult to carry out these safety functions without them and even more challenging to replace them.

### Include GHG Performance in Corporate Incentive Plans

Many oil and gas companies offer incentives to induce workers to perform better and reward those that keep operations running smoothly while maintaining the well-being of their employees. When industry players downsize, as many have during the 2020 pandemic, it is even more important for companies to reward strong performance amid layoffs and corporate restructurings. Already, some companies are linking CEO compensation to GHG emissions.[133] This benchmark should apply to all executives of oil and gas companies and should extend to individual employees within companies that are responsible for emission reductions, including methane leakages.

### Leave Industry Associations That Challenge Climate Science

Oil and gas companies, large and small, must contend with industry association policies that do not always mirror their own stated climate priorities. Rarely are industry associations' climate positions stronger than those of their weakest members. Moreover, smaller, less integrated companies by their sheer numbers can band together and outweigh supermajors in terms of voting power, a dynamic that also can weaken associations' climate positions. Yet when coalitions do not advance strong climate priorities, or when they even deny climate science altogether, they undercut a corporation's public efforts to reduce their GHG emissions. Such policy discordance not only sends mixed signals to the market but also leaves firms and industry associations sometimes working at cross-purposes on climate change.

Companies that are proactive on climate governance should not fund business groups, work with lobbyists, or participate in coalitions that are at odds with their stances. Moreover, every company needs to leave associations that challenge, peddle misinformation on, or deny climate science.

## Offer External Support for the Paris Agreement

Climate agreements only succeed if all actors support their aims and contribute tangibly to their mission. While there are numerous external actions oil and gas companies need to take, they can start with establishing governance proto-cols like science-based targets that align corporate climate plans with the Paris Agreement. Companies also need to advocate for a price on all GHG emissions, not just carbon, including their own lifecycle emissions.

### *Establish Corporate Governance Protocols That Are Supported by Science-Based Targets*

Each company needs to specify how much and how quickly it must reduce its GHG emissions to meet the goals of the Paris Agreement. Embracing science-based targets entails assessing whether corporate action is aligned with current climate science to limit global warming to 1.5 degrees Celsius higher than pre-industrial levels or well below the targeted ceiling of 2 degrees Celsius in tem-perature rises.[134] Companies need to commit to, develop, set, announce, and periodically update new corporate governance protocols to fulfill science-based climate targets.

### *Advocate a Price on Lifecycle GHG Emissions, Including Firms' Own Operations*

Oil and gas companies have been slow to forcefully advocate for the adoption of a price on GHG emissions. Internally, however, many companies assume a shadow price for carbon, which they factor into the economics of new projects. The situation shifted in June 2019, when several major oil companies pledged "to support 'economically meaningful' carbon pricing regimes."[135] This corporate announce-ment followed a personal appeal from Pope Francis for oil and gas companies to help the world deal with the climate emergency at hand.

The details matter. The benchmark for this corporate commitment should specify that the price be placed on a company's Scope 1, 2, and 3 GHG emissions and should expressly include methane and black carbon. Pricing GHG emis-sions fairly involves verifiable corporate GHG levels that prevent leakages and omissions. A tax applied only to end-use petroleum products or $CO_2$ alone is not sufficient. And the use of 100-year global warming potential (GWP) scaling factors for methane and other short-lived climate pollutants is not appropriate. Companies need to properly reflect and pay for emissions from their own pro-duction, processing, refining, and shipping operations.

## Join Forces to Jumpstart a Clean Energy Transition

No company can go it entirely alone to address climate change because oil and gas are versatile commodities that wind their way through complex supply chains as they readily change hands between many industrial actors. While pulling off a clean energy transition requires the participation of all actors, the oil and gas industry must play a leading role in forming partnerships to successfully research, develop, demonstrate, and deploy new technologies.

*Pursue Research and Development to Demonstrate Low-Carbon Technologies*

Historically, oil and gas companies have maintained active research agendas. While many of their R&D efforts have been scaled back in recent years, research is a critical benchmark for gauging whether a company is planning to operate in a low-carbon economy. A key goal is to eventually provide low-GHG substitutes for consumers' many petroleum needs. Beyond an individual company's R&D, meaningful interactions must take place in collaboration with industry, government, academic, and nongovernmental organization (NGO) partners. This includes funding novel low-carbon energy pathways as well as demonstrating and deploying new operations, products, and markets with an eye toward a long-term clean energy transition.

*Convert the Industry's Biggest Challenges into Climate Opportunities*

The oil and gas industry faces technical challenges that could potentially be converted into opportunities. Overcoming these hurdles requires collaborative innovation backed by ample funds for further study to produce breakthroughs. In situ molecular manipulation, for example, involves modifying the contents of a reservoir at its source to avoid removing its carbon altogether.[136] And discovering a way to convert heavy oil to hydrogen underground could eliminate the need for thermal recovery. Likewise, stripping $CO_2$ off acid gas below the surface could eliminate the need for posterior CCS processes. Another opportunity rests with a circular economy that keeps petroleum products in use as long as possible, decreasing the use of raw materials by reducing, recycling, reusing, and removing.[137] Companies are working to ensure that plastics, for example, are always reused, minimizing waste and reducing energy use and GHG levels.[138]

*Play a Leadership Role among Industry Partners*

Oil and gas industry actors tend to be fast followers that resist being the first to renovate their operations and alter their policy positions but quickly follow the leader once their competitors act, resulting in a wholesale sectoral shift. The tide appears to be turning among the European IOCs that are playing a leadership role on climate change, promoting industry best practices for GHG reductions, expanding their climate disclosure, and sharing cross-cutting knowledge to reduce emissions. There is more to be done, however. Leading companies can seek joint climate-related ventures with partners outside the oil industry (like automakers, airlines, and utilities) and work with suppliers to specify emissions reductions in energy procurement contracts and equipment specifications. Time will tell if the other IOCs, NOCs, and INOCs follow suit.

### *Partner with Other Researchers on Low-GHG Solutions*

The OCI+ team has demonstrated that, by partnering with peer, academic, and government researchers, the oil and gas industry can gain access to vast troves of data, talented scholars, and integrated problem-solving skills. Companies cannot afford to scale back their efforts and instead must boost them, especially when oil prices are low. Innovation is the lifeblood of industry, even in mature sectors like oil and gas. The companies that successfully reinvent themselves in the most socially responsible ways are the most likely to thrive in the future.

## *Rethinking Self-Regulation*

In August 2019, the Business Roundtable shook up Wall Street. Corporate CEOs committed to "protect the environment by embracing sustainable practices."[139] Of the nearly 200 industry leaders from Amazon to Xerox who signed on, only 1 in 20 has a direct stake in oil and gas.[140]

This pledge is noteworthy because it breaks with fifty-year-old accepted doctrine that businesses' sole social responsibility is to use their resources to "increase their profits."[141] There is an important caveat, however. Companies must engage in open and free competition without deception or fraud.

Oil industry actors have had a long history of blurring the lines when it comes to deceptive and fraudulent behavior. Robber barons like Rockefeller were not known for their scrupulous business practices. He fundamentally distrusted the free market and openly sought to suppress competition altogether.[142] Secret agreements were his signature move.[143]

Even today, the oil and gas industry is neither fully open nor freely competitive. Much of the world's oil and gas resources are nationalized, and oligopolistic forces influence global prices. For example, in 2020, oil price futures were driven below zero (in other words, sellers would pay buyers to take their oil) because the Russians and Saudis refused to stop pumping crude when the coronavirus pandemic decimated demand. Moreover, negative externalities, like climate change, continue to go unattended. Information asymmetries and data opacity distort every link in the supply chain. Charges have been made that, over the past half century (or longer), ExxonMobil (and other oil companies and their industry coalitions) not only knew their operations clearly contributed to climate change but also funded misinformation to avoid solving the problem.[144]

Perhaps it is unfair to expect the oil and gas industry to regulate itself when it comes to climate change. ARCO's reformulated gas signature move would not have made its mark on the industry without an external regulatory force. Given the climate realities of the twenty-first century, it is time to seriously rethink the carbon part of the hydrocarbon equation. Remaking the oil and gas industry is too big an undertaking for the industry alone to rise to the occasion.

## Defeating versus Partnering

Prevailing wisdom is that the oil and gas industry cannot effectively and durably regulate itself. Most companies are waiting on the sidelines for clear government action. Even companies that admit climate change is a problem place a higher priority on meeting global demand for petroleum products. Policymakers are needed to break this industry-stymied logjam.

A recent Twitter poll posted that two-thirds of participants responded that "defeating" oil and gas companies poses a more successful decarbonization pathway than "partnering" with them.[145] But what does defeating the oil and gas industry mean? Supplying petroleum, which all people consume, is a complex undertaking. Calls for a public takeover of Big Oil is no way to guarantee zero climate risk.[146] To be sure, today's NOCs do not perform better than IOCs.

By partnering with the oil industry instead, governments may be able to better align private interests with the public good. Just as fracking was first developed in a government laboratory before it was commercialized by industry, clean energy innovations that pertain to the oil and gas sector must also be incubated in the public sector. Good coordination and communication between industry and governments has never been more important. As chapter 6 shows, such collaboration underpins a balancing act that has been ongoing since the industry's founding. Now, however, the well-being of the planet is at stake.

# 6

## Governments: Acting in the
## Public Interest

GOVERNMENTS ARE IN a position to contribute to a global energy transition in ways that no other actors can, by crafting a robust regulatory and oversight environment for oil and gas giants to operate in. To date, however, most governments have stumbled and not yet delivered. In 2015, Paris hosted the United Nations (UN) Climate Change Conference, where nearly all nations made their initial pledges on greenhouse gas (GHG) emissions. Despite the achievement that the first global climate accord represented, failure lay in the wings.

Taken together, the commitments that these 196 countries made were woefully insufficient to meet the goal of limiting average global temperatures from rising by no more than 2 degrees Celsius above "preindustrial levels" and the dangerous levels of global warming a major temperature hike would usher in.[1] While half of all nations declared their intention to *increase* emissions through 2030, the agreement furnished no enforcement mechanism to hold countries accountable for their pledges.[2] Moreover, nations did not secure commitments from any oil and gas firms, and therefore, mitigation measures from producing, processing, refining, and shipping oil and gas were essentially missing in action. National commitments were voluntarily being updated in 2020, the first of ongoing five-year nationally determined condition (NDC) revisions under the Paris Agreement. As of spring 2021, just over 100 countries had updated their NDCs.[3] Eighty-five nations are uncertain or have stated that they have no plans to revise their NDCs in 2020.[4] Despite the coronavirus pandemic, atmospheric GHG levels continued to rise in 2020,[5] and it is unclear whether COVID-19 will ultimately disrupt NDC efforts or accelerate them through economic recovery packages that fund low-GHG energy systems and infrastructure investments.[6] If supply-side oil and gas emissions continue to go unaccounted for, the earth will greatly exceed its carbon budget.[7]

To compound matters, even as global governance involving carbon dioxide ($CO_2$), methane, and other GHGs (along with their oil and gas sources) remains feckless, the ways that national, state, and local governments have responded to climate change have been poorly targeted and politically unsustainable.

Climate solutions cannot rest on the shoulders of the public (especially motorists) when the direct target (the oil and gas industry) carries on with business as usual. In 2018, protests erupted in Paris over a long-touted proposed solution: fuel taxes. The yellow vests protesters—*gilets jaunes*, a nickname coined for the safety vests French motorists are required to carry in their cars—wanted to repeal the green tax on diesel. Rural workers who regularly drove long distances said they could not afford higher fuel prices. Although the protest morphed into a larger movement about declining standards of living, it was sparked by ineffective governance. Fuel taxes, especially on working-class people, are regressive, taking a proportionately higher toll on those with lower incomes. Pollution fees and other policies directed at the oil and gas industry are not.

Like with any policy issue, governments are tasked with balancing private interests and the public good in calibrating their responses to rising GHG emissions. To arrest climate change, policymakers, regulators, monarchs, and dictators have to reform the oil and gas industry, and this is a difficult role for them to play because governments are not neutral parties. Oil and gas profits play an outsized role in political decision-making. Petroleum royalty payments and resource rents stock government coffers. Petroleum trade buoys regional economies and is a source of geopolitical bargaining power. Government subsidies work to keep petroleum flowing.

Government researchers are well positioned to study and formulate a safer path forward for oil and gas in a warming world. Chapter 6 discusses prospects of progress on climate policymaking in the face of hard truths about the oil and gas industry. Oversight of and regulations on the international petroleum sector are lacking. Although there are notable exceptions, most nations are not successfully wrestling with oil and gas. While every nation has a role to play, the United States, Canada, Saudi Arabia, Norway, and Russia stand at the fore with their abundant oil and gas supplies. A failure to act forces local communities to pay the price from both petroleum development and the ensuing climate impacts. Beyond governance, the public sector is instrumental to long-term research and development (R&D), the root of most innovation in the oil and gas sector.

## Concerning Global Climate Governance

The challenge of climate change and rising GHG emissions demands concerted policy action.[8] Yet such action is unfortunately hampered by free-riding, when actors benefit from the positive effects of mitigation measures regardless of whether or not they pay their fair share of the costs.

A given government may conclude that it faces few incentives to restrain its own oil and gas sector even though the specter of climate change endangers everyone: if such reforms are perceived as costly and other parties may be willing to foot a good share of the bill, such government officials may be tempted to remain on the sidelines.

Successful climate policy requires deft navigation between competing public and private interests.[9] In 1992, participating countries at the Rio Summit in Brazil adopted the UN Framework Convention on Climate Change (UNFCCC) treaty to prevent "dangerous" human interference with the global climate.[10] The UNFCCC entered into force in 1994 with ratification by 197 parties.[11] Only North Korea, Kosovo, Vatican City, and Taiwan are not currently parties to the UNFCCC.[12]

But the treaty's emission limits are nonbinding and unenforceable. The parties meet annually at gatherings called the Conference of the Parties (COP) to adopt agreements (protocols), assess progress, and renegotiate the UNFCCC's climate objectives. When the 2015 UN Conference on Climate Change was convened in Paris, its participants struck the landmark multinational agreement that invited countries to submit their current emissions trajectories and routinely update their plans (NDCs) to keep future emissions in check to meet future climate targets.[13] Of the 186 NDCs initially submitted,[14] analysis of these submissions finds that nearly one in two nations intends to increase their GHG emissions between 2015, when the agreement was reached, and 2030.[15] Given another shot, countries are generally not upping their ambitions in their 2020 NDC updates.[16] Furthermore, NDCs are voluntary and countries can withdraw at any time, as the United States temporarily did during the Trump administration.

Tallying NDCs is complicated because emissions are calculated according to different baselines that must first be standardized.[17] Even after NDCs are aligned and totaled, a gaping hole remains to meet the agreement's target ceiling of 2 degrees Celsius in temperature rises. Governments need to exert policy pressure to durably reduce GHG emissions. Absent a direct enforcement mechanism, civil society (discussed in chapter 7) is largely tasked with holding countries accountable.

## National Affiliations That Matter

There are meaningful differences between countries' NDCs. There are also differences in their affiliations with membership groups, some of which are more supportive of climate action than others. Nations join forces and form affiliations for different reasons. With respect to energy policy and the environment, economic and political security tend to take precedence over climate action. Figure 6.1 charts the range in various countries' national climate commitments when grouped by their multinational group affiliations. The significance of membership in each of these groups is discussed next.

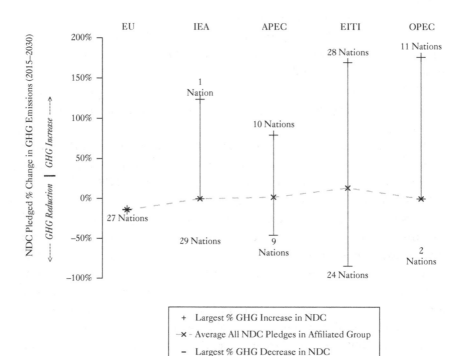

**FIGURE 6.1** NDC Pledges of Countries with Multinational Affiliations

*Note:* Updated NDC updates (starting in 2020) are underway and not included here. Single entry for the European Union because in 2015, all EU nations pledged a 14 percent GHG emission reduction by 2030. APEC, Asia-Pacific Economic Cooperation; EITI, Extractive Industries Transparency Initiative; EU, European Union; GHG, greenhouse gas; IEA, International Energy Agency; NDC, nationally determined contribution; OPEC, Organization of Petroleum Exporting Countries;

*Sources:* L. King and J van den Bergh, "Normalisation of Paris Agreement NDCs to Enhance Transparency and Ambition," *Environmental Research Letters* 14, no. 8 (July 26, 2019); European Union, October 5, 2020, https://europa.eu/european-union/about-eu/countries_en; Organization of Petroleum Exporting Countries, "Member States," https://www.opec.org/opec_web/en/about_us/25.htm; International Energy Agency, "Membership," April 24, 2020, https://www.iea.org/about/membership; Extractive Industries Transparency Initiative, "Members of the EITI Association – 2019-2022," https://eiti.org/files/documents/2019-2022_eiti_members_registry-22_09_2020.pdf; Asia-Pacific Economic Cooperation, "Member Economies," https://www.apec.org/About-Us/About-APEC/Member-Economies.

*Keeping Good Company*

Intergroup dynamics may be a means of promoting good stewardship. Member groups like the European Union and the International Energy Agency (IEA) are composed nearly unanimously of countries that are committed to making GHG cuts. Some of these nations also handle large volumes of oil and gas production, processing, or refining, including Norway, Australia, Mexico, the Netherlands, and Canada. In 2015, these nations pledged NDC reductions that, by 2030, would cut their emissions by 5, 10, 12, 14, and 30 percent, respectively. While the EU nations submitted a relatively-weak unified commitment in 2015 of a 14 percent GHG reduction, in 2021, their mitigation pledge was significantly strengthened to 55 percent.[18] Along with the IEA nations commitments and increased targeting of oil and gas in their 2020 NDC updates, these efforts could help shape climate leadership in other oil and gas nations.[19]

The European Union, with its twenty-seven (post-Brexit) member countries, was founded in 1958 to promote stability in a postwar world.[20] Energy security looms large in the European Union due to Europe's scant oil and gas supplies, a paucity that bolsters the call for energy diversification, integration, and interconnection. Climate mitigation measures may even open up some opportunities. European oil and gas companies are ahead of the pack, as discussed in chapter 5. EU policymakers can creatively engage these international oil companies (IOCs) in their host countries' NDCs, especially when it comes to promoting new net-zero GHG pledges.

Also, it is not surprising that nations affiliated with the IEA subscribe to sustainable energy development and climate security. The IEA, with its thirty member countries, is focused on combating climate change through energy planning, analysis, and policymaking.[21] In 2020, the IEA published analysis targeting the oil and gas industry and what it must do to combat climate change.[22]

*Pushing Associates to Do More*

But not all organizational camaraderie has given rise to lofty aspirations. Countries that are members of the Organization of Petroleum Exporting Countries (OPEC), Asia-Pacific Economic Cooperation (APEC) forum, and Extractive Industries Transparency Initiative (EITI) could rub off on each other in more positive, climate-friendly ways like the EU and IEA members are. Virtually all OPEC members, except for Equatorial Guinea and the Republic of Congo, submitted their first NDCs that projected *increases* in their GHG emissions between 2015 and 2030.[23] These nations remain at odds with the goals of the Paris Agreement. Of these, Iraq is the most concerning with its pledge to increase emissions by 170 percent.[24] APEC nations, with their highly disparate climate commitments, tilt toward increased GHG levels overall. Chile is the outlier, with a nearly fortyfold increase in emissions planned between 2015 and 2030.[25] And although EITI members have pledged some of the largest emission reductions in their NDCs, more EITI members plan to increase rather than decrease their GHGs by 2030.

It is not an easy pitch for these groups to refocus their priorities on climate change. OPEC, with its thirteen member countries in the Middle East, Africa, and South America, was founded in 1960 to stabilize oil markets and provide steady revenues for oil and gas producers.[26] Russia and the ten other nations that have since affiliated themselves with a less formal bloc known as OPEC-plus are not likely to be climate leaders.[27] But this reluctance may be slowly changing. In 2019, OPEC Secretary General Mohammed Barkindo stated that "the oil industry must be part of the solution to the climate challenge . . . and race to lower greenhouse gas emissions."[28] If just a few dominant oil and gas suppliers recommit to reducing their GHG emissions—countries like Saudi Arabia and Qatar, for example, as discussed later—that would contribute mightily to global climate progress.

APEC, a group formed in 1989 that currently boasts twenty-one members, has historically set its sights on economic growth. But with rising sea levels, worsening storms, and the spread of disease threatening the 3 billion combined residents of the APEC members,[29] the bloc's priorities are shifting when it comes to dealing with climate change.[30]

While some members seem to have elevated climate concerns, APEC nations have widely divergent priorities. The various members who put a premium on energy production (like Russia, Canada, and the United States), on climate risks (like Papua New Guinea, Indonesia, and the Philippines), and on broader economic objectives (like China, Japan, and Australia) confound APEC members' ability to exercise collective oversight on matters pertaining to the climate. A major opportunity exists for APEC nations to band together and avoid becoming the last place to unload the most GHG-intensive and highly polluting residual petroleum products, including fuel-grade petcoke and high-sulfur fuel oils.[31]

EITI, with its fifty-two members, was launched in 2002 to promote public resource management and accountability to reduce corruption.[32] In 2019, EITI reiterated its commitment to environmental monitoring and impact disclosure.[33] Similar to the APEC nations, billions of people live in countries that have adopted EITI's transparency standards. However, EITI can do more to promote climate action because over one-half of its member countries have made NDC pledges to increase their GHGs by 2030.

# Climate Oversight Lacking on Oil and Gas Operations

The enticing riches that oil and gas offer can blind nations to their climate responsibilities. Mentions of mitigation strategies by the oil and gas sector are glaringly sparse in essentially all NDCs. Most nations have not specified or sufficiently covered supply-side oil and gas emission reduction measures in their NDCs.[34] This is particularly problematic for national oil companies, where the government exerts direct decision-making power over the petroleum industry. Specifically, in a survey of fifty-seven countries, fewer than half (twenty-one countries) crafted some sort of oil and gas interventions to reduce their production-related GHGs, and eighteen resource-rich countries—including Angola, Colombia, Iran, Kuwait, the United Arab Emirates, Canada, France, and Mexico—stated their intentions to continue oil and gas operations.[35] Others, including Australia, Norway, Russia, and Brazil do not discuss oil and gas in their NDCs. The United States does not mention oil and calls out natural gas only once.[36].

The more oil and gas that a given country supplies, the larger its climate footprint tends to be. Climate pledges from the ninety nations with the largest oil and gas enterprises (each operating over 100,000 barrels of oil equivalent [BOE] per day) have projected an estimated production-weighted GHG *increase* of 18 percent by 2030.[37] A subset of thirteen countries—including the United States, Canada, Russia, and Saudi Arabia—dominate the petroleum sector worldwide, producing and refining nearly 170 million BOE per day of oil and gas. Figure 6.2 plots the relationship between the amounts of oil and gas these nations supply and their projected 2030 GHG emissions levels. Five countries (marked with a negative sign) expressed their intention to reduce their GHG emissions under the Paris Agreement, while the others (marked with a plus sign) plan to increase their GHG levels.

Many countries stated that they intend to increase the share of renewable electricity in their domestic energy supplies. Uptake of renewable energy sources is concentrated in developing countries, however, and not in the major oil and gas supplying nations plotted in the figure. Many nations aim to increase gas use while also reducing methane emissions.[38] But they do not offer sufficient assurances as to how they plan to prevent leakage of this potent GHG, which increases climate uncertainty.

The second round of updated NDCs began in 2020.[39] If past is prologue, countries will not obligate their domestic oil and gas suppliers to shrink their climate footprints. This would validate business projections to spend some $25 trillion on oil and gas infrastructure between 2018 and 2040.[40] Not only do national

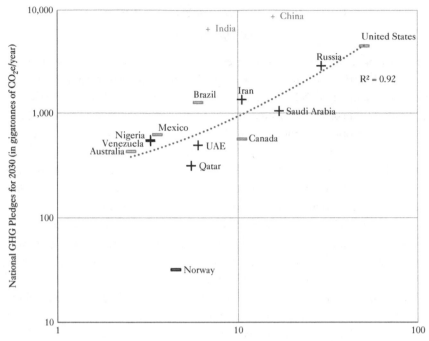

**FIGURE 6.2** Major Oil- and Gas-Producing Nations versus Their NDC Pledged Annual Emissions in 2030

*Notes:* Log-log scale. Countries with "−" pledged reduced GHG between 2015 and 2030; countries with "+" projected GHG increases; China and India are plotted for reference, but they are not calculated into the regression line (polynomial fit) because, while they refine significant amounts of imported oil, they are not major global oil and gas suppliers. BOE, barrel of oil equivalent; $CO_2e$, carbon dioxide equivalent; NDC, nationally determined contribution; UAE, United Arab Emirates.

*Sources:* Author's estimates; L. King and J van den Bergh, "Normalisation of Paris Agreement NDCs to Enhance Transparency and Ambition," *Environmental Research Letters* 14, no. 8 (July 26, 2019); International Energy Agency, "Statistics," https://www.iea.org/data-and-statistics; *Oil & Gas Journal*, "2018 Worldwide Refinery Survey"; US Energy Information Administration, "What Countries Are the Top Producers and Consumers of Oil?" https://www.eia.gov/tools/faqs/faq.php?id=709&t=6; US Energy Information Administration, "Petroleum and Other Liquids," https://www.eia.gov/international/data/world/petroleum-and-other-liquids/annual-petroleum-and-other-liquids-production; and International Energy Agency, "Data and Statistics," https://www.iea.org/data-and-statistics?country=ALBANIA&fuel=Natural%20gas&indicator=Natural%20gas%20final%20consumption; EIA; *International Energy Outlooks (various)*, https://www.iea.org/topics/world-energy-outlook

subsidies by the world's major economies make up a significant portion of these funds, but also governments are saying one thing and doing another.[41]

## National Climate Leadership Prospects on Oil and Gas

The world's high-volume oil and gas value chains (supply plus consumption) are largely a product of around thirty-five countries discussed later. Taken together, these national actors emitted nearly 40 Gt of carbon dioxide equivalent ($CO_2e$) in 2015, just over 80 percent of total reported GHG emissions.[42] Figure 6.3 charts oil and gas supply and demand for every nation. Bars on the right of the figure depict domestic volumes of oil and gas supplied and those to the left depict the volumes consumed. Oil and gas *makers* supply more hydrocarbons than they consume, which permits them to export a significant share of their resources. Oil and gas *takers* consume more oil and gas than they domestically produce and refine. And oil and gas *sustainers* are the rarity that supply and consume relatively similar amounts of oil and gas. In 2018, the United States stood alone as the world's largest oil and gas maker and taker, as shown in Figure 6.3. This distinction continues to hold despite the handful of countries like Russia and China that produce or consume meaningful but much smaller oil and gas volumes than the United States. Most of the rest of the world (150 nations) constitute a relatively small share of oil and gas supply or demand and are only sizeable in the aggregate.

Disaggregating nations' positions can help identify specific strategies for reducing GHG emissions. For example, gas-producing nations will require the most stringent methane rules and regulations, while oil-refining nations need to tightly regulate these industrial complexes.

Although nations (as well as states, provinces, and regions) have historically focused on cutting their oil and gas demand, no government to date has successfully dealt with the climate risks of its supply-side oil and gas activities—extraction, processing, refining, and shipping. While many governments exhibit weak oversight of their petroleum sectors, focusing efforts on the countries with leadership potential could offer the world a possible way forward. Several governments are discussed later, including the potential leaders, as well as where action is plausible and where the greatest climate risks lie.

### Acts to Follow?

National governments are not the only relevant policymaking actors on climate change and the oil and gas supply chain. Many energy and climate policies are implemented by subnational state and provincial governments.

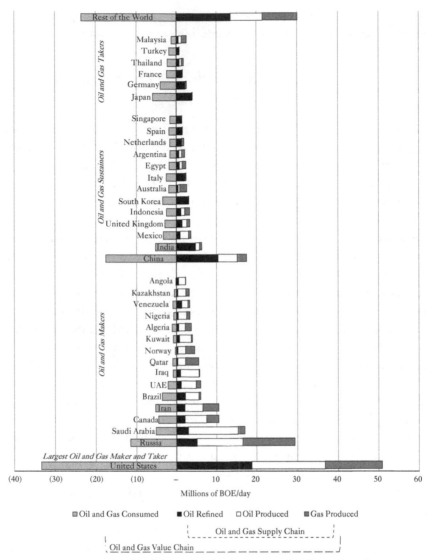

**FIGURE 6.3** National Oil and Gas Actors by Category and Volumes (2018)

*Notes:* Consumption volumes noted in parentheses to differentiate them from volumes in the oil and gas supply chain. Sustainers, production and consumption volumes within 1 billion BOE/day of one another; Maker, production greater than 1 billion BOE/day, but consumption is significantly less; Taker, consumption greater than 1 billion BOE/day, but production is significantly less; Maker and Taker, production and consumption volumes greater than 1 billion BOE/day, but not within this range of one another. US Energy Information Administration data do not distinguish between oil and gas industry consumers and all other consumers. BOE, barrel of oil equivalent; UAE, United Arab Emirates.

*Sources:* International Energy Agency, "Data and Statistics," https://www.iea.org/data-and-statistics; US Energy Information Administration, "What Countries Are the Top Producers and Consumers of Oil?," April 1, 2020, https://www.eia.gov/tools/faqs/faq.php?id=709&t=6; and *Oil & Gas Journal,* "2018 Worldwide Refining Survey."

To merely govern is not enough. Effective climate governance is indispensable.[43] Although no countries have perfected climate leadership, a few notable cases are worthy of mention. Canada, Norway, Australia, and the state of California, for example, have ample oil and gas resources (including unconventional supplies). They have also undertaken efforts to address climate change. Looking to the future, these governments could lead the way to provide greater resource transparency, target the oil and gas supply chain in their climate policymaking, and further bolster their strong R&D capabilities. R&D is especially crucial for developing and demonstrating the innovations in industrial processes needed to reduce the climate risks emanating from their oil and gas enterprises. Table 6.1 summarizes these nations, their affiliations, their NDC pledges, and their best respective courses of action.

### Canada: Where Innovation Could Pay Off

Canada, one of the world's largest gas producers and the holder of massive reserves of oil sands, could be an act to follow if the nation ceases political infighting on climate change. National policies to price carbon, promote clean electricity, and make buildings as energy efficient as possible rank Canada as a climate leader.[44] But when it comes to oil and gas supplies, the country's succession of elected leaders has tended to vacillate between protecting the environment and maximizing commercial energy development. Market forces prevail over climate policy as Canada struggles to prevent oil and gas spigots from turning on whenever global prices rise.

Nevertheless, given its strong research capabilities, Canada is well positioned to leapfrog others. For example, to the extent that methane is closely regulated,[45] Canada's natural gas supply could shape up to be one of the world's cleanest.[46] And future underground production of gas from oil sands—a process (discussed previously) designed to sequester the carbon in oil sands while producing only gas—holds out the promising possibility that Canada could ultimately become the world's top climate leader, bar none.[47]

### Norway: Upping Its Game

Norway staked out a leadership claim on oil and gas management decades ago. In 1972, the country's oil industry regulator adopted a Ten Commandments of sorts: commandment number five required oil industry development to protect nature and the environment, and number six made flaring Norway's offshore exploitable gas unacceptable.[48] Such environmental oversight remains a priority for offshore oil and gas development, and gas flaring is still prohibited except on an emergency basis.[49] In 1991, Norway was the first to introduce a carbon tax to

Table 6.1 Improving Climate Governance of Oil and Gas among Global Leaders

| State Actor | Oil and Gas Status | Affiliations | Total Supply[a] (MBOE/day) | 2015 NDC | Best Courses of Action |
|---|---|---|---|---|---|
| Canada | Maker | IEA, APEC | 10.5 | −30% | End political wavering; strengthen climate change priorities; increase R&D to advance in situ production innovations |
| Norway | Maker | IEA, EITI | 4.5 | −5% | Become lowest-GHG producer; fund oil and gas innovation with sovereign wealth fund; protect the Arctic |
| Australia | Maker | IEA, APEC | 2.5 | −10% | Prioritize climate change in resource development plans; reduce GHGs from LNG |
| California | Taker | [b] | 2.3[c] | −40%[d] | Collect and publicize oil and gas data; convert refining infrastructure to green hydrogen; conduct joint R&D with Canada |

[a] Includes total volume of domestic oil and gas production plus refined oil in MBOE per day.

[b] California is indirectly in the IEA and APEC as part of the United States, but it does not enjoy full membership benefits.

[c] California refines 1.9 million bpd of oil and produces 0.4 million BOE per day of oil and gas.

[d] California is not required to submit an NDC to the UN, but the legislature passed a law called AB 398 in 2017 setting a new GHG target of at least 40% below the state's 1990 emissions by 2030.

APEC, Asia-Pacific Economic Cooperation; EITI, Extractive Industries Transparency Initiative; GHG, greenhouse gas; IEA, International Energy Agency; LNG, liquefied natural gas; MBOE, million barrels of oil equivalent; NDC, naturally determined condition; R&D, research and development; UN, United Nations.

*Sources:* Author's calculations; United Nations; L. King and J. van den Bergh, "Normalisation of Paris Agreement NDCs to Enhance Transparency and Ambition," *Environmental Research Letters* 14, no. 8 (July 26, 2019); California Energy Commission, "California's Oil Refineries," https://www.energy.ca.gov/data-reports/energy-almanac/californias-petroleum-market/californias-oil-refineries; US Energy Information Administration, "California," July 15, 2021, https://www.eia.gov/state/data.php?sid=CA

prompt the oil industry to reinject (not dump) and electrify offshore rigs using renewable hydroelectric sources.

What is Norway doing now to shrink its oil and gas climate footprint? The country's carbon pricing scheme has not reduced its petroleum production, gas production is on the rise, and Norway is considering venturing into more sensitive ecosystems where an estimated two-thirds of its undiscovered hydrocarbon resources reside.[50] As discussed in chapter 5, Norway's own state-owned oil company (Equinor) emits far more GHG emissions on foreign soil,[51] which are not accounted for by the nation itself even though these revenue streams enrich its sovereign wealth fund.[52] How exactly its sovereign wealth is invested and spent,[53] whether Norway ventures deeper into the Arctic, and how aggressively Equinor reduces its emissions in other nations will determine if Norway remains a global climate leader that other nations can follow.[54]

### Australia: Canary in the (Gas) Mine

Australia has wavered on climate protection over the past decade. But its recent title as the world's largest liquefied natural gas (LNG) exporter could squash its future role as a climate leader.[55] Australia is a resource-rich nation that is well situated geographically to export its oil and gas (and coal) to energy-hungry Asian neighbors. This situation strengthens the position of the nation's fossil fuel lobby and explains its relatively small 10 percent NDC climate reduction goal.[56]

However, the increase in severe droughts, record-breaking heatwaves, and destructive wildfires has left Australians growing impatient for their national leaders to strengthen the country's climate governance. All eyes are on this ecological disaster–prone nation, which is already accommodating millions of migrants displaced by rising sea levels, water and food shortages, armed conflicts, and natural disasters.[57] Australia is the proverbial canary in the coal mine. How will its government play its oil and gas (and coal) hand to mitigate global warming in the decade ahead?

### California: Refocusing Climate Success on Oil and Gas

California produces, imports, and refines millions of barrels a day of some of the world's most GHG-intensive oil.[58] Yet, curiously, the oil and gas sector has not been the primary focus of the state's renowned climate leadership. Instead, California has focused its efforts on curbing the GHG emissions of the transportation sector, adopting regulations, like the Zero-Emission Vehicle mandate (ZEV), which promotes electric vehicles (EVs) and gave rise to Tesla,[59] and the low-carbon fuel standard (LCFS), which reduces the carbon intensity of California's pool of transport fuel.[60] And in 2020, the government announced plans to prohibit the sale of new gasoline cars by 2035.[61]

Meanwhile, oversight of the state's fifteen refineries and their climate foot-prints has been too lax.[62] Much more needs to be done to convert California into an overall climate leader whose achievements can be replicated elsewhere. The state needs greater data transparency on the oil it produces, refines, and imports so that supply-side GHG emissions can be fully assessed. The state needs to adapt its refining infrastructure to produce and use carbon-free green hydrogen.[63] And joint government R&D will be key: there are natural affinities between California's and Canada's heavy oil resources. Pooling research funds and dem-onstrations could offer the two potential partners tangible climate gains toward a clean energy transition.

## Where Greater Government Action Is Possible

All governments can—and should—more prudently oversee their oil and gas supplies in the future. The nations with powerhouse petroleum sectors that matter most in this regard include the United States, Brazil, and Mexico. Moreover, Saudi Arabia, the United Arab Emirates, Qatar, and Oman have the capacity to become frontrunners on climate mitigation. If these national gov-ernments, plus a host of states and provinces, commit to reducing the GHG emissions of their supply-side oil and gas operations, they could lower global climate risks given the large volumes of hydrocarbons they control. Table 6.2 summarizes these nations, their affiliations, their NDC pledges, and their best respective courses of action.

### The United States: Hitting the Oil and Gas Lottery, Struggling on Climate Progress

As of 2014, the United States has held the world titles for being the single-largest oil producer, gas producer, and oil refiner.[64] This notable record has endured ever since the shale oil and gas production grew in the 2010s, and no other coun-try has ever claimed this trifecta. This may seem like the perfect recipe for self-sufficiency, but it is not—even though the country's on-the-books oil and gas supplies outpace domestic consumer demand. Despite a favorable energy balance sheet, the United States imported 40 percent of the oil it refined in 2019 and exported roughly one-quarter of the oil it produced.[65] Likewise, by the end of 2019, the United States produced one-third more gas than Americans demanded, yet the country still imported 9 percent because infrastructure is lacking in cer-tain locations to ship gas domestically, while building infrastructure to export gas has a higher profit margin.[66] The future concern is that the United States' ongo-ing quest for energy security, real or perceived, will block climate progress—both at home and abroad.

Table 6.2  Improving Climate Governance of Oil and Gas among Powerhouse Actors

| State Actor | Oil and Gas Status | Affiliations | Total Supply[a] (MBOE/day) | 2015 NDC | Best Courses of Action |
|---|---|---|---|---|---|
| United States | Both maker/taker | IEA, APEC | 50.9 | −31% | End political wavering; strengthen climate change priorities; align federal agency efforts; convert refining infrastructure to green hydrogen |
| Saudi Arabia | Maker | OPEC | 17.0 | +63% | Reduce NDC; publish durable climate plan; use renewables in oil and gas operations; minimize flaring; commercialize CCUS |
| Brazil | Maker | IEA | 6.0 | −10% | Regain climate leadership role; develop low-GHG deepwater methods; maintain low-GHG biofuels production |
| UAE | Maker | OPEC | 6.0 | +53% | Reduce NDC; use renewables in oil operations; increase R&D to advance in situ production innovations |
| Qatar | Maker | OPEC | 5.5 | +69% | Reduce NDC; reduce systemwide methane leakage; reduce GHGs from LNG |
| Mexico | Sustainer | IEA, APEC, EITI | 3.6 | −10% | Strengthen climate priorities; reduce methane leakage; adopt California's and Canada's heavy-oil low-GHG innovations |

(continued)

Table 6.2  Continued

| State Actor | Oil and Gas Status | Affiliations | Total Supply[a] (MBOE/day) | 2015 NDC | Best Courses of Action |
|---|---|---|---|---|---|
| **Alberta** | Maker | [b] | 6.3 | −30%[b] | End political wavering; strengthen climate change priorities; increase R&D to advance in situ production innovations |
| **Texas** | Maker | [c] | 14.4 | n/a | Low-methane fracking; zero-routine flaring; tighter MRV; convert refining infrastructure to green hydrogen |
| **Other US States**[c] | Makers | [c] | 8.6 | n/a | Low-methane fracking; zero routine flaring; tighter MRV; employ methane feebates |
| **Japan** | Taker | IEA, APEC | 4.0 | −23% | Convert refining infrastructure to green hydrogen; purchase low-GHG certified oil and gas; increase low-GHG biofuels production |

[a] Includes total volume of domestic oil and gas production plus refined oil in MBOE per day.

[b] Total supply includes how Alberta set a GHG reduction target of 30% below 2005 levels by 2030. Alberta is indirectly in the IEA and APEC as part of Canada, but it does not enjoy full membership benefits.

[c] The US states with major gas fracking operations, including New Mexico, Colorado, North Dakota, and Pennsylvania. Texas and other US states are indirectly in the IEA and APEC as part of the United States, but they do not enjoy full membership benefits

APEC, Asia-Pacific Economic Cooperation; CCUS, carbon capture, utilization, and storage; EITI, Extractive Industries Transparency Initiative; GHG, greenhouse gas; IEA, International Energy Agency; LNG, liquefied natural gas; MBOE, million barrels of oil equivalent; MRV, measuring, reporting, and verification; NDC, naturally determined condition; OPEC, Organization of Petroleum Exporting Countries; R&D, research and development; UAE, United Arab Emirates.

*Sources:* Author's calculations; United Nations; L. King and J. van den Bergh, "Normalisation of Paris Agreement NDCs to Enhance Transparency and Ambition," *Environmental Research Letters* 14, no. 8 (July 26, 2019).

US political parties and their priorities constantly change over time, making climate progress a stop-and-go affair.[67] For example, the country dramatically withdrew from the Paris Agreement just a few years after leading the charge to draft it at the 2015 UN Climate Change Conference. Regulatory agencies have recently been working at cross-purposes. For example, as the Department of Interior opens up protected federal lands to oil and gas development and the Environmental Protection Agency (EPA) momentarily backtracks on methane rules and then picks them up again, National Aeronautics and Space Administration (NASA) researchers are establishing the world's preeminent Carbon Monitoring System (CMS) using satellites and other measurement and modeling methods to track oil and gas (and other) sources of global warming.[68] Like others, the United States can make great strides on climate change by focusing efforts on methane and applying its deep scientific knowledge in the oil and gas sector.

### *Other US States to Follow*

Aside from California, there are other oil- and gas-endowed US states to watch, including Texas, New Mexico, Colorado, North Dakota, and Pennsylvania. Each is engaged in fracking. All have inadequate governance structures and data collection systems that predate fracking and heightened climate concerns. None of them has entirely mastered managing the climate super-pollutant methane, which is inadvertently leaked and purposefully released throughout the value chain.[69] Fracking infrastructure and management practices need to be renovated now to significantly reduce the GHG intensity of oil and gas supplies from these states. And if states pursue massive infrastructure buildouts, including pipelines, LNG terminals, chemical plants, and other oil and gas systems, their investment plans will have to be reconciled with US climate targets and mitigated using carbon capture and storage (CCS) or other techniques to avert a dangerous increase in global temperatures.

Texas's fate and fortunes have long been tethered to oil and gas supply chains.[70] Still, the unbridled pace of development that occurred prior to the coronavirus pandemic was alarming—even by Texan standards. New Mexico, which shares the Permian Basin with Texas and also produces hydrocarbons from its Four Corners region, has recently emerged as a top oil-producing state. Although significant production occurs on federal lands that state policymakers cannot directly control, New Mexico's governor signed an executive order in 2019 for a comprehensive clean energy transition strategy.[71]

Colorado is adopting landmark methane rules and making up for prior mistakes on oil and gas governance by concurrently granting and denying permits for new oil and gas drilling.[72] North Dakota continues to experience clashes over building new pipeline infrastructure to move its (and Alberta's) oil bounties to distant markets in North America and beyond. Strict limits on methane flaring and venting could help North Dakota reduce GHG emissions from its oil production.[73] In the case of Pennsylvania, which went from mining coal (in the 1700s) to striking oil (in the 1850s) to fracking gas (in the 2000s), the state is an energy powerhouse that is unlikely to ever get off hydrocarbons. Given its abundant resources, Pennsylvania policymakers must overhaul their regulations so that the climate intensity of their oil and gas operations statewide are as low as best practices allow. How these other US oil and gas states tackle—or mismanage—their climate responsibilities will be telling. Success in one place can spawn more success in others, especially in basins that span state borders.

### Mexico and Brazil: Protecting the Climate during Political Shifts

While Mexico and Brazil pledged GHG reductions in 2015, this pair of oil- and gas-rich nations has undergone political transitions since the Paris Agreement was signed. The Mexican government has not been able to offer sustained attention to climate change. Although economic development is a top priority, other considerations like access to international finance, local pollution, and reputational pressures are driving Mexico's oversight on GHG emissions.[74] As a result, the Mexican government continues to struggle to deal with the adverse climate impacts of its predominantly heavy oil production.

Brazil was an early leader on global warming, hosting the historic Rio Summit in 1992, setting ambitious GHG reduction goals in 2009, and submitting one of the region's more ambitious NDCs in 2015. But in 2019, Brazil's current president, Jair Bolsonaro, citing fiscal constraints, refused to host the annual follow-on Conference of the Parties (COP 25) meeting, which had to be moved to Spain.[75] Brazil's oil and gas sector supplies roughly double that produced by Argentina, Venezuela, or Colombia.[76]

Setting aside the thorny question of whether Brazil should curtail development of its oil riches, the country could markedly lift its climate status by cutting GHG emissions if it were to adopt similar offshore drilling and flaring techniques that producers use in the Gulf of Mexico.[77] Increasing the production of biofuels, both ethanol and biodiesel, and integrating these gains into the petroleum industry could be a lift to the leadership standing of Brazil's oil sector.[78] Latin America's oil- and gas-rich nations look to Brazil for technical support. As such, the fate of the continent's climate potential may largely rest in Brazil's hands.

### The Middle East: A Need for Greater Climate Progress

Climate change could seriously disrupt life in the Middle East, one of the hottest regions on Earth.[79] Average temperatures here have already risen more than

2 degrees Celsius since preindustrial times.[80] Mass migration poses real security risks, and rising sea levels threaten to damage oil and gas supply infrastructure.[81]

Yet major oil and gas producers have failed to act decisively so far. Saudi Arabia has not exerted climate leadership—far from it.[82] Currently listed as a "very low performer," ranking sixty out of sixty one, Saudi Arabia lags in the Climate Change Performance Index, a collaborative nongovernmental organization (NGO) assessment tool that surveys countries' efforts on climate change.[83] In 2015, Saudi Arabia pledged a 63 percent *increase* in its GHG emissions by 2030.[84] And in 2019, Saudi Arabia disputed the findings of the global scientific community when it attempted to block the Intergovernmental Panel on Climate Change's (IPCC's) consequential move to lower the global climate target to 1.5 degrees Celsius temperature rise.[85] Today, volatile global oil prices, regional unrest, and future reductions in oil demand have Saudi Arabia questioning its economic and political future.

While the climate intensity of its oil production and refining is one of the lowest globally, because it produces a huge volume of oil, Saudi Arabia has a massive annual carbon footprint, some 215 million metric tonnes reported in 2015.[86] Getting off oil is not a viable option for Saudi Arabia, at least not in the near term. Instead, the nation is pursuing other avenues: using renewables to power gas production, minimizing flaring, developing CCS and utilizing the captured carbon, and investing in R&D.[87] But the country's self-proclaimed "ultra-clean energy" efforts are not enough. Saudi Arabia needs a robust climate plan, one that squarely addresses its oil sector, transparently tracks emissions, and funds R&D to achieve progress toward a durable low-GHG energy transition.[88]

Like Saudi Arabia, Qatar, the United Arab Emirates, and Oman are major global oil and gas suppliers with insufficient pledges that would *increase* their GHG emissions in 2030 by as much as 69 percent.[89] While they are not on the right path, their climate leadership on oil and gas matters.[90] Gas-rich Qatar is the largest global emitter of GHG emissions per capita when its huge volumes of exported LNG are included. While Qatar is adding natural gas capacity faster than solar energy, it announced plans to capture and store millions of tons of carbon from its LNG facilities and reduce methane leakage across its gas value chain.[91]

Meanwhile, the oil-rich United Arab Emirates is planning to use solar energy for up to 30 percent of its energy needs by 2030, including to power its own oil operations.[92] And Oman is planning major solar installations, including some to generate clean steam for its depleting oil fields.[93] In the sunny Arab Gulf, these efforts could significantly cut oil sector emissions in the short term. But over the long term, these oil and gas nations' economic ambitions must be complemented by loftier climate goals.[94] Otherwise, they will be relegated to the ranks of other precarious laggards on climate change.[95]

### Alberta: Capable of So Much More

Alberta is the province primarily tasked with slashing the climate impacts of Canada's heavy oil. Political wrangling has gone on for decades; one premier tightens the reins, and the next loosens them. Carbon taxes are adopted and then rescinded. Research funds flow and then dry up. Infrastructure is planned and then abandoned. The province—with its costly oil and long distances to market—is very sensitive to oil market conditions, which can rapidly accelerate (or halt) development cycles. For example, a nine-year project to extend oil sands mining was abandoned in 2020 when investors raised political, economic, and environmental concerns. Such on-again, off-again plans hamper Alberta's climate progress.[96]

Studies comparing Canada's oil and gas sector have found that Alberta has historically underreported its methane emissions by as much as 50 percent.[97] Still, Alberta has at least committed itself to tighter methane regulations.[98] Given its advanced R&D capabilities through government laboratories in partnership with local universities, Alberta could become a climate leader if it ultimately develops underground methods for extracting hydrogen and keeping the carbon buried. But oil-rich Alberta will never be a climate leader if it freely extracts its hydrocarbon resources using current methods, regardless of government restrictions.[99]

### Japan: The Preeminent Oil and Gas Taker

Japan's innovative capacity should not be overlooked, as this nation has little in the way of a domestic oil and gas supply but a large energy appetite. While biofuel substitutes for oil and gas remain a focus, Japan's refineries (which process nearly 4 million barrels of crude per day) together with its significant imports of petroleum products underscore that Japanese policymakers need to selectively do business with suppliers that certify low-GHG oils.[100] The government is working to assess the climate impacts of future crude oil production and has expressed interest in using the Oil Climate Index + Gas (OCI+) in its policymaking efforts.[101]

## Where the Greatest Climate Risks Lie

The rest of the world's oil and gas majors—Venezuela, Nigeria, and Russia—are climate laggards that pose serious climate risks. While, at present, China and India are not major oil and gas suppliers beyond their own borders, they could make meaningful climate progress if they lower the GHG intensity of the significant volumes of oil they refine. Table 6.3 summarizes these nations, their affiliations, their NDC pledges, and their best respective courses of action.

Table 6.3 Improving Climate Governance of Oil and Gas among High-Risk Nations

| State Actor | Oil and Gas Status | Affiliations | Total Supply[a] (MBOE/day) | 2015 NDC | Best Courses of Action |
|---|---|---|---|---|---|
| Nigeria | Maker | OPEC, EITI | 3.3 | +26% | Reduce NDC; zero routine flaring; reduce systemwide methane leakage; reduce GHGs from LNG; improvements from technical assistance |
| Russia | Maker | OPEC⁺, APEC | 29.3 | +12% | Reduce NDC; zero routine flaring; reduce systemwide methane leakage; avoid developing in the Arctic; strengthen climate change priorities |
| Venezuela | Maker | OPEC | 3.3 | +70% | Reduce NDC; strengthen climate change priorities; adopt California's and Canada's heavy-oil low-GHG innovations |
| Iran | Maker | OPEC | 10.5 | +48% | Reduce NDC; reduce systemwide methane leakage; zero routine flaring |
| India | Sustainer | IEA | 6.3 | +229% | Reduce NDC; convert refining infrastructure to green hydrogen; purchase low-GHG-certified oil and gas; enforce petcoke and residual bans |
| China | Sustainer | IEA, APEC | 17.5 | +24% | Reduce NDC; convert refining infrastructure to green hydrogen; purchase low-GHG-certified oil and gas; reduce GHGs from LNG; enforce petcoke and residuals bans |

APEC, Asia-Pacific Economic Cooperation; EITI, Extractive Industries Transparency Initiative; GHG, greenhouse gas; IEA, International Energy Agency; LNG, liquefied natural gas; MBOE, million barrels of oil equivalent; NDC, naturally determined condition; OPEC, Organization of Petroleum Exporting Countries.

*Source:* Author's calculations; United Nations; L. King and J. van den Bergh, "Normalisation of Paris Agreement NDCs to Enhance Transparency and Ambition," *Environmental Research Letters* 14, no. 8 (July 26, 2019).

*Nigeria, Russia, and Venezuela: Mismanaging Their Vast Resources*

Nigeria is the only nation in the world that reportedly made initial commitments to both wind down its oil and gas production and reduce production-related emissions.[102] However, the government provided no details, calling its climate commitment into question. Worse yet, since 1990, it is reported that Nigeria's GHGs have increased by 270 percent, a troubling trend that the government deserves help to reverse.[103]

As the world's second-largest oil and gas producer and refiner,[104] Russia is planning to vastly expand its oil and gas development in the Arctic. Hundreds of billions of dollars in investments are being tapped to create a major new Arctic company (Vostok Oil). Efforts to underwrite a massive new Arctic hydrocarbon development present a clear warning that climate progress cannot be left only to national leaders.[105] International governance will be required to safeguard such fragile ecosystems. Moreover, if petroleum resources with the lowest carbon intensities (like Saudi stocks, for example) are replaced by Russia's GHG-intensive Arctic oil, the planet's climate will pay the price.

Unlike Russia, which uses oil and gas to reinforce its geopolitical power, several other nations struggle to manage the climate impacts of their vast resource endowments due to political instability and civil unrest.[106] Venezuela, Iran, and Nigeria also depend on oil and gas to prop up their economies. But these nations, with NDCs projected to increase emissions by up to 70 percent, have not proven stable enough to balance their energy and climate goals.[107]

Their different types of unconventional oil and gas lead to assorted climate problems. Venezuela's heavy oil requires solutions along the same lines as Canada's wares. The light oil, condensate, and gas in Nigeria and Iran face challenges from excess methane emissions like those of Qatar—or Texas. For example, it is reported that temperatures in the already sweltering nation of Nigeria are 12 degrees Celsius (22 degrees Fahrenheit) higher in the vicinity of gas flares.[108] These nations will require technical assistance to reduce the climate risks of their resource endowments, but first they have to stabilize their political scenes and improve on governance.

*Taymyr: The Climate Bomb You Haven't Heard of*

Few have heard of Taymyr, a northern Russian province located at the mouth of the Arctic Ocean. Here, land temperatures have already hurdled past a perilous point by about 4.5 degrees Celsius since 1960.[109]

Despite these alarming numbers, Taymyr is also where Russia is planning to build fifteen new towns, a pair of airports, thousands of kilometers of pipelines and electrical lines, and a seaport to sail a fleet of top ice-class tankers. The goal is for Vostok Oil, a newly established Arctic division of the major Russian energy company Rosneft, to export millions of tons of northern Ural tundra oil.[110] India is a big investor in the venture, and Western investors are also expected to participate. Such projects could push the global climate over the edge. Russia cannot be trusted to safely govern in ways designed to mitigate climate change and protect its far northern regions. Instead, financing for such projects needs to come under greater scrutiny, so potential investors—Japanese, Indian, and Western backers—understand that developing this region's oil and gas supplies poses inordinately high climate risks.[111]

*India and China: Greening Oil and Gas Supplies through Refining*

While they do not supply vast volumes of oil and gas to the rest of the world, India and China loom large when it comes to safeguarding the climate. These major oil and gas consumers have been busy building out their oil-refining capacities, where they stand in second and third place after the United States.[112] They have been pursuing oil and gas purchase agreements and joint ventures with other nations to fill up their growing refining capacity.

How these nations supply their energy will have huge climate implications. In 2015, India and China respectively pledged 229 and 24 percent *increases* to their already massive GHG footprints by 2030.[113] But then in 2020, China pledged to zero out its GHGs by 2060 and to transform its economy away from fossil fuels.[114] India would benefit by following China's lead, if for no other reason than it would help clean up their dangerously polluted air.[115]

To achieve this, these economic giants could do several things to reduce their supply-side oil and gas GHG footprints and bolster their leadership credentials. China and India can preferentially trade with low-GHG producers; require greater oil and gas data transparency from their trading partners; ban imports of bottom-of-the-barrel, dirty petroleum products like petcoke and high-sulfur fuel oil; and invest in R&D to transform the refining sector beyond twentieth-century techniques. These supply-side measures could make a significant dent in global GHG emissions.

# Local Communities Pay the Price

For all this focus on how major countries worldwide can curb GHG emissions, the effects of environmental problems like smog, polluted water, and climate change are predominantly felt locally. Cities, suburbs, and rural communities feel the effects of various links in the oil and gas supply chain from daily operations, intermittent upsets, and horrific accidents. They also experience the fallout of climate disasters, like storm surges, droughts, fires, electricity outages, and pandemics. Often, local governments are well positioned to provide oversight in ways no other government actor can. But reducing GHG emissions is rarely their top priority—or actual charge. Climate resilience is the immediate focus in many localities. To follow are a few of the many stories about the clash between oil and gas supplies and climate change told by local journalists and countless residents.[116]

## Oil and Gas Hot Spot

Houston has one of the highest per-capita levels of GHG emissions in the United States.[117] The city is also extremely hot and getting hotter.[118] And Houston also took the brunt of a polar vortex that created a deep freeze in Texas in the winter of 2021.[119] But despite being inundated by storms and floods from record-breaking rainfall and despite suffering repeated triple-digit temperatures on an annual basis and recently without heat, power, water, and phone service for a full week,[120] Houston is not taking climate change seriously enough. Sometimes dubbed the "energy capital of the world,"[121] Houston is jam-packed with hydrocarbon infrastructure: refineries, gas plants, chemical facilities, pipelines, tank farms, company headquarters, and one of the country's busiest ports.[122] Rather than tightly govern the oil industry, Houston has the nation's highest-emitting refineries, has issued new permits for major LNG terminals, and is planning to expand its port.[123]

The area's climate footprint is as wide as it is large. Neighboring Beaumont, dubbed "cancer alley,"[124] is home to dozens of chemical plants.[125] And more facilities are in planning for the Gulf Coast region around Houston.[126] Beyond $CO_2$, chemical facilities pump out other powerful GHGs—methane and volatile organic compounds (VOCs)—by the ton. In addition to air pollution, these toxins continue to contaminate water and elevate risks of explosions and fires.

Even though the Texas Oil and Gas Association claims that local companies are committed to a future of lower emissions, Houston and Beaumont are climate laggards. Houston's first climate action plan does not even mention the oil and gas industry.[127] But there are numerous steps the two cities can take to reduce the GHG emissions in this sector, including improving reporting and transparency, requiring companies to conduct R&D on reducing their climate risks, imposing an industrial emissions fee on carbon and methane,[128] and retooling the plastics sector.[129]

## Piling Up Petcoke

Petcoke also continues to be a looming concern. In 2013, black heaps of the stuff three stories high piled up along a river in Detroit, Michigan.[130] It took some digging for the community to figure out that the substance was petcoke left over from Canadian-refined heavy oil. Since the petcoke was too dirty to secure a permit to be burned in US power plants, Marathon's oil refinery transferred it to Detroit Bulk Storage, which handed it off to Koch Carbon to be shipped down the Mississippi River, through the Gulf of Mexico, and across the Pacific Ocean to Asia.

In 2017, Detroit's municipal government issued a local ordinance requiring that petcoke piles be covered to prevent particles from being released into the air. But in 2019, Detroit's petcoke problem was again in the news, with Marathon requesting a variance to store their petcoke uncovered.[131] This chain of events calls into question whether oil and gas companies can effectively self-regulate to protect residents located near their operations. When hazards arise, locals are instructed to say something if they see something. As the case of Detroit and its petcoke shows, this mantra does not always come to the rescue.[132]

## Gas-Induced Earthquakes

One of the world's largest gas fields lies underneath Groningen, the Netherlands, and has been producing for about sixty years.[133] Long before that, windmills powered the region. With vast stores of its natural gas removed, Groningen's land is now sinking—and the earth beneath is shaking. Medieval churches are being ruined, homes are collapsing, ceilings are crumbling, and chimneys are falling. Costs to repair dwellings have run into the billions.[134] Meanwhile, the government is being forced to tear down buildings and halt gas production. This example further shows that the negative externalities of the oil and gas sector can profoundly upend the lives of local communities.

## *In Search of Public Sector Innovation*

The climate challenges that oil- and gas-supplying nations and localities face do not have simple solutions. Hydrocarbons are so thoroughly enmeshed in economies and political structures that surgical approaches are required to disentangle the two. Solving these problems will take deep study. In addition to academia (discussed in chapter 7), the actors best suited to the task at hand are government research laboratories, national academies of sciences, and public agencies.

### Twenty-First-Century Government Research

As one journalist observed about the power of public interest knowledge, "we've seen projections from non-profits and [private] research outfits before, but now comes the official government confirmation."[135] Tracking data, making projections, and exploring new systems—these are the special abilities wielded by experienced and far-seeing government researchers who have long time horizons and more durable budgets.

Historically, public sector innovation has had a long history of expanding and strengthening the oil and gas supply chain. Ludvig Nobel (Alfred's brother of Nobel Prize fame) is credited with creating the Russian oil industry, operating research laboratories that invented the oil tanker in the late 1880s. Over the course of a century, government researchers worldwide have contributed several oil- and gas-related advances to foster energy security. In the early 1980s, US government scientists undertook research to hydraulically fracture shale deposits, a technique that industry then commercialized to great effect.[136] Government scientists in the United States, Japan, and elsewhere remain focused on unlocking methane hydrates, the most plentiful supplies of natural gas worldwide.[137]

But not all government researchers concentrate on extending the life of oil and gas supplies. National laboratories have engineered processes that produce crude oil from harvested algae and other nonfossil energy feedstocks.[138] Global space agencies like NASA are collecting and studying satellite data to better attribute climate change to fossil fuels and other factors.[139] The German and Dutch governments are studying the production of green hydrogen from their offshore wind plants.[140] Net-negative GHG technologies like CCS and other climate-engineering approaches are also being investigated in government labs.[141]

The best and brightest minds are being called to the fore to research the low-GHG energy transition as the world shifts from energy security to climate security. Government scientists can unlock vast troves of data for analytics and can model complex energy systems. This knowledge is essential for developing a coordinated supply-side approach that serves as an oil and gas roadmap for investors, policymakers, and civil society.

## Building Back Better

Calls for investments in infrastructure keep growing louder. But such plans need to mitigate and adapt to global warming. For example, US President Joe Biden's plan to "build back better" rests on an economy that is stronger than the one prior to the start of the 2020 coronavirus pandemic. Accomplishing such an arduous task will entail investing in infrastructure, innovation, manufacturing, clean energy, and more.[142] Billions would need to be devoted to research incubators focusing on climate change solutions.

Such public investments in science and technology must be reinvigorated to reverse the major decline in relative spending worldwide (excluding China) over the past two decades[143]—lowered by a factor of four in the United States alone, as a percentage of both total outlays and gross domestic product (GDP) since 1960.[144] Major climate-friendly innovations in the petroleum sector are long overdue, and no other actor has the wherewithal to dig deeply into precommercial, cutting-edge R&D than government laboratories. In collaboration with industry researchers and civil society actors (academics and expert retirees), government laboratories and agencies have the know-how and staying power to develop a clean energy transition strategy.

The United States alone has seventeen national energy laboratories that translate basic science into innovation, and I have benefited from working with many of them.[145] This includes the Advanced Research Projects Agency-Energy (ARPA-E), the national government's advanced research project agency focused on energy. ARPA-E scientists are beginning to study a transition strategy for decarbonizing oil refining.[146] This effort and others like it could be further enhanced by the creation of a new federal agency, the Advanced Research Projects Agency-Climate (ARPA-C), an advanced research project agency focused on carbon and climate change.[147]

Such efforts do not stop at US borders. For example, Saudi Arabia has exponentially increased its R&D spending over the past two decades and launched research centers worldwide. Across the globe, government energy R&D spending has remained relatively flat and grew by only 3 percent in 2019, with the most funds spent in Europe, the United States, and China. The good news is that roughly 80 percent of these funds went to low-GHG technologies, such as CCS and green hydrogen—which took off in 2020.[148] However, government R&D tends to wax and wane amid economic cycles, a tendency that poses risks to the development of clean energy technologies, especially in countries that intend to grow their oil and gas sector in the coming decades.

## Getting Governments to Act

The sinking city of Groningen is a stark reminder of how little even experts know about the unintended consequences of oil and gas development. Crumbling homes foreshadow what the world will face as the effects of climate change intensify. Drilling, extraction, processing, and shipping of petroleum products are hazardous to people and ecosystems, especially as the quantities involved increase.

Ideally, market forces are supposed to help countries and the world reach and maintain a desirable equilibrium between the production and consumption of products like petroleum and prevent undesirable side effects, like negative environmental externalities.[149] Visible environmental fallout, like piles of petcoke, garner public attention and are still difficult to remedy. Invisible GHGs, like methane, are easily ignored.

Climate change and other negative externalities attributed to the oil and gas sector set up markets to fail. Special tools are needed when this happens, as discussed in chapter 8. Getting governments to act, however, involves pressure from civil society, as seen in chapter 7. NGOs, academia, the media, and the public play important roles, alerting others to the harmful aftershocks that accompany oil and gas resources. Many governments foresee an increased role for gas in their energy repertoires in the years to come. Civil society will be instrumental in holding governments and corporations accountable and helping societies around the world tackle this and the other manifold challenges that climate change poses.

# 7

# *Civil Society: Driving Change*

WHILE ADDRESSING CLIMATE change will undoubtedly require changes in supply-side business practices by the industry titans trading in petroleum and more consistent scrutiny and visionary leadership from governments national and local, these actors sometimes need to be spurred to action. This is where civil society in all its forms can come in.

Our Oil Climate Index + Gas (OCI+) research group was invited to give a seminar at the Bill Lane Center for the American West at Stanford University in April 2019.[1] Oil and gas factor prominently into the lands of western North America, including Texas's Permian Basin, Alaska's Arctic resources, Alberta's oil sands, and Mexico's vast stores of hydrocarbons. We shared results from modeling California's heavy oils through the OCI+ and discussed the prospects for reducing their elevated GHG emissions.

After the event concluded, an attendee who works for a nonprofit environmental group approached me. We discussed how increased data transparency could further guide Californian policymakers, both in the short term to clean up their complex oil and gas supplies and in the long term to structure the managed decline that Governor Gavin Newsom had recently announced.[2] This conversation led to further study and preparation of an internal report. Once armed with new information, the nongovernmental organization's (NGO's) workers crafted a policy proposal and approached potential legislative sponsors. In early 2020, a bill was introduced in the California Assembly.

The Know Your Oil bill—AB 3217—would provide long-overdue information disclosure from California's petroleum producers and refiners about the makeup of the complex oil they produce and process.[3] This policy could be adopted nationally and internationally.[4] This experience is a prime example of how civil society, the third sector (distinct from government and for-profit companies), can influence government actors to adopt policies that direct the petroleum industry to change its business-as-usual practices.

This chapter features the mixed cast that makes up civil society—NGOs, philanthropic foundations, academia, think tanks, the media, and the public at large.[5] These actors disseminate information, conduct advocacy, provide accountability, and participate in social movements, roles in which private and public actors do not excel.

Civil society has scored historic wins, such as pressuring the auto industry to make more fuel-efficient, cleaner cars and compelling electric utilities to invest in renewable energy sources. But, when it comes to the petroleum industry, these actors put too little energy into altering the oil and gas supply chain. Instead, they have been laser-focused on cutting consumers' demand for petroleum products. This blind spot is likely the result of too little impactful supply-side research and public outreach, the connective tissue that strengthens the outcomes that civil society drives.[6]

To shrink the climate footprints of petroleum producers, refiners, and shippers—especially those trading in unconventional oil and gas—NGOs must expand their capacity to model oil and gas, weigh GHG trade-offs, and hire staff with private sector experience. The media must focus less on reporting oil price fluctuations and more on reshaping the industry's future. Academics must collaborate with each other and with government researchers to connect deep knowledge to policymaking. Think tanks must delve more deeply into models like the OCI+ and integrate its findings into their own studies. Philanthropic foundations must take care to fund systems analysis and support partnerships with the public and private sectors. And activists must become more technically versed on oil and gas processes and industry practices. If civil society actors can expand and sharpen the tools available to them, they can play a critical role in successfully transforming the oil and gas sector in a warming world.

## Consumers versus Suppliers

Civil society has long focused on urging consumers to cut back on petroleum products while overlooking the substantial climate footprints of the producers of such products. In 1991, four environmental NGOs joined forces to publish a major report, *America's Energy Choices*.[7] I was part of the team that conducted the analysis. We envisioned a "climate stabilization" future that demanded half as much total energy economy-wide, replaced most oil with biomass, and furnished more renewables than natural gas. The report detailed nearly 100 national policies and pointed out federal agencies that could be tasked to implement them. This plan was designed to make the twenty-first-century United States more productive and competitive while cutting expenses and air pollution for all Americans.

Looking back, what we advised was rational at the time. In 1973, the world suddenly appeared to run out of oil when the Arab nations ceased exports to the West. Hence, the environmental NGOs launched in conjunction with Earth Day were convinced that oil and gas supplies would soon peak and run out. The logical next step was to conserve energy and pivot to cleaner alternative sources. But that isn't at all how things turned out.

No one, myself included, expected oil and gas supplies to resurface in abundance. By 2010, when I began my OCI+ research at the Carnegie Endowment, fracking, oil sands, and other abundant unconventional oil and gas resources were emerging. While this sea change fundamentally altered my thinking on how we must wrestle with petroleum systems themselves, many of my civil society colleagues have struggled to grasp this massive shift. Instead, most have doubled down on demand-side strategies, failing to appreciate that their actions now need to incorporate supply-side oil and gas approaches.

As civil society actors continued to push efforts to cut petroleum use, oil consumption increased 50 percent and natural gas use doubled between 1990 and 2019.[8] Civil society actors keep riding the third wave of environmentalism, urging citizens to "think globally, and act locally."[9] But tactics are changing. The new "environmentalism of everyday life" would have us reconfigure energy systems and material flows.[10] This shift could engage civil society to help rebuild a new zero–greenhouse gas (GHG) oil and gas industry supply chain, but that is only possible if we can move beyond the fixation that primarily consumers can solve the world's climate-related oil and gas problems.

## Consumers under a Microscope

It is a wide-held misconception that consumers are sovereign and can control markets by shifting their demands. While this is certainly true in the aggregate, it will take billions of consumers—from private citizens to commercial entities—acting in concert to meaningfully reduce oil and gas emissions. Table 7.1 details the variety of petroleum consumers and estimates their emissions contributions in terms of their share of total lifecycle GHGs for different oil and gas assets.

Such granularity is essential because the numbers can shift dramatically depending on what assumptions are used about the petroleum product in question and how it is used. For example, motorists buying gasoline may account for as little as 7 percent of the total emissions of a barrel of oil or as much as 42 percent, depending on what kind of oil it is. Likewise, utilities may account for as

Table 7.1 Sample Ranges in Share GHG Emissions by Minimum and Maximum Consumer Responsibilities

| Oil and Gas Consumers | Crude Oil | | Natural Gas | | Sample Oil and Gas Averages[a] |
|---|---|---|---|---|---|
| | Min. | Max. | Min. | Max. | |
| **Motorists** (gasoline) | 5% | 28% | <1% | 29% | 13% |
| **Air Passengers and Freight** (jet fuel) | 3% | 19% | <1% | 12% | 8% |
| **Truckers** (diesel) | 3% | 22% | <1% | 5% | 6% |
| **Industry, Businesses, Homes** (fuel oil) | <1% | 12% | <1% | 2% | 2% |
| **Industry, Shipping, Power** (residual fuels, LNG, petroleum coke)[b] | 2% | 36% | <1% | 12% | 9% |
| **Petrochemical Industry** (NGL feedstocks)[b] | <1% | 28% | <1% | 41% | 8% |
| **Homes and Businesses** (LPG for heat, stoves)[b] | <1% | 2% | <1% | <1% | <1% |
| **Power and Industry** (natural gas)[b] | <1% | 43% | 11% | 72% | 21% |
| **Everyone** (asphalt, solvents, lubricants, sulfur, solvents, and waxes)[b] | ? | ? | ? | ? | ? |

[a] Sample set includes 28 oil and gas resources in OCI+ Beta (excluding United Arab Emirates $CO_2$ Enhanced Oil Recovery estimates) and sum of these average values amount to ~68% (all consumers' responsibility).

[b] Industry consumer, including companies that may also make up the oil and gas sector such as LNG shippers and petrochemical plants.

*Notes:* Minimum and maximum shares are calculated for different resources and are not additive; assumes 20-year GWPs (100-year GWPs may differ, but generally represent similar min. and max. GHG ranges).

$CO_2$, carbon dioxide; EOR, enhanced oil recovery; GHG, greenhouse gas; LPG, liquefied petroleum gas; NGL, natural gas liquid; OCI+, Oil Climate Index + Gas; UAE, United Arab Emirates.

*Source:* Author's estimates using OCI+ Preview Web Tool, 2020.

little as 13 percent or as much as 100 percent of the total emissions in a barrel of oil equivalent (BOE) of gas.[11]

Those barrels of oil that produce less gasoline in effect transfer the responsibility for the remaining GHGs onto the shoulders of many other disparate consumers. Simply driving less will not necessarily cut all the carbon out of a barrel of oil. Conversely, some barrels of oil produce more high-GHG residual fuels, while other oil and gas assets produce more petrochemical feedstocks. In these cases, reducing the climate footprints of hydrocarbons (especially unconventional ones) implicates not ordinary individual consumers but industrial consumers, especially shippers and the oil and gas industry itself.

## Refocusing on the Supply Chain

Civil society actors have successfully taken on and overhauled supply chains in the past. Activist Ralph Nader forced the auto industry to make safer cars. Instead of telling citizens not to drive, Nader's NGO (the Center for Auto Safety) detailed how automakers could remake cars with seat belts, collapsible steering columns, padded dashboards, and, eventually, air bags.[12] Likewise, the International Council on Clean Transportation—an NGO focused on cutting vehicle-generated pollution—caught VW and other automakers cheating on their emissions tests and got them to retool their diesel engines and improve diagnostic systems.[13]

Civil society actors have also shown a propensity to focus on supply-side changes on other occasions involving the manufacturing and utility sectors.[14] Consider the example set by an NGO leader named Amory Lovins. Instead of telling chemical plants and data centers to turn off, Lovins's NGO (Rocky Mountain Institute, now renamed RMI) detailed how they could retrofit pipes, pumps, ducts, fans, and drive systems to be more energy efficient.[15]

It is puzzling why the same supply-side expertise is missing in the petroleum sector even though its emissions can account for a major share of GHG levels. Too few civil society actors are working to change oil and gas operations and the sector's business model. One notable exception is the International Energy Agency (IEA).[16] Instead of assuming that civil society can or should destroy petroleum demand, the IEA recognizes that oil and gas are critical in today's economy and that these supplies will be necessary in the future. The organization's position has evolved over time. When the IEA was first established in the wake of the 1974 shocks to the global oil supply, the NGO promoted demand reduction as a primary solution, an idea that was reinforced by high oil prices at the time.

But today, with an ample petroleum supply and low prices, it is curious why civil society actors cling to their old playbook to destroy demand for petroleum products. Three-quarters of Americans surveyed want to see higher emissions standards imposed on industry.[17] Public concerns (paired with mitigation strategies identified by the OCI+) bode well for the eventual tightening of industry standards that can lead to supply-side innovations in the oil and gas sector. There will always be a need for liquid and gaseous energy, and the petroleum industry must develop and commercialize low-GHG ways to supply these resources. Sustainability must be a top priority with the goal of reducing emissions and waste throughout the supply chain. Civil society has many tools at its disposal to help advance that goal, starting with information.

## Civil Society Approaches

Civil society actors have essentially four ways to advance their aims, including disseminating information, conducting advocacy, providing accountability, and participating in social movements. Each of these approaches encompasses numerous strategies, which civil society actors are well practiced at using.

### Disseminating Information

In the oil and gas sector, industry actors gather an abundance of technical minutiae—data from wells, refineries, and shipping operations worldwide. Systems data become commodities in and of themselves. Often these data are not publicly available and can only be purchased at exorbitant prices. Even then, problems linger as this information can be incomplete, outdated, or not well-suited for cross-comparisons.

The lack of information transparency hinders civil society's ability to see problems, recognize trends, and analyze solutions in the petroleum industry. This paucity of solid information leads to erroneous assumptions about oil and gas homogeneity that result in a tendency among activists to treat them all the same in terms of their climate impacts, a mentality that can lead to missed opportunities for reducing GHG levels.

### Educating the Public

Public education is critical for ensuring that citizens understand complex systems, like oil and gas production, that are dynamic and require ongoing vigilance. The knowledge involved is often highly technical and not readily consumed by the public. Instead, know-how needs to be translated into common terms and reshaped into concrete policy and public calls to action. For example, when oil and gas fracking emerged a decade ago, it stoked widespread public fears about increasing GHG emissions and other environmental impacts. We now know that global climate risks from fracking depend on tightly managing gas leakage (mainly methane) at all stages, whereas the regional risks to water supplies, air quality, induced seismicity, and other local hazards from these novel techniques require ongoing study.

NGOs—working internationally, nationally, and at the grassroots level—assume most of the responsibility for educating the public. They connect academic and think-tank research to policymakers and the public. Even well-funded NGOs with philanthropic support, however, rarely have the expertise and budgets to acquire knowledge gained through government grants and privately funded studies. NGOs often borrow from academic research to craft their public messages and policy strategies.

Public education is one of the most important tools that civil society actors like me and the other OCI+ researchers have at their disposal to address damaging environmental externalities that harm the public and hamper markets. Through ongoing research, publications, events, and outreach facilitated by NGOs and the media, public audiences worldwide have learned lessons about lifecycle GHG emissions in the oil and gas sector.[18] Online learning platforms cite OCI+ studies, and high schools have used our research in AP Environmental Science classes.[19] Our most global public education efforts to date include a presentation at the 2015 United Nations (UN) Conference on Climate Change in Paris,[20] two "Ask Me Anything" (AMA) online question-and-answer (Q&A) sessions through Reddit,[21] and a website devoted to OCI+ lessons learned.[22]

## *Diving in Deep*

Researchers at universities worldwide have probed deeply and focused narrowly on various technical and policy particulars that encompass all elements of climate change and energy policy. Think tanks are also diving deeply into these issues. Academics are some of the most valuable sources of new information. Academic journals (the currency used to promote academics) are peer reviewed, funding sources are specified, and conflicts of interests are disclosed. Analysis conducted and models developed by academics tend to be more transparent than the analytical tools used in the private sector and more readily available than those used in the public sector.

At its core, the OCI+ is an academic collaboration between researchers from three universities—Brown, Stanford, and the University of Calgary—who have spent their careers at dozens more academic institutions and collaborated widely with other scholars around the world. Over nearly a decade, the team has engaged undergraduate, graduate, and postdoctoral students and formed various forms of academic collaboration to widely spread knowledge.

*Scientific Assessments*

Groups of experts, like those affiliated with the Intergovernmental Panel on Climate Change (IPCC), provide scientific information so that governments can create sensible climate policies.[23] While the IPCC is an organization of governments, its assessment reports are written by volunteer scientists from academia, think tanks, and NGOs. Rather than conduct their own research, these civil society actors review thousands of scientific papers on climate change that are published each year and report on where experts agree and where further research is needed.

This open and transparent system of collecting and evaluating information informs governments, provides key input into international climate negotiations, and furnishes equally important and trustworthy information for civil society actors. For example, the IPCC's special report on the impacts of global warming that would ensue from temperature rises of 1.5 degrees Celsius above preindustrial levels continues to be used by numerous NGOs to further their own outreach and advocacy on climate action.[24] In their report, the IPCC highlights the need for a rapid drop in the use of oil and gas, with the stated goals of curbing emissions by 37 and 25 percent, respectively, below 2010 levels by 2030.[25]

*Giving Expert Input*

Opportunities for civil society actors to inject their knowledge into the political process can arise in many ways, such as through government testimony, in national academy studies, and on blue-ribbon panels. These formal occasions during which experts are invited to participate carry weight and remain on the record forever, reaching countless readers and listeners at home and abroad. Such fact-gathering efforts are often undertaken before governments tackle new problems, a fact that further heightens their influence.

I have engaged in several of these valuable public education efforts over the years. In 1992, I participated in an automotive fuel economy study that eventually led to tightening vehicle standards decades later.[26] In 1998, the Partnership for a New Generation of Vehicles pressed for research and development (R&D) on advanced automotive technologies.[27] Numerous National Academy of Sciences (NAS) efforts have focused on climate change research, but these have never centered on mitigating the climate impacts of the oil and gas industry. Instead, when it comes to studying petroleum systems, the focus has been on reducing accidents.[28]

More recently, OCI+ research has been featured in congressional testimony. These particular hearings focused on whether to lift the ban on US crude oil exports, an energy policy imposed in the 1970s.[29] While there was general agreement in these hearings about the newfound abundance of oil, what was particularly striking was the degree to which the congressional members as well as the other experts that testified alongside me did not know about the growing heterogeneity between different types of global oils. In terms of policymaking, different types of oils have different economic value and different climate impacts, variables that decision makers were not fully aware of. Moreover, when one of the other congressional witnesses testifying alongside me remarked that he was unaware of the impact of oil heterogeneity on climate change, I was reminded that not all experts are knowledgeable about everything. This blind spot has been common in my civil society discussions over the years.

*Calling on Citizen Scientists*

The US government piloted a grassroots effort in 2014 to engage citizen scientists.[30] In 2016, Congress adopted the Crowdsourcing and Citizen Science Act to facilitate broad public participation in innovation through open, voluntary collaboration between the public, federal agencies, and national laboratories.[31] This arrangement, whereby citizen scientists formulate research and solve complex problems, is well suited to help address the climate impacts of the oil and gas sector. Citizens' expertise is complemented by the recent wave of oil industry retirees who worked in environmental divisions and have first-hand experience mitigating emissions and devising energy transition strategies.

Beyond such engagement, the rapid rise of smartphone ownership has put powerful, miniaturized computers in the public's hands. Apps have been developed, for example, for citizen scientists to track the onset of seasons and report weather conditions. Drones are flying overhead to collect data. And emerging technologies, such as automated vehicles armed with cameras, lasers, and GPS systems, can be equipped as roving sensors to monitor real-time pollution levels.[32] Tomorrow's pedestrians, cyclists, and motorists may find themselves sleuthing environmental problems while they move about. Ultimately, real-time monitoring data collected by civil society actors could be input into the OCI+ model to update its emissions estimates and track changes in emissions over time.

## Conducting Advocacy

Those who possess unique knowledge—or have experience navigating the maze of rules that govern lawmaking—are well positioned to influence decision makers. While most professional lobbyists have previously served in government, a cadre of civil society actors also advocates from time to time. Whether the group involved is an NGO or a coalition of private citizens, the ability to petition governments is a cornerstone of democracy. But the line between interested individuals and parties with (often monetary) special interests can be easily blurred.

The more robust and transparent the data used to advocate for decision makers to take certain actions, the better equipped they will be to assess trade-offs.

### Telling a Better Story

In 1990, while working as a transportation analyst for the Union of Concerned Scientists, I advocated for vehicle fuel efficiency legislation introduced by Democratic Senator Richard Bryan (Nevada) and Republican Senator Slade Gorton (Washington). Staffers from other environmental NGOs joined me in these congressional visits as we fielded questions about the bill.[33] We knew the vote would be close, and the bill was eventually pulled from consideration because it was a few votes short of the sixty needed to limit debate.[34]

Civil society advocates are most effective when they are duly informed and tell the whole truth. During one key visit, legislative staff asked how much fuel would be saved if the bill were enacted. Before I could answer, my colleague chimed in citing a grossly inflated figure, as if corporate average fuel economy (CAFE) standards applied to all vehicles and not just the sliver of new cars and trucks sold each year. I remained silent. But as we exited the Hart Senate Office Building, I spoke up about his wrong answer. I expected him to thank me for pointing out the mistake. Instead, he replied, "I know. But I tell a better story." I then responded, "I will never advocate for a policy alongside you again." And, staying true to my word, I never did.

While it is not uncommon to push the boundaries when trying to influence decision makers, doing so can ultimately weaken the appeal of the advocate's message. This is why professional lobbyists—those paid to promote the private sector—are among some of the least trusted individuals in society.[35] Public distrust of many big businesses, like oil industry actors, further underscores the importance of unbiased, transparent assessment tools to fill the knowledge vacuum.[36] Such knowledge enables civil society actors to participate in what would otherwise be a one-sided decision-making process.

### Updating Your Story

Civil society actors tend to hold onto issues tightly and stick to their story. Since my involvement thirty years ago, environmental NGOs have advocated for CAFE standards on the grounds that they are the biggest step to combat global warming.[37] Such resolve can be an asset in a world with manifold concerns yet short public attention spans. But even the most dogged advocates should challenge themselves to update their arguments.

Justification and trade-offs for a particular action can change over time. For example, when CAFE standards were first adopted in 1975 after a major oil supply shock caused long lines at gas pumps, US oil imports were on the rise, and supply security was a growing concern. Figure 7.1 charts the imbalance between 1993 and 2013 when US oil imports exceeded domestic production. As minivans entered the market and light trucks replaced cars, gasoline consumption rose. So at the time, NGOs' argument for CAFE standards was justified: higher fuel economy could help manage the growth in US oil consumption. CAFE standards were finally tightened in 2007. Oil consumption fell 15 percent by 2013, but oil production was also on the rise due to fracking and other unconventional oil supplies.[38]

At this point, the story changed from an energy security issue to a climate risk problem. In a world of oil abundance, reducing gasoline consumption is no longer the pat solution it is cracked up to be. Refining a barrel of oil makes a wide array of petroleum products that trade globally. When an American motorist use one less gallon of gasoline than before, this will not remove one gallon of oil from the global economy. Instead, US refiners will either export this gallon of gasoline or process it into a different product, such as diesel or fuel oil for use domestically or abroad.[39] In other words, as long as plentiful oil is used to supply wide-ranging demand for various petroleum products worldwide, consuming less gasoline in the United States could shift GHG emissions but not eliminate them.

### *Convening Industry Retirees*

I joined Chevron in 1982, after back-to-back oil crises had spurred development of alternative (and ultimately unconventional) energy sources. The company amassed young, talented engineers who came of age during the environmental movement of the 1970s. Solving pollution problems was our calling. Although I left the company after five years, many of my colleagues remained and eventually began to retire in the 2010s. (This was the case at other oil companies as well.) I had the idea to reach out to industry retirees thinking that those with strong environmental ethics would be concerned about climate change. What started out as a couple of like-minded folks turned into a dozen. Together, we analyzed and promoted low-GHG petroleum strategies.[40] Having first-hand industry knowledge is a powerful enabler for both information and advocacy.

There are other such groups too, like the Citizens Climate Lobby (CCL), a grassroots advocacy group that educates and trains citizens to advance climate policies and engages industry and other retired energy professionals, including George Shultz (former secretary of state) and Steven Chu (former secretary of energy) who serve on CCL's advisory board.[41] From fossil fuel subsidies to carbon fees, CCL advocates for policies that could be further refined using results from the OCI+ model.[42]

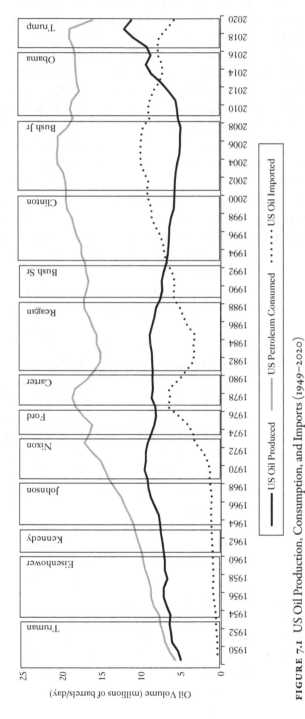

**FIGURE 7.1** US Oil Production, Consumption, and Imports (1949–2020)

*Sources:* Author's estimates using data from the US Energy Information Administration, including Field Production of Crude Oil, Imports of Crude Oil, and Product Supplied of Finished Petroleum Products, https://www.eia.gov/.

## Providing Accountability

Civil society serves as a watchdog, calling for accountability in the public and private sectors. The work does not end when problems are studied and laws and regulations are enacted. Various civil society actors stand guard to make sure government obligations are fulfilled and that industry actors comply with regulatory requirements. There are numerous ways that civil society can demand accountability, including by monitoring and reporting progress, advising stakeholders, and coordinating through partnerships. The media (both traditional outlets and social media) also plays an important role in accountability through the stories it tells its audiences.

### *Reading the Daily News*

While readers cannot swear by every word the media prints, journalists play the most visible civil society role in holding other actors accountable for their climate impacts.[43] Drawing on the insights of academics, think tanks, NGOs, and their own research, reporters post the vast majority of stories that examine and scrutinize the climate impacts of the oil and gas industry. Major outlets—like the *New York Times*, NPR, and the *Guardian*—frequently feature climate reporting. Targeted media organizations also remain hypervigilant on climate change, including Inside Climate News, Climate Nexus, ClimateWire, and GreenBiz, and blogs like *DeSmog* and *Heated*.[44] A slew of other outlets weigh in too.

Unfortunately, journalists can inadvertently minimize the oil and gas industry's direct role in global warming. These outlets along with academic journals sparingly report on the specific climate problems stemming from oil and gas operations. Climate news tends to focus on consumer action, maybe because it is more relatable than supply-side oil and gas GHG emissions. For example, the media often conflates all consumers (including industrial consumers), citing that overall petroleum consumption accounts for the majority of oil and gas GHG emissions. This chiding suggests that citizens are most responsible for mitigating GHG levels from the petroleum sector. But this is misleading. Differentiating end users' climate responsibilities, as detailed in Table 7.1, suggests that the world's more than 1 billion motorists, on average, may account for less than 20 percent of total lifecycle emissions from a barrel of oil (and even less in a BOE of gas). That means the other 80 percent of GHGs are spread out among the oil and gas industry itself (supply-side GHGs), an array of industrial consumers, and the public at large.

The media could help other civil society actors understand that, like a game of Jenga, petroleum products cannot be pushed out, one by one, over time to get the world over its reliance on oil. Instead, displacing entire BOEs of oil and gas and replacing them with decarbonized alternatives requires a technically and economically feasible systems approach—featuring both supply-side and demand-side strategies. The media is well positioned to convey this information.

## Monitoring and Reporting GHG Emissions

The constant flow of GHGs being pumped into the atmosphere risks serious environmental damage, much like a bathtub that is not draining. Monitoring emissions and their buildup has been improving over time. While industry-reported emissions were once the main source of data, other forms of global monitoring are increasingly the norm. Together, monitoring and reporting offer a powerful pair of tools for ensuring accountability.

Monitoring GHGs often involves civil society groups in partnership with government and industry actors. For example, a new NGO, Carbon Mapper, was established in 2021. This unique multi-stakeholder consortium consists of public (NASA-Jet Propulsion Lab and the California Air Resources Board), private (Planet), nonprofit (RMI), academic (University of Arizona), and philanthropic (High Tide Foundation and Bloomberg Philanthropies) partners and is on the cutting-edge of monitoring methane and carbon dioxide from oil and gas and other superemitting sources.[45] And in 2014, the Environmental Defense Fund launched the Methane Detectors Challenge, a multistakeholder partnership to enable oil and gas companies to detect and fix methane leaks in real time.[46] When civil society groups form coalitions—and invite industry and government partners to participate—there is a greater chance that innovative new practices will be implemented. Advanced technologies are no longer restricted to the deep pockets of industry leaders and governments. Civil society actors have at their disposal an increasing array of smart devices, including satellites, to undertake the next wave of technologically driven environmentalism.[47]

Emissions reporting remains vital to solving the puzzle of climate change, and the validity of the data is important. Civil society depends on GHG emission inventories to benchmark progress and identify mitigation targets. Such oversight is so critical that the United Nations has published a public handbook on the subject.[48] Reported data, once verified and made public, can enable certification standards for differentiating oil and gas GHG emissions. Like in the cases of fair trade coffee or sustainable palm oil, certification standards set by NGOs—such as low-methane natural gas supplies[49]—can be used in consumer advertising and by markets to monetize social impacts. Valuable emissions certificates for the oil and gas sector could also be traded alongside commodities themselves. Ultimately, this information could be used to identify the emissions intensities of consumer fuels to help consumers differentiate between more and less eco-friendly varieties of commodities like gasoline at the pump.

## Tracking Climate Disclosure and Compliance

Assessment tools are important accountability benchmarks for gauging how well prepared corporations are for tackling climate change. Companies are continually updating their climate ambitions and disclosing their GHG emissions, but these findings are rarely presented on an equivalent, readily comparable basis. (The same concern applies to the nationally determined contributions [NDCs] that countries have put forth under the Paris Agreement).[50] Companies mix and match their GHG accounting methods, sometimes disclosing the emissions over which they have direct operational control and sometimes reporting the share of emissions that they directly profit from (regardless of operational control). Both of these measures are critical in assessing corporate accountability, and all companies in the oil and gas sector should be required to disclose both metrics in the future.[51]

One civil society group called the Transition Pathway Initiative (TPI), staffed by researchers at the London School of Economics, is undertaking side-by-side comparisons of corporate climate pledges and assessing whether these commitments are aligned with the Paris Agreement.[52] As the public's climate expectations rise, companies add in new elements and think in more long-term ways. For example, corporate disclosures are increasingly featuring Scope 3 emissions and 2050 targets. All companies need to go further. The TPI cannot substantiate current corporate claims of "net zero" emissions or alignment with the Paris Agreement's targeted ceiling in temperature rises of 1.5 degrees Celsius. More transparent disclosure is required to help investors track companies' progress.

High-quality management is a lead indicator for corporate climate leadership. The influence of CEOs who prioritize the global climate can help accountability trickle down both within their own companies and among their business partners. The TPI calls out ten steps that companies can take to increase the accountability of the oil and gas industry and help investors assess their climate progress. These steps include (1) adopting a standardized approach, (2) committing to "net zero" emissions, (3) incorporating all Scope 3 emissions, (4) clarifying carbon capture and storage (CCS) and their emissions' offsets, (5) establishing total GHG as well as GHG per BOE targets, (6) detailing the role renewable sources play, (7) paying executives and employees who meet climate targets, (8) partnering with their industrial customers, (9) detailing Scope 3 targets, and (10) acknowledging climate change as a key management proposition.[53]

## Forming Coalitions

Very few deep-rooted societal problems can be fixed by any single actor alone. This is especially true for managing the powerful GHG methane. Progress on

reducing methane emissions has been painfully slow. In the face of barriers, advocacy coalitions were formed. To cite one example, a collaborative organization
called the Climate and Clean Air Coalition (CCAC) is a civil society–led collection of public and private actors aimed at demonstrating and overseeing systematic reductions in methane emissions.[54] This task involves developing guiding
principles across gas value chains, improving the accuracy of methane emissions
data, advocating sound policies and regulations on methane emissions, and
increasing industry transparency. Given that oil and gas are part of the energy
resource mix that will shape the future, it is imperative to improve the accountability and environmental performance of key industry actors.[55]

Likewise, the World Bank has been holding the oil sector accountable for
its gas flaring, the 160-year-old wasteful practice of burning off unwanted gases
throughout the petroleum supply chain. The World Bank's Zero Routine Flaring
by 2030 initiative, established in 2015, calls for gases to be reinjected, used onsite,
or gathered and marketed. In 2019, Saudi Aramco endorsed the initiative, joining
over eighty international oil companies (IOCs), national oil companies (NOCs),
governments, and NGOs worldwide.[56] The Visible Infrared Imaging Radiometer
Suite (VIIRS) government satellite, which visualizes flaring from space, has been
instrumental for tracking progress and holding parties accountable.[57]

### Challenging on Legal Grounds

NGOs and other civil society actors frequently file lawsuits when environmental damage takes place or rules are rolled back or unenforced. Often concerns
rest on whether the oil and gas industry has too much authority to determine
its own project impacts and whether oil and gas firms are doing enough to protect the public. Civil society actors and activists routinely use the courts to hold
the industry and regulators accountable and prevent future ecological and atmospheric damage, which is fundamental to environmentalism—litigation is "in
their DNA."[58] From blocking offshore drilling to challenging the Keystone XL
Pipeline (discussed later), NGOs parlay scientific studies into evidence to hold
polluters accountable.

Civil society actors in the United States and elsewhere are increasingly liti-
gating petroleum companies' contributions to global warming. Children,[59]
fishermen and farmers,[60] philanthropists,[61] and others have notably joined
forces with states and cities to demand climate accountability through various
courts.[62] For example, in 2015, various civil society actors released reports claiming ExxonMobil knowingly misinformed the public about climate science.[63] In
such instances, plaintiffs argue that companies should pay to repair infrastructure
that suffers climate damages and incur renovation costs to bolster future climate
protection out of the profits they reaped from taxpayers over the decades. And

in 2019, seven environmental NGOs filed a lawsuit on behalf of 17,000 Dutch citizens against Shell on the grounds that the company is "threatening human rights" by aiding and abetting climate disasters. In 2021, the court ordered Shell to drastically reduce its GHG emissions.[64] These cases pitch the industry's financial motives against civil society's wellbeing and moral resolve.

## Participating in Social Movements

When civil society actors strongly object to the actions of the public or private sector, they can assemble in protest.[65] At times, governments can undermine these efforts by force and other means.[66] For example, in 2020, at least three US states passed laws criminalizing protests against "critical" fossil fuel infrastructure.[67]

Protesting the climate impact of the oil and gas industry is a global phenomenon. Petroleum infrastructure (pipelines and platforms) is a common flashpoint. So too is drilling, especially when newer techniques (like fracking) or fragile ecosystems (like the Arctic) are involved. Industry accidents precipitate outrage, as does the very existence of big oil companies, like ExxonMobil in the United States and Shell in Europe (the Netherlands).[68] Revealingly, a Google search of "oil and gas protest" returns some 50 million results.[69]

Such activism, however, does not always involve gathering in person, blocking roads, hanging from bridges, or shrouding equipment with banners.[70] Recent campaigns are underway to empty the deep corporate pockets that otherwise invest in fossil fuels. These civil society–led efforts involve the divestment of various endowments and other funds that profit directly or indirectly off revenues generated by oil and other fossil fuels.

### *Blocking Infrastructure*

Markets struggle to make a sound case for new oil and gas infrastructure because demand fluctuates, suppliers are wont to overbuild, and government oversight is often politicized. Projects are costly and involve multiple permits, making infrastructure an ideal target for civil society activists whose protests can delay, derail, or significantly alter industry plans.[71] Success rests on tactics that reconcile with short- and long-term consumption patterns. As long as petroleum demand persists, all infrastructure cannot be endlessly blocked.

Take the case of the Keystone XL pipeline.[72] By the late 2000s, US demand for oil sands was up and projected to triple or quadruple between 2009 and 2030.[73] But landlocked Alberta did not have enough pipeline capacity to get more of these unconventional hydrocarbons to market.[74] As industry actors sought permits to expand the pipeline, questions arose about what Alberta-based suppliers planned to put in the pipeline. Details about the unconventional oil that would flow through nearly a dozen US states along the pipeline's route were sparse. Citing environmental concerns, the Obama administration blocked the pipeline as it signed the Paris Agreement because oil sands had higher GHG emissions and allowing them to flow into the United States would shift a share of Canada's carbon burden to its southern neighbor.[75]

Protests can delay the actions of industry players, but they can also have unintended (and sometimes harmful) consequences. In the case of Keystone XL, because pipeline capacity was lacking, diluted bitumen was exported by rail instead—a less safe and higher-emitting option.[76] In 2020, the Trump administration approved the Keystone XL pipeline just as demand for oil sands sank amid reduced mobility and shuttered economies during the lockdowns in response to the global coronavirus pandemic. The Trump administration's regulatory greenlight was not a definitive all clear to move ahead.[77] In fact, President Biden changed course and cancelled Keystone XL, a reversal that many states are legally challenging.[78] Low oil prices do not foster increased oil sands production, which means that, even if it is eventually permitted, customers could be slow to fill the expansion in pipeline capacity—a temporary victory for Keystone XL protesters.[79] But infrastructure plans can be put on hold only for building to resume down the line, which makes it hard for protesters to claim certain victory.

### Investment Movements

Because it is so apparent that money talks, civil society actors are increasingly using the power of their purses to exert influence over the oil and gas industry. Investment movements target financial actors such as banks, endowments, companies, and their shareholders. The bottom line is to apply normative and societal pressure to convince oil and gas firms and their investors to incorporate environmental, social, and governance (ESG) factors into their investment decisions. Divestment campaigns that remove investment capital from oil and gas companies are a popular tactical climate investment strategy. Other options include impact investing to provide capital to ESG firms or stranding assets by devaluing them and converting them to worthless liabilities.[80]

ESG-designated funds have been amassed with the idea of generating returns that are good for people and the planet. Since the Paris Agreement, various investors have made commitments to divest some $11 trillion in assets out of fossil fuels.[81] Philanthropic organizations and universities are in various stages of divesting their endowments, but not all of these actors agree that oil and gas divestment is a workable strategy.[82] Harvard University, for one, has been holding out and announced plans to decarbonize its investment portfolio.[83]

It is worth noting that investors need transparency to get the results they desire. For example, when activist investors pressed Marathon Petroleum to reduce its GHG emissions, the refiner moved to spin off and sell its Speedway gas stations to international firms—a move that shifts responsibility to other actors but does not ultimately put fewer emissions into the atmosphere.[84] NGOs and philanthropic foundations leading the divestment charge can get caught up in the drawn-out process that takes years to completely overhaul an institution's oil and gas investment holdings.[85] Although private actors—such as Bloomberg, Dow Jones, and others—have developed tools to track divestment efforts, detailed analyses are not publicly available or readily updated in a way that would provide constantly up-to-date scorecards economywide on the performance of the oil and gas sector.[86] This highlights the need for civil society actors to have access to open-source analytic tools like the OCI+ to assess different oil and gas assets' GHG emissions in order to fine-tune their divestment and investment strategies.

### Shareholder Activism

Each spring, publicly traded oil and gas companies hold annual meetings for their shareholders.[87] In the United States, investors owning shares worth at least $2,000 can offer a company resolution, as long as it follows guidelines issued by the US Securities and Exchange Commission.[88] (The bar for such resolutions is lower in Europe, where regulators cannot prohibit the placement of shareholder resolutions.) As an alternative to divestment, some investors are pursuing resolutions to press companies to lower their climate risks.[89]

Over the past couple of decades, shareholder resolutions have increasingly focused on climate accountability and risks. Coalitions of institutional investors are forming. For example, Climate Action 100+ (a bloc whose more than 450 investors collectively manage over $40 trillion in assets) is driving climate action through direct engagement with corporations.[90] And Ceres, a US NGO that participates in Climate Action 100+, tracks shareholder resolutions filed by members of their investor network in a comprehensive database.[91] Some hedge funds are getting involved too. In 2019, TCI vowed to vote in opposition to corporate climate resolutions when companies do not reveal their emissions levels.[92] And in 2021, an activist investor voted its shares to successfully insert grassroots activists onto ExxonMobil's board.[93]

European oil and gas majors have been more responsive to shareholder pressure than their American counterparts.[94] In 2018, Shell's shareholders persuaded the company to pledge emissions reductions from the products it sells. Likewise, BP's shareholders voted to require the company to align its policies with the Paris Agreement. One of the best uses of shareholder resolutions is to broadly require corporate data transparency from companies that are notoriously opaque about their current practices and future plans. Disclosing details about oil and gas assets under development relates directly to projected GHG emissions and helps assess whether a company may (or may not) be profitable under future climate policy-making scenarios.

### Mentoring Youth Activists

Intergenerational issues with existential risks, like climate change, are lightning rods for young activists who are concerned about the condition of the world they will inherit.[95] Many young people are fed up with the status quo. In a recent US national poll, two out of three teenagers surveyed said that the oil and gas industry causes problems rather than solves them.[96] They want action. The prominent Swedish teenage climate activist Greta Thunberg, for example, famously initiated a "school strike for climate" and sailed from Stockholm to New York to address a 2019 UN Climate Conference.[97] Thunberg wants governments and businesses around the world to move faster to cut carbon emissions.

Such bold goals require serious educated thought, as showcased at Oxford University in February 2020. Students occupied the bursar's office and demanded that the college *immediately* divest its shares in oil and gas companies. In response, the bursar made the students a counteroffer saying that, while he could not arrange any divestment at short notice, he could immediately arrange to shut off the college's gas-powered central heating. The counteroffer concluded with the line "please let me know if you support this proposal."[98] Certainly, the bursar was being provocative. But he was also instructing the students to think clearly when he explained, "It is all too easy to request others to do things that carry no personal cost to yourself."[99] This experience was hopefully a wake-up call that may encourage such students to become highly educated about the complexities of oil and gas, especially supply-side dynamics, and labor judiciously to change the course of this mighty sector.

Likewise, Harvard College has engaged with students and faculty over a work-able path to achieve net-zero GHG emissions. When some students and faculty members demanded that Harvard divest its $40 billion endowment from fossil fuel holdings,[100] university management countered with its preference to deploy the university's unique strengths—teaching, research, and operations—to "prepare for and accelerate the necessary transition to a fossil fuel-free economy."[101] In 2020, Harvard embarked on a five-year transition to enhance its investment portfolio's transparency and produce comprehensive data and methodologies to assess its own lifecycle carbon footprint. This effort is not unlike a country's preparation of an NDC under the Paris Agreement. Harvard's success in developing a credible climate plan, one that openly and durably supports net-zero emissions, will help guide other civil society actors as well as countries around the world. A deep understanding of oil and gas markets—and why altering the supply side is more crucial in the short term than cutting demand—will be crucial as young people assume future leadership responsibilities in private companies, government posts, and civil society organizations.

## Coordinating Civil Society

The varying skill sets of different civil society actors are well suited to different approaches. Some require more knowledge and others require more action to reach their ends. Connecting these various levers offers a multipronged approach, as shown in Figure 7.2. Using the OCI+ as an example, academics and think tanks deeply study the oil and gas industry's GHG emissions to develop new models and solutions. These data in turn make their way into news stories as journalists report on a given model's findings, creating the need for greater accountability. NGOs then read these reports, write their own reports, and advocate public and private decision makers to increase the industry's transparency, alter its behavior, and make new rules. And protesters assemble, as the yellow vests have in Paris, to weigh in and apply pressure for systematic change.

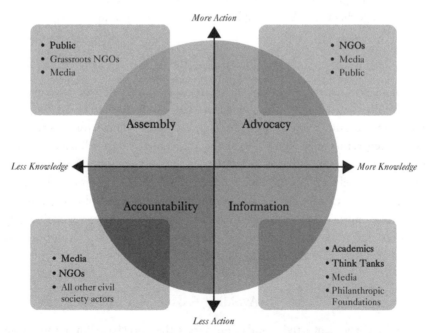

**FIGURE 7.2** Civil Society Coordination by Levers, Actors, and Capacities

*Note:* The central civil society actors that tend to lead in each capacity are bolded. NGOs, nongovernmental organizations.

*Source:* Author's assessment.

## Strengthening Civil Society

The forte of civil society is its massive size, underpinned by billions of citizens worldwide who are free to unite. Unlike industry, these actors are not bound by trade secrets and antitrust measures. Unlike governments, they are not wedded to certain leaders and bound by procedural red tape. When NGO activists, professors, media mavens, and philanthropists band together, civil society can be formidable. This hive approach injects a mighty force that can overpower the private sector and government interests that ordinarily dominate decision making in the oil and gas industry.

Still, there are limits to civil society's capabilities that must be addressed for civic actors to be highly effective. Attending to their weaknesses can make civil society actors stronger, especially as they look to be engaged on climate issues over the long haul. More intellectual honesty is needed about the massive size of the task ahead. If fossil fuel substitutions were as simple as many civil society actors profess, then the world would have been off petroleum decades ago. And if companies could transition as readily as civil society actors claim, these enterprises would have moved beyond petroleum already. Civic actors should continue to hold governments and companies to high standards, but they must also readily acknowledge that the task is a long-term one that offers few quick or easy fixes.

### Herding Cats

Managing civil society actors and channeling their energy in productive ways is notoriously challenging.[102] This unenviable task can be likened to herding cats. However, successfully mitigating the oil and gas sector's climate impacts requires civil society actors to be highly disciplined, organized, and coordinated. Instead, many of these actors struggle to gauge success, shed staff, and fundamentally reshape themselves when necessary. Studies find that successful civil society engagement on climate change and corporate social responsibility depends on constantly improving institutional design and capacity.[103] The better civil society balances clear goals with institutional diversity, innovation, and evolution, the more likely these civic actors are to ultimately succeed.[104]

### Avoiding Fads

As vital as the work of activists and nonprofits is, the world has a limited attention span and limited funds for underwriting the large number of civil society actors that are active worldwide. The good news is that civil society invites participation, much of it voluntary (though reliance on volunteers can contribute to the coordination challenges of herding proverbial cats as discussed earlier).

On any day of the week, a new protest is waged, a new NGO coalition forms, a new academic study is released, or a new media story is published. This eagerness to act sets up a fierce competition for bandwidth and money. One civic organization's relentless push to grab attention and pull purse strings, whether those of philanthropic foundations, government agencies, private companies, or ordinary citizens, can sideline others' efforts. The desire to chase bright and shiny objects or new, momentarily trendy fads can subvert real solutions. The climate-forcing capacities of the oil and gas supplies that simultaneously underpin the world economy require nuts-and-bolts systematic approaches to craft enduring change.

## Capturing Co-benefits

Whether the societal problem activists are combating is deadly air pollution, abject poverty, police brutality and discrimination against racial minorities, or other forms of social injustice, many climate activists resist embracing other real-world issues.[105] Nevertheless, there is (or can be) intersectionality between climate change and other societal concerns, and looking for solutions with multiple benefits can reduce barriers to change. According to the IPCC, such climate mitigation "co-benefits" can involve "the mitigation of air-pollution impacts, energy-supply security (by increased [clean] energy diversity), technological innovation, reduced fuel cost, employment and reducing urban migration."[106] This is especially true in the case of methane, which is both a potent GHG as well as an air pollutant that sickens people and shortens lives. The United Nations highlighted these co-benefits for reducing methane in a hallmark 2021 report.[107]

Climate advocates should not focus too narrowly on the energy policy of climate change, because their success is deeply tied to addressing other societal inequities.[108] For instance, the world will have a hard time addressing climate change until climate-minded activists acknowledge that the burden of curbing GHG emissions cannot fall disproportionately on already poor and marginalized populations who may have less latitude to shoulder the costs of greener energy solutions. Civil society actors cannot afford to be focused too narrowly on climate advocacy as if no other social ills matter. This is especially critical in the Global South where elevated risks of premature deaths from air pollution and energy insecurity compound other forms of destruction from climate change. With tailored regional inputs, the OCI+ can be adapted to model co-benefits like air and water pollution.

## The Trappings of Membership Groups

Earth Day commemorations in 1970 rallied tens of millions of citizens and boosted membership in environmental groups worldwide. Civil society groups surged in size by nearly 40 percent between 1969 and 1972, many of them operating via memberships.[109] When individuals participate in these ways, they pay dues and provide opinions. While membership-based NGOs offer democratic governance, this organizing model can come with less advantageous trappings, especially as their agendas increase in complexity and require greater strategic thinking. Studies show that, when civil society organizations reach a certain size, the ability of their members to shape the NGO diminishes as they invest in professional expertise to carry out their work.[110] The more technical the issue (such as modeling lifecycle GHG emissions in the oil and gas sector), the more general memberships in civil society groups can be a hindrance rather than a conduit to knowledge. For civil society to carve out climate positions and set goals for the oil and gas sector, such groups must complement nonexpert, action-oriented individuals with technically trained staff.[111]

## New Knowledge for Civil Society Actors

It is an uncontested fact that burning fossil fuels injects GHGs into the atmosphere, emissions that build up, blanket the earth, and increase its temperatures. Scientists have known this for centuries, and policymakers have wrestled with a response for decades. Industry actors continue to manufacture and market climate-forcing oil and gas commodities. Civil society actors are trying to stop them, while also pressuring governments to act.

Yet the planet itself will be the final arbiter of how much ecological damage is too much. Since this is hard for the public to adjudicate, it means that civil society's job is never done. The bad news is that civil society actors have no sure way to strand oil and gas and permanently keep the carbon such resources generate in the ground despite the harm they cause.[112] They also cannot make laws or change industry operations of their own accord. Nor can they credibly generate an actual red line—a demand that industry actors may not cross: the point of no return is up to Mother Nature herself.

The good news is that civil society actors have sheer numbers and dogged resolve in their corner. Moral suasion is on their side. And, if they so choose, the full complement of these actors, forming broad alliances, can embolden civic activism.[113]

However, civil society actors must be extremely careful not to impose more harm than good when using their full arsenal. For example, those promoting low-carbon biofuels can inadvertently cause destruction of forests that exacerbates climate change and create food insecurity. And past efforts to shift to more fuel-efficient diesel cars led to an increase in air pollution that indirectly results in more climate forcing. Such trade-offs are not uncommon.

As such, one thing civil society needs more of is information. Their effectiveness ultimately rests on knowledge about facts that are constantly changing. This is certainly the case regarding the dynamic global structures and natural and manmade systems that both climate change and oil and gas markets operate within.

Open-source, user-based tools like the OCI+ are at civil society's full disposal. Eliciting more oil and gas data transparency, NGOs and others can help facilitate model updates and expansions. Such progress can provide civil society actors with more forward-looking information in a virtuous cycle.

In the meantime, what we are doing now is not working. The oil and gas industry, with its large and variable climate footprints, still dominates global energy supplies. The oil and gas industry experiments on the atmosphere with each pound of carbon dioxide, methane, and other GHGs that it emits. Civil society actors counter by experimenting with oil and gas markets each time they propose changes to the petroleum sector. Ultimately, it will fall to governments to judge how to balance private interests with public goods. With these challenges in mind, we must construct a plan. This task is covered in the book's next, and final, chapter.

# PART III

## The Pathways

# 8

## Moving Forward: Oil and Gas Climate Solutions

TACKLING THE PLANET'S climate problems in a credible way will require a novel mix of demand-side and supply-side solutions, an adaptive approach that even some of the world's most ardent, well-intentioned proponents of climate mitigation efforts do not seem to fully grasp. The Obama White House hosted a climate summit in the fall of 2015 after eighty-one major companies signed on to the American Business Act on Climate Pledge in support of that year's United Nations (UN) Climate Change Conference in Paris.[1] I spoke on an industry panel alongside representatives from Walmart, Procter and Gamble, and Alcoa. The other panelists offered up their solutions, which ranged from using suppliers with lower greenhouse gas (GHG) emissions to cutting operational carbon intensity through material and process innovations.

I reported on the core findings of the Oil Climate Index + Gas (OCI+) and how the petroleum sector could mitigate climate change, pointing out that heterogeneous oil and gas resources have wide-ranging climate footprints and that these differences are large enough to matter. If we shift our mindset to assume that oil and gas are here to stay, we can actually minimize their supply-side emissions now instead of imagining that they will go away.

The panelist next to me whispered, "I had no idea." Surprisingly, some of the most savvy people in government, civil society, and industry have expressed similar sentiments. This reaction keeps coming up in seminars with faculty and students, academic collaborators, government scientists, public agency analysts, legislative staff, policy analysts at global nongovernmental organizations (NGOs), foreign officials, security experts, foundation officers, think tank experts, and countless journalists. Even petroleum company representatives are taken aback by the large variances in the GHG emissions of various types of oil and gas and what this means for climate change.

But not knowing is no excuse for inaction—not when it comes to climate change or any other existential threat. The first step to heightening transparency about the climate impacts of oil and gas involves getting the OCI+'s insights about climate change and GHG emissions into stakeholders' hands. That is the reason I wrote this book.

While it is essential, knowledge alone will not reliably shrink the climate footprints of oil and gas. In reality, there is no single strategy that any one actor can execute to turn the planet's climate around. Instead, like the vexing arcade game whack-a-mole, blocking the oil and gas supply chain in one place with one tool can cause hydrocarbons to re-emerge elsewhere rather than disappear altogether. Such tactics do not result in the net reductions the planet needs: they just lead to shell games in which companies pawn off emissions on each other while overall GHG levels continue to rise. For example, the historic breakup of Standard Oil in 1911 (a piecemeal approach) did not prevent oil and gas—with its attendant climate impacts—from resurfacing elsewhere. This event merely converted the industry into its present, more convoluted and intractable structure.

Successfully solving the oil and gas sector's climate problem requires a surgical approach, one that makes near-term improvements while formulating a long-term vision. In this final chapter, I summarize four guiding principles for twenty-first-century oil and gas systems. I then lay out pathways for reducing the supply-side climate footprints of oil and gas. These small-scale, incremental 2 percent solutions involve modest reforms that add up over time as we strive to make complex oil and gas systems more environmentally sound. The book concludes by envisioning a twenty-first-century end game for decarbonizing this critical economic sector.

## Guiding Principles on Climate Change for the Twenty-First Century

If the petroleum sector vanished today and rematerialized in 2030 in a dangerously warmer world, what would the oil and gas industry look like and what environmental, social, and governance (ESG) directives would it operate by? The four following guiding principles are meant to help answer this question. All climate mitigation solutions should check one or more of these boxes.

### Principle #1: The Public Has a Right to Know

Most ordinary citizens are unaware that no two oils or gases are the same in terms of their GHG emissions. Some otherwise-equivalent barrels of these resources can emit three times more total lifecycle GHG emissions than others, as discussed in chapters 3 and 4. There can be as much as a tenfold difference between the lowest and highest levels of supply-side GHG emissions generated by the production, processing, refining, and shipping of a given barrel of oil or cubic foot of gas depending on its type. Regardless, officials assign an average GHG emission factor of 432 kilograms of carbon dioxide ($CO_2$) to a barrel of oil, while a barrel of oil equivalent (BOE) of natural gas is reported to emit 308 kilograms of $CO_2$.[2]

Markets cannot function efficiently when such different types of oil or gas are essentially treated the same or when their highly variable GHG emissions are hidden. Opaque, incomplete, and missing information leads to excessive emissions and squandered opportunities to reduce GHGs. Greater data transparency on oil and gas is needed to inform the decision-making necessary to reduce climate risks for consumers, homeowners, workers, manufacturers, investors, insurers, and other members of society.

## Principle #2: Prices Should Reflect True Societal Costs

Climate change will continue to impose damage and costs on societies and ecosystems around the world, with a price tag that many economists believe is currently underestimated.[3] While oil and gas emit the majority of global GHGs, these emissions are not reflected in their prices. Plans put forth to date to price carbon fail to reflect the full scope of the petroleum sector's true societal costs.[4] As such, these artificially low costs do not take into account all GHGs like black carbon, do not accurately reflect the added warming caused by short-lived climate pollutants like methane, and do not fully tally lifecycle emissions throughout the oil and gas sector.[5] The economic cost caused by an additional ton of both short- and long-term climate pollutants being released into the atmosphere converts into a variable cost per barrel, depending on which oil or gas is used.[6] Beyond petroleum-based fuels, every petroleum product including plastics and other petrochemicals must be priced to account for the GHGs emitted from cradle to grave.

## Principle #3: Industry Must Be Responsible for All Supply-Side GHGs

It is difficult to track the GHGs from the multitude of industry actors connected through complex, convoluted, and changing corporate structures and joint venture business arrangements. Studies reveal that there are no independent sources of data on oil and gas emissions, and all government datasets rely primarily on self-reported statistics.[7] Pollutants and byproducts are routinely leaked and not fully accounted for, including methane, black carbon, $CO_2$, volatile organic compounds, petcoke, and single-use plastics.[8] Leakages occur when infrastructure does not function properly or when operators wantonly dispose of unwanted byproducts.[9] Holding producers, oil refiners, gas plant operators, shippers, traders, and all other industry actors responsible for all supply-side GHG emissions can help improve operations, system designs, government oversight, and public safety.

## Principle #4: Policymakers Should Place Supply-Side Oil and Gas on a Safe Climate Path

Oil and gas emissions need to be reduced to keep the earth's temperature from rising beyond a ceiling of 1.5 to 2 degrees Celsius above preindustrial levels. The Intergovernmental Panel on Climate Change (IPCC) traces various pathways for keeping emissions in check.[10] The most aggressive scenario envisioned to date calls for dramatically lowering oil and gas demand by 87 percent and 74 percent, respectively, by 2050—excising roughly four out of five barrels of petroleum from the economy.[11]

This is unrealistic. Demand-side policies and voluntary supply-side measures will not be enough. Industry and civil society can take voluntary actions like aligning future business plans with the principles enshrined in the Paris Agreement. But the onus for decarbonizing oil and gas systems falls on the shoulders of governments, especially in the case of hydrocarbon assets with the highest GHG emissions. A climate path well below 2 degrees Celsius cannot simply proceed with business as usual, using twentieth-century petroleum supplies. Public-private-nonprofit collaborations can advance monitoring, research and development (R&D) and the commercialization of GHG-intensive industry operations and petroleum product substitutes. Government incentives can also help the public and industrial consumers use low-GHG petroleum products.

## *Pursuing New Pathways*

The progression from simpler to more complex oil and gas supply chains over the past century necessitates greater transparency, smarter decision-making, and stricter policy guidance. Demand-side actions such as urging consumers to drive less and use carbon-free fuels will no doubt be necessary, but these measures will not provide sufficient climate protection by midcentury. Consequently, supply-side oil and gas climate strategies must feature prominently in any realistic plans to reduce near-term GHG emissions in line with the Paris Agreement.

Five main mitigation tools are available to governments, industry, and civil society actors. These include (1) publicly monitoring, reporting, and verifying lifecycle GHG emissions; (2) formulating voluntary agreements and binding regulations; (3) enacting sanctions and prohibitions; (4) conducting R&D and pursuing technology transfer; and (5) offering financial incentives and disincentives. Table 8.1 presents dozens of potential mitigation policies that private, public, and NGO actors can employ to harness each of these tools. Using the OCI+, actors can input particular oil and gas data to determine which strategy has the best chance of excising the most GHGs. Such analyses are a prudent way to factor climate impacts into supply-side decision-making in the oil and gas sector.

These pathways are not mutually exclusive. In fact, strategies can—and should—be mutually reinforcing. For example, greater data transparency through more robust oversight can inform focused standards, smarter incentives, and critical R&D that leads to successful technology transfer.

## The Sum of 2 Percent Solutions

A comprehensive fix to the problem of climate change would be ideal if such an option were viable. Unfortunately, such a silver bullet does not exist—one rarely does. Even though no single approach will quickly turn the titanic oil and gas industry around, there are numerous elegant approaches for tackling the tough climate challenges imposed by supply-side oil and gas GHG emissions. Such small cuts can add up to sizable results—what I refer to as the *sum of 2 percent solutions*.

The sections that follow detail what policymakers, industry, and civil society can do. The more these actors coordinate, the more impactful the outcome will be.

### What Policymakers Can Do

Government actors are discussed first because there is no substitute for policy action to motivate the wholesale changes required to decarbonize supply-side oil and gas enterprises. Most industry actors will not invest in near-term climate mitigation without government leadership (or the real threat of leadership fueled by civil society actors). There are dozens of government policies that merit consideration. Table 8.2 identifies which of the four guiding principles apply to fifteen promising government-led strategies.

## Table 8.1 Sample Oil and Gas Supply-Side GHG Mitigation Strategies

| GHG Mitigation Tools | Government | Industry | Civil Society |
|---|---|---|---|
| Public Monitoring, Reporting, and Verifying (MRV) Lifecycle GHGs | Collect open-source oil and gas data; mandate auditable company GHG reports; update GHG inventories; expand and integrate remote sensing and expedite public reporting | Submit and update corporate climate plans; create a new reporting framework for oil and gas supply-side GHGs | Demand near real-time MRV for oil and gas GHGs; establish low-GHG certification for oil and gas; tap the expertise of retirees from industry and government |
| Voluntary Agreements and Binding Regulations | Update oil and gas GHGs in NDCs; tightly regulate methane; coregulate GHGs and air pollutants | Decommission targeted oil and gas assets; commit to zero routine gas flaring, fugitives, and venting | Advocate for Know Your Oil regulations; incorporate supply-side oil and gas emissions in updated NDCs |
| Financial Incentives and Disincentives | Price oil and gas sector lifecycle GHGs; establish a methane feebate on the oil and gas industry | Incentivize top-down corporate climate mitigation; stop financing high-GHG projects | Promote incentives to safely decommission and redevelop oil and gas properties |
| Sanctions and Prohibitions | Ban residual fuels and waste byproducts; prohibit operations in fragile ecosystems | Apply future production quotas to high-intensity GHG oil and gas assets | Tailor divestment decisions to lifecycle GHG emissions; push endowments to adopt GHG investment criteria |
| R&D and Technology Transfer | Update GWPs; facilitate clean energy tech transfer; promote R&D for nonenergy petroleum products; research climate engineering impacts | Integrate renewables into oil and gas systems; accelerate research on net-zero emission technologies (NETs) | Conduct R&D on refining for a clean energy transition; facilitate tech transfer of open-source GHG tools and remote detection systems |

GHG, greenhouse gas; NDCs, nationally determined contributions; R&D, research and development.
*Source:* Author's assumptions.

Table 8.2  Government Strategies to Reduce Supply-Side GHG Emissions*

| Government Supply-Side GHG Strategies | Principle #1: The Public Has a Right to Know | Principle #2: Prices Reflect Social Costs | Principle #3: Hold Industry Responsible | Principle #4: Position Sector on Safe Climate Path |
|---|---|---|---|---|
| Mandate auditable company GHG reports | x | | x | x |
| Update GHG inventories | x | | x | x |
| Ban residual fuels and waste byproducts | | x | x | x |
| Collect open-source oil and gas data | x | | x | |
| Expand and integrate remote sensing and expedite public reporting | x | | x | |
| Update supply-side projections of oil and gas GHGs in countries' NDCs | x | | | x |
| Tightly regulate methane | | | x | x |
| Coregulate GHGs and air pollutants | x | x | | |
| Price oil and gas sector lifecycle GHGs | | x | x | x |
| Establish a methane feebate on the oil and gas industry | | x | x | |
| Prohibit operations in fragile ecosystems | x | | | x |
| Update GWPs | | x | | x |
| Facilitate clean energy tech transfer | | | x | x |
| Accelerate R&D for nonenergy petroleum products | | | x | x |
| Research climate engineering impacts | x | | | x |

GHG, greenhouse gas; NDCs, nationally determined contributions; R&D, research and development.

*Note:* These strategies are not ranked in a particular order since the majority advance two guiding principles.

*Source:* Author's assumptions.

### Mandate Auditable Oil and Gas Company GHG Reports

Climate change is increasingly viewed as an existential financial risk for firms that requires closer tracking. Producing auditable reports that detail oil and gas companies' respective climate risks and decarbonization plans could help reduce the most significant accounting errors. Similar to how the Securities and Exchange Commission (SEC) traces and mandates financial disclosures in the United States, a government agency can train its attention on incomplete or fraudulent GHG reporting by oil and gas companies. Currently, GHG-emitting companies do not report their ESG risks at all, or they do so inconsistently.

Efforts are underway to manage these risks in Europe. In 2019, the European Union established new rules for investors to incorporate harmonized ESG risks into their business procedures.[12] Establishing an auditable reporting system would permit regulators to challenge lacking or faulty ESG data. The OCI+ could aid such efforts by quantifying and ranking oil and gas assets by their individual climate risks.

### Update GHG Inventories

The current oil and gas GHG inventory reports and nationally determined contributions (NDCs)[13] submitted by governments under the Paris Agreement are woefully inadequate.[14] One glaring example is oil refining. The US Greenhouse Gas Reporting Program shows constant emissions over the past decade even though major investments and capacity additions were made in this timeframe to process heavy, dirty crudes.[15] Without accurate inventories, it is impossible to set and meet global climate targets. Existing flaws were exposed, for example, after a 2012 methane leak detected remotely in New Mexico when inadequate inventory data led to underreported GHG emissions by nearly a factor of four.[16] The IPCC refined its 2006 guidelines for national GHG inventories in 2019 so as to help improve inventory accuracy.[17] This is a particularly germane exercise for oil and gas because the sector significantly changed through the 2010s due to fracking and liquefied natural gas (LNG) shipment. Improved and updated inventories will facilitate OCI+ modeling and satellite-based GHG measurement and attribution studies.

### Ban Residual Fuels and Waste Byproducts

Wealthier nations pay a premium for more refined petroleum products with fewer contaminants, while the degraded, less refined residuals with the highest levels of GHGs and air pollutants are cast off to poorer nations.[18] Countries are reacting to this state of affairs, however, citing results from OCI+ analyses.[19] For example, in 2016, China banned petcoke imports from the United States, which then shifted to India until that country's Supreme Court banned the burning of high-sulfur petcoke nationwide.[20] In 2019, China followed suit, outlawing the use of high-sulfur petcoke from domestic and imported sources.[21]

Similar dumping concerns involve high-sulfur bunker fuels used by maritime vessels traveling in open seas where no single country governs (the global commons). In 2020, the International Marine Organization initiated a ban on the use of heavy fuel oils exceeding 0.5 percent sulfur.[22] Plastic straws and single-use plastic bags are also targets of national bans.[23] For bans to be successful, there need to be market-ready, affordable, low-GHG substitutes available; suppliers cannot use displaced feedstock for other products; and rules must be strictly enforced.[24] There also must be proper disposal options in the short term when refiners are not able to shift operations to adjust these product slates.

### Collect Open-Source Oil and Gas Data

Decision makers need oil and gas data to determine which assets to turn off in a managed ratcheting down of some parts of the hydrocarbons industry and which to turn back on after the global coronavirus pandemic temporarily destroys demand.[25] Policies mandating transparency provide knowledge to governments and civil society. For example, US legislation, the Know Your Oil Act, was introduced in 2015 and again in 2017 to authorize federal agencies to collect and publish oil and gas data.[26] Similar state legislation was introduced in California in 2020 to create public libraries with production and refining data.[27] And work is underway to expand the legislative language to include gas.

If enacted, these policies would help the OCI+ model better identify cost-effective climate mitigation strategies for global oil and gas assets. Several articles in the Paris Agreement specify the need for transparency and public access to information to support the UN's Sustainable Development Goals (SDGs). Efforts to digitize big data in the oil and gas sector could further enable climate transparency.

### Expand and Integrate Remote Sensing and Expedite Public Reporting

Independent, nearly real-time data collection is required to verify GHG monitoring and reporting. As such, remote sensing equipment operated by government space agencies, private firms, academic researchers, NGOs, and other actors is on the rise. Redundancy helps such sensors avoid missing hard-to-detect GHG emissions blocked by clouds, released in geopolitically sensitive regions, discharged intermittently, or below equipment detection limits. Multiple devices can be used to correct erroneous reports and prevent the sale of misleading GHG data.[28] This calls for a remote sensing ecosystem approach, such as advanced by Carbon Mapper, a collaboration of National Aeronautics and Space Administration (NASA)/Jet Propulsion Laboratory (JPL), the California Air Resources Board,

Planet, the University of Arizona, the Rocky Mountain Institute (RMI), the High Tide Foundation, and other philanthropies.[29]

For example, in November 2019, remote sensors picked up elevated methane and benzene levels followed by high flaring in the southern United States: the anomalies turned out to be a blowout in a Texas Eagle Ford gas well. By triangulating remote-sensing devices along with the OCI+ model, experts estimated that the well blowout emitted nearly 100 times more methane than would have been released from normal operations.[30] Sensing a gas plume is not enough. Questions remain about its methane content and the intermittency of emissions, questions that require additional analysis. Integrated diagnostics will be increasingly important for evaluating and mitigating supply-side oil and gas GHG emissions.

### Update Supply-Side Projections of Oil and Gas GHG Emissions in Countries' NDCs

Countries report and update their GHG emissions under the Paris Agreement through the NDCs they submit.[31] Governments are tasked with accurately reporting GHG levels and mitigation measures for all petroleum sources in their jurisdictions. However, in their initial NDCs, the majority of countries did not expressly indicate mitigation measures for their oil and gas supply chain emissions. The NDC updates posted beginning in 2020 need to increase their ambition and include more specificity on reductions in the GHG emissions from oil and gas production, processing, refining, and shipping operations.

These voluntary submissions should include not only $CO_2$ but also other GHGs—methane, nitrous oxides, black carbon (particulate matter [PM]), volatile organic compounds (VOCs), nitrogen oxides, carbon monoxide, and fluorocarbons. Peer-reviewed studies using the OCI+ models offer insights through cross-national supply-side GHG comparisons that identify highly variable emissions that can be readily mitigated.[32]

### Regulate Methane Tightly

Global methane emissions are at all-time global highs, and leakages throughout the oil and gas sector are primary drivers of this trend.[33] As such, policymakers are beginning to regulate short-lived climate pollutants (SLCPs) like methane. In North America, for example, the Obama administration set out to plug fugitive emissions from leaky equipment, while Mexico adopted regulations to curb methane, and Canada regulated production facilities. Meanwhile, Canadian provinces added gas capture to drilling permits and set limits on methane venting.[34] However, a broader science policy framework is needed to manage methane effectively.[35] Global rules are warranted on the grounds of both the climate-affecting and regional pollution that methane generates. Whether methane (along with other air toxics) are released through fugitive emissions, venting, or flaring, such practices not only exacerbate climate change but also can cause severe illnesses or even be lethal.[36]

### Coregulate GHGs and Air Pollutants

Global climate change worsens local air pollution and vice versa. These dual pollution problems sicken and kill many millions each year from heart disease, lung cancer, respiratory infections, heat exhaustion, and the spread of viruses, especially among poor and marginalized populations. Addressing one problem while inadvertently worsening the other can be deadly.[37] For example, in the 1990s, the European Union adopted tax breaks on the purchase of diesel cars because they are more fuel efficient and mitigate climate change. However, diesel fuel emits more PM, nitrogen oxides, and air toxins than gasoline and requires additional emission controls. As diesel car sales took off, VW and other European automakers rigged their cars to fraudulently pass tests, jeopardizing millions of citizens in the European Union and throughout Asia.[38]

Governments cannot afford to focus on one problem at a time when it comes to protecting an atmosphere where pollutants know no boundaries. Moreover, decision makers in highly polluted regions can use air pollution as a political wedge to address climate change while also improving air quality.

### Price the Oil and Gas Sector's Lifecycle GHGs

We are all already imposing a costly price on carbon as the globe warms through higher air conditioning bills, infrastructure damages, agriculture losses, insurance premiums, and health care costs. But this effective tax is indirect and contains only costs—no benefits. Pricing GHG emissions directly provides ancillary benefits through lower emissions, as long as all responsible parties pay their fair share. A smart tax like this could use the OCI+ to assess fees on hydrocarbons producers, processors, refiners, shippers, and consumers based on their specific GHG responsibilities.[39] Applying carbon taxes only on gasoline (and other end-use petroleum products) places an unfair burden on motorists and misses the remainder of supply-side GHG emissions.

Effective carbon pricing must explicitly apply to the oil and gas industry's Scope 1 and Scope 2 GHGs. And, if tackling all lifecycle carbon dioxide equivalent ($CO_2e$) GHG emissions is too massive an undertaking at the outset, a price could be placed initially on only methane emissions.[40] Instead of involving all actors, the fee would be targeted on this powerful SLCP and would charge a subset of industry actors—oil and gas suppliers, dairies, and landfills—and a few select consumers.

*Establish a Methane Feebate on the Oil and Gas Industry*

An alternative to charging fees, which are purely punitive, involves feebates—a self-financing system of fees and rebates that not only penalizes poor performance but also rewards good behavior.[41] Feebates have already been applied to motor vehicles to motivate consumers to purchase more fuel-efficient cars and trucks. As governments worldwide struggle to adopt a carbon fee or tax, they can instead focus first on a single SLCP in a single sector.

Establishing a methane feebate on the oil and gas sector would create a disincentive for industry operators that discharge methane and offer rebates to those that conduct routine methane monitoring, reporting, and verifying; do not routinely flare; minimize fugitives; and eliminate venting. Fees can be used to pay for rebates, independent certification, and other program administrative expenses, imposing no direct cost on the public to reform industry practices and foster innovation in the abatement of methane emissions.

*Prohibit Oil and Gas Operations in Fragile Ecosystems*

Some ecosystems are more fragile than others and carry outsized climate risks, including the Arctic and Antarctic, permafrost-covered lands, boreal forests, peat lands, rainforests, and other movement- and flood-prone areas. When oil and gas are present, oil and gas industry actors often exert economic pressure to develop these areas.[42] The OCI+ estimates the effect that land use in these areas has on GHG emissions.

The risks vary depending on the landscape. In the Arctic, burning diesel and residual fuels emits black carbon that accumulates on permafrost, heating it up, melting it, and then releasing stored pockets of $CO_2$ and methane.[43] The abrupt thawing of frozen ecosystems can cause an outpouring of emissions and cause land to suddenly sink (or subside), damaging pipelines and other infrastructure that further damages sensitive ecosystems.[44] Likewise, developing oil and gas operations in rainforests disrupts and dries out biomes that naturally store $CO_2$.[45] Although ranching and farming may convey large risks, installing and operating oil and gas infrastructure in these sensitive ecosystems is also debilitating. Blanket prohibitions are challenging, especially in nations where the rule of law is difficult to enforce. But outright prohibitions may ultimately be necessary to prevent runaway climate change.

*Update Global Warming Potentials*

Global warming potential (GWP) is a handy way to sum up different GHGs with different radiative forcing properties and report their varying emissions on a standardized $CO_2e$ basis. In 1990, the IPCC laid out GWP assumptions for different GHGs, and as scientists uncover more data, the calculus of GWPs continues to evolve in their assessment reports (ARs).[46] For example, in 1990, methane's 100-year and 20-year GWPs were 21 and 63, respectively. These values were updated in 2005 (AR4) to 25 and 72, and again in 2013 (AR5) to 36 and 86. They are due to be updated yet again (AR6) in 2021–2022.[47]

The current GWP approach masks the true behavior of potent SLCPs like methane. Researchers are developing a brand-new GWP metric—$CO_2$ warming equivalent ($CO_2$-we). This method will theoretically account for SLCP emission reductions more accurately than current methods in carbon budgets alongside long-lived climate pollutants.[48] Nevertheless, many governments, companies, and civil society actors are still using outdated AR4 GWPs from 2005, which significantly undercount the climate-forcing potential of GHGs and the value of SLCP mitigation. Using current GWPs is especially important in the oil and gas sector, where methane and black carbon are quite prevalent.

### Facilitate Clean Energy Tech Transfer for Oil and Gas

Major climate innovations in the petroleum sector are long overdue. In the oil industry, budgets for clean energy R&D and technology transfers amount to miniscule fractions of pennies on the dollar, a tiny share of capital expenditures. Consequently, low-GHG breakthroughs in oil and gas systems have not occurred. The same is true of public R&D spending on energy, which has remained flat as a share of gross domestic product since the 1980s.[49] Discovering and inventing low-carbon energy technology requires continued support for R&D and technology transfer.

For example, the creation of a new US federal agency—which could be called the Advanced Research Projects Agency–Climate (ARPA-C)—has been floated.[50] The idea would not be to duplicate the efforts by the Advanced Research Projects Agency–Energy (ARPA-E), which focuses on non–fossil fuel energy breakthroughs. Rather, in tackling climate change head-on, ARPA-C could tackle oil and gas R&D, an area that ARPA-E has largely avoided. The abundance of US shale resources calls for an agency dedicated to reducing the GHG footprint of the oil and gas sector. Public R&D paired with technology transfer in conjunction with private and NGO sector collaboration will be vital for a successful clean energy transition.

### Accelerate R&D for Nonenergy Petroleum Products

A million-mile car battery is under development that could wean the transportation sector off gasoline fuel.[51] This breakthrough follows decades of research aimed at boosting motorists' confidence in the reliability of electric vehicles (EVs). Battery research underscores the need for similar R&D efforts to replace all of the other commodities derived from a barrel of oil, including other fuels and nonenergy petroleum products. Accelerated R&D is needed to replace the complex hydrocarbons that make up petrochemicals with green electrons. This is a tall order for lubricants, waxes, solvents, sulfur, and a plethora of other petrochemicals that are the backbone of many industries.

The search is on for ready replacements, especially low-GHG substitutes. For example, US and Canadian national laboratories are studying how to turn continuous supplies of excess, cheap renewable electricity (and their enabling electrons) into carbon-neutral petrochemicals and fuels.[52] Such theoretical pioneering approaches can involve chemical processes that build up $CO_2$ and water rather than break down and refine hydrocarbons to manufacture industrial inputs. The quest for such renewables will likely proceed slowly.[53] Nonetheless, the fact that many GHG emissions stem from unexpected corners (and byproducts) of the petroleum industry means that such solutions are necessary: without such progress, other efforts to curb GHG emissions will fall far short of the world's stated climate mitigation goals.

*Research Climate Engineering Impacts*

The earth's temperature is rising with concomitant risks for environmental protection, public health, economic fallout, and geopolitical security. In response, scientists in the public and private sectors are exploring engineering approaches that would involve altering and recalibrating the planet's climate system, known as climate engineering or geoengineering. Governments are partnering with international oil companies (IOCs), for example, in Canada, California, and Switzerland, to advance direct air capture (DAC).[54] And oil-producing nations like Saudi Arabia and the United States are blocking efforts to examine the risks of DAC and other technologies that affect the atmosphere, oceans, and land.[55] As climate engineering gains traction, there need to be greater (not less) transparency on efforts underway, more international dialogues on the progress achieved to date, and reliable assessments on its feasibility, risks, and benefits.[56] To succeed, geoengineering will modify the earth's climate, whether by scrubbing $CO_2$ out of the air, blocking sunlight by injecting chemical particles into the atmosphere, thinning clouds, accelerating the earth's carbon uptake, or seeding oceans. Prudent public research, governance, tracking, disclosure, and coordination of climate engineering is necessary, and ARPA-C, discussed earlier, could provide agency oversight on climate engineering as well as other forms of climate mitigation.[57]

# What Industry Can Do

In an ideal world, governments would take hold of the reins on climate change mitigation. But multinational climate governance to date has not been an outright success. Moreover, government leadership is fraught when it comes to oil and gas supply-side GHGs because these resources pad (and in some cases dominate) governments' budgets and further a nation's geopolitical clout. Nevertheless, even a hint of forthcoming climate policy action, as discussed previously, could motivate some companies to act and become tomorrow's zero-GHG energy providers.[58] For those companies that choose to seriously mitigate their climate footprints, there are ample opportunities to reduce their supply-side oil and gas emissions. Their motivations for doing so can span from outdoing competitors to upgrading their own ESG rating. Each company has its own calculus, using various methods (like shadow pricing their GHG emissions) to determine their respective climate liabilities in the face of changing markets, government action, or civil society pressure. Investors, industry consultants, assurance agents, and other industry actors abide by their own set of climate priorities and methods. Table 8.3 identifies a host of recommended industry strategies, especially ones that satisfy the maximum number of overriding principles possible.

### Accelerate Research on Net-Zero Emissions Technologies

All technological advances are not equally effective at reducing lifecycle GHG emissions. Net-zero emissions technologies (NETs) require further research to evaluate their effectiveness before they are widely deployed.[59] In many cases, it is better to use ancillary pockets of GHGs that the oil and gas industry produces anyway rather than harvest additional, naturally occurring GHG supplies solely to use them for ostensibly more ecofriendly petroleum operations. For example, removing $CO_2$ from underground caverns (where nature has already stored it) and using it for enhanced oil recovery (EOR) does not qualify as a low-GHG carbon capture and storage (CCS) approach. Nor does developing an ultra-sour gas field to strip off its $CO_2$ and inject it in EOR operations. Conversely, DAC technologies that capture $CO_2$ directly from the air to manufacture synthetic transport fuels returns at least some of the captured $CO_2$ into the atmosphere through its lifecycle and could be considered a NET.

Ongoing research is needed to guarantee that CCS, DAC, and other similar techniques certifiably result in net-negative emissions—removing more lifecycle GHG emissions than they produce. To reverse global warming, NETs will require international buy-in and be employed on a planetary scale while safely storing $CO_2$ without unintended societal consequences.

### Integrate Renewables into Oil and Gas Systems

To at least partially offset their traditional dependence on the massive energy inputs of their own byproducts and intermediaries to fuel their operations, oil and gas companies are slowly substituting renewable energy sources in their supply chains in various ways. These methods include employing renewables to

Table 8.3  Industry Strategies to Reduce Supply-Side GHG Emissions (Ranked by Guiding Principles)

| Industry Supply-Side GHG Strategies | Principle #1: The Public Has a Right to Know | Principle #2: Prices Reflect Social Costs | Principle #3: Hold Industry Responsible | Principle #4: Position Sector on Safe Climate Path |
|---|---|---|---|---|
| Accelerate research on NETs | x | x | x | x |
| Integrate renewables into oil and gas systems | x | | x | x |
| Create a new reporting framework for oil and gas supply-side GHGs | x | | x | x |
| Apply future production quotas to high-intensity GHG oil and gas assets | | x | x | x |
| Commit to zero routine gas flaring, fugitives, and venting | | | x | x |
| Decommission targeted oil and gas assets | | | x | x |
| Submit and update corporate climate plans | x | | | x |
| Stop financing high-GHG projects | x | x | | |
| Incentivize top-down corporate climate mitigation | | x | x | |

GHG, greenhouse gas; NETs, net-zero emission technologies.
*Source:* Author's assumptions.

make the hydrogen, steam, heat, and electricity they use onsite for oil and gas operations. For example, Aera (a joint venture with ExxonMobil that Shell is now exiting) plans to install the largest concentrated solar array in California for steam-injection EOR.[60] And Equinor is installing the world's largest wind farm in the North Sea to power its oil and gas platforms and millions of homes.[61]

Full integration of renewables into oil and gas systems will not only reduce the climate footprint of their operations but also help the industry capitalize on clean energy sources, diversify its portfolios, and cross-train its workforce for future deployment in the renewables sector. That said, in the notoriously boom-and-bust petroleum industry, successful integration of renewables will require uninterrupted private financing from companies and their investors.[62]

### Create a New Reporting Framework for Supply-Side Oil and Gas GHG Emissions

Oil and gas operations are unlike other industrial entities in that they use vast quantities of their own petroleum products to manufacture the commodities they sell. The tight (often vertical) integration of this diverse and mature industry means that companies are at once oil and gas consumers *and* producers. Existing GHG accounting and reporting protocols do not suit the oil and gas sector because they miss GHG emissions from a variety of sources including byproducts (like petrochemicals and petcoke) and ancillary operations (like hydrogen production and distribution).[63] Incomplete reporting makes it impossible to compare corporate climate risks.

This point is especially germane for investors as they develop low-GHG financial products, services, and indices. For example, in 2019, Goldman Sachs announced that it would restrict financing to new oil exploration and production in the Arctic and adopt strict lending policies for fossil fuel companies.[64] The European Investment Bank announced an end to its multi-billion-euro financing for new oil, gas, and coal projects after 2021.[65] And the hedge fund TCI is applying pressure on asset managers that do not require public GHG reports from the corporations they invest in to change their ways, an indication that financial traders are increasingly concerned about how climate change affects their portfolios.[66] Tailoring GHG reporting and certification to suit supply-side oil and gas GHG emissions could close the gap on emissions leakage and better target the operations that pose the highest climate risks.[67] An example of such a voluntary corporate certification program that could also be adopted by governments is the methane intelligence quotient (MiQ) standard that differentiates natural gas production and midstream emissions based on their varying methane emissions.[68]

*Apply Future Oil and Gas Production Quotas to High-Intensity GHG Assets*

Geopolitics fueled the six-month embargo of Middle Eastern oil to the United States and Western Europe in 1973 in retaliation for supporting Israel during the Yom Kippur War. Today, disruptions in the Middle Eastern oil trade tend to be economic in nature. For example, in 2020, Saudi Arabia and Russia orchestrated voluntary oil production cuts to prop up petroleum prices that had collapsed when demand waned during the coronavirus pandemic and ensuing economic lockdown. Such market collusion is not uncommon, especially by the twenty-three countries in the Organization of Petroleum Exporting Countries-plus (OPEC+) coalition that controls about half the world's oil production.[69]

From an environmental perspective, the same tactic could ultimately work to address climate change by systematically shutting down (or not turning back on) assets that generate the highest GHG emissions.[70] If migration, rising sea levels, hunger, and political unrest mount due to global warming, Saudi Arabia and others could get more serious about combating climate change. They would then have the market power to orchestrate a race to the top that favors producers and refiners, like themselves, that offer products with the lowest GHG emissions.

*Commit to Zero-Routine Flaring, Fugitives, and Venting*

With methane emissions at an all-time high, efforts are getting underway to stop routine gas flaring and venting and plug fugitive emissions from well completions, workovers, liquids unloading, corroded pipelines systemwide, and many other aspects of oil and gas operations.[71] In 2018, for example, the Nigerian National Petroleum Corporation announced a new strategy by which it would utilize all of its gas and prohibit routine flaring by 2020.[72] But ample progress has not been made. In 2021, Nigeria was named as one of seven countries with the most continuous flaring worldwide.[73]

Likewise, in 2019, several big oil and gas companies withheld support for the rollback of US methane emissions standards over concerns that noncompliant operators could provoke even tighter, more prescriptive regulations down the line.[74] At least eleven companies signed on to new guiding principles for methane that improve the accuracy of methane emissions data, advocate sound methane regulations, and increase corporate transparency.[75] Recent efforts underway through industry collaboratives, such as the Methane Guiding Principles, are geared toward mitigating methane in the oil and gas sector.[76] However, there is a limit to voluntary oversight, according to Shell's former CEO.[77] Although methane is money and leak-free systems make business sense, the industry still requires clear guidelines to level the playing field between industry actors worldwide, large and small, operating in different geographic settings.

### Plan to Decommission Targeted Oil and Gas Assets

Petroleum systems are not designed to operate forever. Over time, the GHG emissions generated by a given oil field or gas field tend to rise as their hydrocarbon assets age and are depleted. The lifetimes of these petroleum assets can be cut short by accidents (such as fires) and incidents (such as storms and rising sea levels) that can cause irreparable damage. Detailed plans to decommission damaged, depleted, and old fields and facilities rather than sell them to another company can prevent high-GHG operations from merely trading corporate hands after their usefulness and profitability erode or end.

This insight applies to wells, production sites, tank batteries, processing plants, refineries, shipping terminals, pipelines, and other petroleum facilities. For example, successful efforts include the United Kingdom's Brent production platforms (the highest-GHG asset currently modeled using the OCI+), which are in the process of being decommissioned after some forty years in operation. But the climate would have benefitted if Brent had been decommissioned a decade earlier than it actually was. Likewise, a bankruptcy judge moved to decommission (rather than rebuild) an old Philadelphia refinery that was damaged in a 2019 fire.[78] Periodic economic downturns, pandemics, and other exogenous forces shine a spotlight on oil and gas assets whose climate costs have come to outweigh their energy benefits. Planning ahead to safely decommission and decarbonize these legacy assets is consistent with climate pledges and low-GHG energy strategies.

### Submit and Update Corporate Climate Plans

Corporate climate pledges have become the norm since the 2015 Paris Agreement was signed. Take, for example, BP's pledge to offset its $CO_2$ emissions by 2050 or Shell's ambition to become a net-zero energy business by that same year.[79] Without transparent, standardized, routinely updated climate plans, however, these pledges are just fleeting words that governments, NGOs, and investors struggle to assess and compare.

A credible climate plan contains certain steps that, if missing, courts may weigh in and order.[80] First, each company formalizes its public climate commitment to reduce its GHG emissions. Next, operational and equity emissions are inventoried, mitigation measures are developed, and reduction targets are set to maintain compliance with the Paris Agreement. Going forward, on an annual basis, inventories are adjusted, targets are updated, and future GHG levels are forecast. Financial disclosures are provided to detail capital investments in oil, gas, nonfossil fuels, and other projects and report on spending on low-carbon R&D. Finally, companies can engage with low-GHG suppliers, remain active in low-carbon joint ventures with private and public actors, and adopt industry best practices. Armed with credible corporate climate plans, NGOs can track companies' actions to assess whether a given petroleum company is positioned to meet the targets under the Paris Agreement.[81]

### Stop Financing High-GHG Oil and Gas Projects

Companies have historically received funds from numerous investors to finance their projects. Investors are increasingly shying away from funding oil and gas capital expenditures. For example, Lloyd's Banking Group[82] and JP Morgan Chase (reportedly the oil industry's bank of choice) announced they would no longer finance Arctic oil and gas projects.[83] And BlackRock is planning to broadly steer its funds away from fossil fuels and into sustainable investments.[84]

These are wise moves on climate grounds. But investors, reinsurance companies, and risk firms can do even more to strategically defund future oil and gas projects with the highest GHG emissions. The data needed to make such climate determinations exists. Larger oil and gas companies use shadow prices to account for climate policies and determine if a given project will be economically viable under a GHG price regime. Investors should obtain shadow pricing data to inform their decision-making, and if these data are not available, investors can use the OCI+ to estimate the climate impacts of certain oil and gas projects. Such a targeted exit by investors from oil and gas funding can lead to greater climate benefits in the short term than a broad-based approach that selects only a few assets to strand.

### Incentivize Top-Down Corporate Climate Mitigation

The oil and gas industry is a top-down business with many ancillary divisions and byzantine financial considerations. Managing climate change has historically not been a priority in companies' centralized plans, which use financial incentives to change the industry's mindset and culture. New incentives for climate mitigation are coming to light, however. For example, in 2016, Shell announced plans to link executive pay to short-term carbon emissions targets, and in 2019, BP announced it would link the bonuses of its employees to reduction targets for GHG emissions.[85]

Other companies are focusing corporate incentives on specific operations, such as Chevron's move to create a scorecard for reducing methane emissions and flaring intensity, a move that would determine new ways to gauge incentive pay for its thousands of employees. Since a significant share of GHG emissions are the direct responsibility of oil and gas operators yet indirectly under the watchful eye of such companies' CEOs, companies can use top-down corporate incentives to more reliably reduce their GHG emissions.

## What Civil Society Can Do

Social movements are designed to change people's views, press for government action, and compel industry decisions. Civil society actors have many ways to bring about desired change. Reforming the global oil and gas sector via such societal pressure requires technical and economic knowledge, systems-level thinking, and long-term time horizons. Table 8.4 identifies several top strategies for civil society to consider, especially ones that maximize the number of guiding principles that they satisfy.

### Advocate for "Know Your Oil and Gas" Regulations

More oil and gas data will likely result in more petroleum studies performed and climate intelligence acquired by civil society actors in academia and the NGO community. The assumptions that underpin these actors' respective research and advocacy activities must be updated to keep pace with new knowledge about the highly heterogeneous nature of hydrocarbon resources worldwide.

The ability to differentiate these resources' diverse GHG emissions can lead to more nuanced policy analysis and public campaigns. For example, legislation to promote oil and gas transparency advanced by RMI (formerly Rocky Mountain Institute), the Natural Resources Defense Counsel and Communities for a Better Environment can have impactful multiplier effects. Using an oil with half as many GHG emissions per barrel in a car that is twice as fuel efficient would result in reduced emissions by a factor of four for each mile driven. Whether civil society is developing emissions offsets for future petroleum projects or lobbying to prevent oil and gas development in the Arctic, civic actors can quantify oil- and gas-specific climate benefits with the OCI+, leveraging these insights to develop more persuasive policymaking arguments and put the onus on industry to provide greater data transparency in the future.

### Establish Low-GHG Certification for Oil and Gas

Another workaround civil society can use to buttress policymaking is through certification programs that differentiate whether or not ESG elements apply to a given commodity, as has been done for fair-trade coffee and sustainable palm oil.[86] Such a program for low-methane oil and gas could certify that operators in the supply chain tightly manage and do not inadvertently or purposefully leak methane. NGOs like RMI and Resources for the Future are working to certify and shift the market to low-emissions gas.[87]

The OCI+ can support certification by modeling global gas assets and creating a rubric that quantitatively estimates any operation's methane leakage, flaring, venting, and combustion emissions.[88] Factoring a fair emissions premium into

Table 8.4 Civil Society Strategies for Reducing Supply-Side GHG Emissions

| Civil Society Supply-Side GHG Strategies | Principle #1: The Public Has a Right to Know | Principle #2: Prices Reflect Social Costs | Principle #3: Hold Industry Responsible | Principle #4: Position Sector on Safe Climate Path |
|---|---|---|---|---|
| Advocate for **Know Your Oil** regulations | x | x | x | x |
| Establish low-GHG certification for oil and gas | x | x | x | x |
| Demand near real-time MRV for oil and gas GHGs | x | | x | x |
| Tap the expertise of retirees from industry and government | x | | x | x |
| Tailor divestment decisions to lifecycle GHG emissions | x | | x | x |
| Urge endowments to adopt GHG investment criteria | x | | | x |
| Incorporate supply-side oil and gas emissions in updated NDCs | | | x | x |
| Promote incentives to safely decommission and redevelop oil and gas properties | | | x | x |
| Conduct R&D on refining for a clean energy transition | | | x | x |
| Facilitate tech transfer of open-source GHG tools and remote detection systems | x | | x | x |

GHG, greenhouse gas; MRV, monitoring, reporting, and verifying; NDCs, nationally determined contributions; R&D, research and development.
*Source:* Author's assumptions.

natural gas market prices could extend beyond civil society actors and require a commitment to a universally accepted methodology and implementation scheme by buyers, sellers, certifiers, regulators, and industry organizations.[89]

### *Demand Near Real-Time Monitoring, Reporting, and Verifying for Oil and Gas GHGs*

It is not enough to have decent laws on the books if industry actors do not abide by them and instead opt to pay fines or if rules are not duly enforced. In such cases, civil society needs to invoke other tools. One option is for civic actors to establish their own monitoring, reporting, and verifying (MRV) programs operating alongside government programs. For example, Climate TRACE,[90] an alliance of civil society actors supported by Google, uses artificial intelligence to rapidly and simultaneously analyze various GHG monitoring data, a multi-NGO initiative aimed tracking emissions from all sectors in all countries, with RMI using the OCI+ to provide oil and gas emissions.[91]

Extending rapid monitoring to the oil and gas sector will present new data challenges owing to the industry's constellation of resources, various actors, global reach, market power, and geopolitical sway. These complications can be remedied through close coordination between these NGO efforts, government scientists (such as NASA's Carbon Monitoring System [CMS] network), Carbon Mapper, Methane SAT, and other remote sensing instruments. Such multi-dimensional, multi-stakeholder GHG tracking system can help resolve monitoring disagreements and reduce market confusion and industry meddling.

### *Tap the Expertise of Retirees from Industry and Government*

A retirement wave has hit the oil and gas sector in recent years, and another one is coming.[92] The share of baby boomers, who accounted for 19 percent of the oil and gas workforce in 2015, is projected to plummet to 7 percent by 2025.[93] Given record-low oil prices in 2020, slated retirements could be compounded by a flood of layoffs.[94] This is good news for civil society organizations that desire to enlist technical professionals with experience in the petroleum industry to join the cause of combating climate change.[95] Similarly, scientists retiring from national laboratories can share their expertise with civil society.[96] Technical projects like Climate TRACE and science policy NGOs could benefit greatly from the involvement of retiring scientists from the oil and gas industry and government laboratories, experts whose main goal is to hand down a positive legacy.

### *Tailor Divestment Decisions to Lifecycle GHG Emissions*

Even the most ardent advocates prodding investors to divest from the oil and gas industry will have a hard time maintaining an investment portfolio with absolutely no oil and gas assets baked into it. Removing IOCs from portfolios may seem like an obvious, straightforward play, but what if one company's assets have a lower GHG footprint than another? And what about natural gas suppliers? Then there is the question of airlines, big-box stores, food companies, and pharmaceutical firms that all require petroleum products in one form or another to serve their customers' needs.

Rather than write off entire swaths of the economy, divestment campaigns can parse oil and gas assets held by different companies and pressure them to shift their portfolios to truly low-GHG resources and operations. For example, certain oil and gas assets, such as LNG projects, remain favorable with investors, while others (including deepwater oil) have not regardless of their actual respective GHG footprints.[97] Civil society investor initiatives like Climate Action 100+ can work with companies to move beyond disclosing corporate climate risks to dissecting and differentiating them.[98] Efforts are underway to bring financial institutions' investment activities into alignment with climate goals.[99] The ability to quantify lifecycle GHG emissions at the asset level using the OCI+ provides civil society actors with the technical details they need to target pressure points, assess trade-offs, and prevent unintended consequences.

### Urge Endowments to Adopt GHG Investment Criteria

Those who manage public and private endowments seek to maximize returns. They also tend to be underresourced when it comes to their climate initiatives.[100] They tend not to have clear-cut criteria for cutting their given portfolio's GHG emissions. Minimum standards could help rationalize investors' decision-making, replace blunt determinations, and direct divestment decisions.

Minimum standards like those for food or medicine could be set for GHG investments. For example, it is not uncommon for civic actors to call out a particular resource type, like oil sands, and simply demand that investors divest from any company that manages such assets, regardless of whether a particular firm's GHG emissions are or are not lower than those of other oil and gas assets.[101] The New York State Common Retirement Fund has set out numerous criteria for making such evaluations, such as investing in companies that have a track record of reducing their GHG emissions rates year on year and divesting from companies that have no climate policy in place.[102] Endowment managers could benefit from a structured approach, one that both educates them and employs incentives to encourage them to develop and implement low-carbon investment plans.

### Incorporate Supply-Side Oil and Gas Emissions in Updated NDCs

Given that the United Nations is in charge of collecting countries' updated NDCs under the Paris Agreement, civil society has an opportunity to engage countries on their supply-side oil and gas GHG emissions. The first round of NDCs submitted after the 2015 UN Climate Change Conference in Paris largely missed the potential to drastically reduce emissions.

In addition to stressing demand-side reductions of transportation fuels and increased renewable energy supplies, there is a parallel conversation that civil society actors can foster about countries' oil and gas production, refining, and shipping activities to make sure that they are covered in GHG inventories and slated for climate mitigation measures.[103] One way to accomplish this is for civil society actors to appeal to nations with the lowest supply-side oil and gas GHGs—including Norway, Saudi Arabia, and Qatar—and engage them in a conversation at future follow-on summits.

### Promote Incentives to Safely Decommission and Redevelop Oil and Gas Properties

Decommissioning and removing oil and gas assets offers companies little to no return on their investment. But permanent shutdowns and abandonments can have major public benefits for civil society, including significant reductions in GHG emissions. For example, once the United Kingdom's Brent oil field is entirely decommissioned, it is estimated that over one tonne of $CO_2e$ will be curbed for each BOE produced by a low-GHG gas asset.[104] This assumes that high-GHG volumes would be replaced by another field somewhere else that can produce oil and gas with a smaller carbon footprint. But if petroleum demand is also cut, even more GHGs can be reduced.

Beyond GHG savings, the environmental impacts and future uses of decommissioned properties are hotly debated topics, and civil society actors are best suited to influence procedures and redevelopment decisions.[105] In the case of Brent, Greenpeace activists are calling for zero oil waste to be dumped into the sea, which is a possible risk because even abandoned oil and gas wells can leak methane and other emissions in the future. Oil and gas decommissioning plans should include financial incentives for companies to set aside part of their proceeds to entirely eliminate future climate and other environmental risks—and in some cases they could even redevelop properties for public benefit—when outdated projects reach the end of their useful lives.

### Conduct R&D on Refining for a Clean Energy Transition

The future is highly uncertain for global oil refiners that have tight profit margins and limited technical flexibility to shift their operations. Regardless, experts typically associate oil with only gasoline when, in reality, crude comprises thousands of other commodities too.[106] The 2020 coronavirus pandemic dealt an uneven hand to refining operations, slashing demand for jet fuel and gasoline, barely affecting diesel fuel, and radically shifting petrochemical feedstocks from car seats and pipes to personal protective gear and food containers.[107] While not all oil companies plan to abandon refining, pessimistic oil demand projections along with calls to decarbonize have Shell and BP shedding their petroleum assets.

Over the long term, however, demand for petroleum products like lubricants, solvents, asphalt, fertilizer, and medicines will persist. The prospects of refining in a low-carbon future are understudied but crucially important.[108] This calls for academics (along with national academies of engineering and other disciplines) to ramp up R&D on refining innovations that drastically reduce GHG emissions while pursuing benefits from petroleum products that do not have non–fossil fuel substitutes.

### Facilitate Technology Transfer of Open-Source GHG Tools and Remote Detection Systems

Amid the ongoing quest to invest in R&D for mitigating climate change, part of the puzzle is expanding the OCI+ model by adding global gas assets, expanding fugitive emissions modules, and updating refinery process calculations. Other academic efforts complement the OCI+. For example, Harvard, CalTech, the University of Arizona, and the Colorado School of Mines are each involved in exploring the potential of remote GHG detection and developing open-source tools to input data they generate into models like the OCI+.

Academics are also working with various NGOs including the World Bank, Environmental Defense Fund (EDF), and RMI to scrutinize GHG emissions in the oil and gas sector.[109] And civil society actors like the United Nations–led Climate and Clean Air Coalition (CCAC) have voluntarily come together around the shared goal of reducing SLCPs to protect the climate and improve air quality. Philanthropic foundations and governments facilitate civil society research through grantmaking and other funding sources. Given how little funding the industry devotes to climate and air quality R&D, GHG mitigation is one area where civil society can lead as long as ample funds are available.

## The Endgame

Where will all these noble efforts ultimately lead? The world has been wedded to oil and gas for nearing 150 years. Petroleum supplies sprung forth in the late 1800s. Over the course of the twentieth century, the industry took off owing to a series of technological advances, which most recently included the newest marginal barrel of oil and gas obtained through hydraulic fracturing and horizontal drilling.

Oil and gas have undeniably helped give rise to the modern trappings, innovations, and conveniences that characterize life in the twenty-first century and fueled globalization. Every economic sector, from transportation to power and agriculture to industry, has been advanced by hydrocarbons. The public surely benefited from the petroleum age in a multitude of ways. And the oil and gas industry and their host governments have also profited handsomely supplying these resources worldwide. But as oil and gas volumes have risen, so too have the societal disruptions they unleash. The world desperately needs an endgame for the gravest fallout of petroleum's heyday: climate change.

A pervasive sense of desperation has the world clinging to simple solutions: we *must* stop producing and forgo using oil and gas, the conventional thinking goes. As desirable as such outcomes may be, they are difficult to deliver due to powerful and dynamic market forces that create new states of equilibrium in response to constant fluctuations in supply and demand. Absent other, far-reaching changes, for some individuals to use less petroleum would merely lower prices and cause someone else to use more (sometimes far afield), a trend that then tightens the petroleum supply and causes prices to rise back up, spurring more production.

Breaking this cycle is only made harder by the way that major world powers like the United States, Russia, and Saudi Arabia peddle oil and gas for financial and political gain. For their part, consumers are not sovereign actors that can turn off oil and gas spigots, especially since petroleum products feed all manufacturing and make up nearly everything ordinary consumers use every day. Lastly, GHGs pollute the climate regardless of where they are emitted, so reshuffling the petroleum deck into different byproducts sold in other places merely perpetuates global warming without tangibly addressing the underlying problem.

Rather than striving to eliminate oil and gas altogether, governments, companies, and civil society actors alike should embrace tailored strategies to minimize supply-side oil and gas GHG emissions. This task entails shifting our mindset and acknowledging three key facts: oil and gas are here to stay, they are becoming more heterogeneous, and the operations employed to extract, convert, and deliver these resources require greater attending to.

Although demand for petroleum products remains highly uncertain due to automation, pandemics, or economic volatility, it is highly unlikely that the world will transition to a 100 percent oil- and gas-free economy over the next thirty years, the period in which climate change is projected to impose high costs and carry dire consequences. The GHG footprints of existing supplies of oil and gas must shrink in the near term. Such actions can bridge the gap to a clean energy future, one that is carbon-free and more equitable, flexible, and resilient.

This is a tall order. The endgame for oil and gas in a warming world involves numerous pathways taken by different actors in different nations over many years.

It requires shifting petroleum demand increasingly to green hydrogen and renewable electricity while figuring out how to make the oil and gas supplies remaining in the global economy generate net-zero emissions.

Now is the time to redress the prevailing opacity that has plagued the oil and gas sector with greater transparency about diverse hydrocarbons and their varying climate trade-offs. To this end, the OCI+ can provide ongoing climate insights to guide all actors as they use all available tools to develop the necessary short-term and long-term strategies. Using such tools will help the world make measurable gains *now* rather than postpone progress as the industry's large GHG footprint grows even larger. As daunting as the task is, there is clearly room for well-tempered optimism and gradual progress if the petroleum industry, government regulators, and civil society work in tandem and decisively act on the creative supply-side solutions this book has sought to offer.

# *Notes*

INTRODUCTION

1. Point Arguello, California, was considered second only to Prudhoe Bay, Alaska, in terms of its oil in place—up to 3 billion barrels. See Draft Environmental Report prepared by Arthur D. Little, July 9, 1984, https://play.google.com/books/reader?id=Lio3AQAAMAAJ&hl=en&pg=GBS.PA1

2. For an extensive history of the 1970s oil crises and how they shaped politics, see Meg Jacobs, *Panic at the Pump: The Energy Crisis and the Transformation of American Politics in the 1970s* (New York: Hill and Wang, 2016).

3. Thomas W. Lippman, "Oil War Fought on U.S. Shores Approaches an Uneasy Armistice," *Washington Post*, April 9, 1993, https://www.washingtonpost.com/archive/politics/1993/04/09/oil-war-fought-on-us-shores-approaches-an-uneasy-armistice/dcf5e143-09b3-4057-a6f7-3e739a613901/

4. On January 29, 1969, Union Oil Company's (UNOCAL) offshore Platform A experienced a blowout that continued leaking for months into the ocean floor through existing faults and fractures. See Kate Wheeling and Max Ufberg, "'The Ocean Is Boiling': The Complete Oral History of the 1969 Santa Barbara Oil Spill," *Pacific Standard*, November 7, 2018, https://psmag.com/news/the-ocean-is-boiling-the-complete-oral-history-of-the-1969-santa-barbara-oil-spill

5. Jon Hamilton, "How California's Worst Oil Spill Turned Beaches Black and the Nation Green," *NPR*, January 28, 2019, https://www.npr.org/2019/01/28/688219307/how-californias-worst-oil-spill-turned-beaches-black-and-the-nation-green#

6. "October 28. 1980 Debate Transcript," Commission on Presidential Debates, https://www.debates.org/voter-education/debate-transcripts/october-28-1980-debate-transcript/; *Offshore Leasing: Key to U.S. Oil Security*, Heritage Foundation, September 19, 1985, https://www.heritage.org/trade/report/offshore-oil-leasing-key-us-energy-security. Oil security is not simply a domestic issue; it also plays a

major role in foreign policy. See Jeff Colgan, *Petro-Aggression: When Oil Causes War* (Cambridge: Cambridge University Press, 2013).

7. The Reagan Ranch, Rancho del Cielo, which is now owned and operated by a foundation, is located less than 10 miles as the crow flies to Gaviota. See https://reaganranch.yaf.org/

8. In 1980, Jackson Browne built a home in Hollister Ranch, an exclusive, secluded coastal parcel in Santa Barbara. (A couple of years later, Chevron sought permits to build an oil and gas pipeline through Hollister Ranch.) For decades, he played dozens of concerts for Santa Barbara activists who were fighting oil development (among other efforts). See https://www.independent.com/2018/08/02/interview-jackson-browne/. In 1985, Bonnie Raitt teamed up with Jackson Browne, and together they performed at least two "oil initiative" concerts in Santa Barbara. See Bonnie Raitt Benefit History, 1980s, https://www.bonnieraitt.com/content/benefit-history-1980s. Also see Get Oil Out!, https://getoilout.org/

9. See pp. 34 (selective catalytic reduction [SCR] to mitigate nitrogen oxide emissions), 300 (Get Oil Out! comments), 301–2, 319, 340, 343 (Hollister Ranch comments) in the Response to Comments prepared by A. D. Little for Santa Barbara County, et al., "Point Arguello Field and Gaviota Processing Facility Area Study and Chevron/Texaco Development Plans EIR/EIS," October 11, 1984, https://play.google.com/books/reader?id=5CE3AQAAMAAJ&hl=en&pg=GBS.RA2-PP2

10. US Department of Interior, "Pacific Summary Report/Index," *OCS Information Report*, MMS-86-0060, 1986, https://play.google.com/books/reader?id=4CFPAAAAMAAJ&hl=en&pg=GBS.PP2

11. See Bureau of Ocean Energy Management map tracking oil and gas pipelines from Pt. Arguello's three platforms to landfall and then onshore to the Gaviota plant: https://www.boem.gov/sites/default/files/oil-and-gas-energy-program/Leasing/Regional-Leasing/Pacific-Region/Leasing/pacific-ocs-map.pdf

12. Plains Exploration and Production Company, "Revisions to the Platform Hidalgo Development and Production Plan to Include Development of the Western Half NW/4 of Lease OCS-P 0450," submitted to US Bureau of Ocean Energy Management, October 2012, https://www.boem.gov/sites/default/files/about-boem/BOEM-Regions/Pacific-Region/DPPs/5C1---2012-10-Platform-Hidalgo-DPP-Revison-NW-4th-of-Lease-OCS-P-0450.pdf

13. In 2020, oil futures temporarily sank to negative prices in response to the global shutdown during the coronavirus pandemic, and real-time prices fell to similarly low levels as 1999. See "Crude Oil Prices – 70 Year Historical Chart," *Macrotrends*, https://www.macrotrends.net/1369/crude-oil-price-history-chart

14. "Chevron Sells 26% Stake in Oilfield," *Reuters*, July 6, 1999, https://www.nytimes.com/1999/07/06/business/chevron-sells-26-stake-in-oilfield.html; "FCX Completes Acquisition of Plains Exploration & Production," News Release, Freeport-McMoRan, May 13, 2013, https://investors.fcx.com/investors/

news-releases/news-release-details/2013/FCX-Completes-Acquisition-of-Plains-Exploration--Production-Co/default.aspx

15. The Plains All American Pipeline was a joint venture with Chevron, Texaco, and Sun Oil. See US Securities and Exchange Commission filing: https://www.sec.gov/Archives/edgar/data/1070423/000110465916100030/a15-24557_110k.htm

16. For a map of interconnecting Line 901 (that ruptured) and Line 903 and the oil and gas production assets they serve, see http://curious.kcrw.com/wp-content/uploads/2016/06/energy_division_map_0816.png. Note that the 24-inch Line 901 was constructed in 1987 along the Gaviota coast to service the crude produced by several oil companies' offshore platforms. Both the pipeline that ruptured (Line 901) and Point Arguello's adjoining pipeline (Line 903) had to be insulated so the viscous, heavy oil would remain heated to make it flow. However, thermally insulated pipelines are known to have aggressive corrosion risks. Investigators found Line 901 to have corrosion similar to that on cars when the paint bubbles and flakes off. Ultimately, moisture migrated along the pipeline (Line 901) to the lowest local elevation point and created an electrochemical corrosion cell that ruptured. Regulators found that "on coated, insulated and buried pipe, the 'corrosion product' grows and remains in close proximity to the pipe steel . . . similar to the type of corrosion on vehicles, in which the corrosion under bubbled paint can be easily flaked off." See US Department of Transportation, "Failure Investigation Report," May 2016, https://www.phmsa.dot.gov/sites/phmsa.dot.gov/files/docs/PHMSA_Failure_Investigation_Report_Plains_Pipeline_LP_Line_901_Public.pdf

17. Kristen Hays, "Regulators Tell Plains to Purge Shut California Oil Pipeline," *Reuters*, November 13, 2015, https://www.reuters.com/article/plains-all-amer-pipeline-california/regulators-tell-plains-to-purge-shut-california-oil-pipeline-idUSL1N1382UV20151113

18. California State Lands Commission, Bureau of Ocean Energy Management (BOEM), and Bureau of Safety and Environmental Enforcement (BSEE), *A Citizen's Guide to Offshore Oil and Gas Decommissioning in Federal Waters Off California*, 2019, https://www.slc.ca.gov/wp-content/uploads/2019/07/IDWG-Decomm-Guide-FINAL-July-2019.pdf

19. Steve Coll, "Getting the International Oil Companies to Address Climate Change," Unconventional Oil Symposium, Carnegie Endowment for International Peace, January 9, 2013, https://carnegieendowment.org/2013/01/09/getting-international-oil-companies-to-address-climate-change-pub-50588

20. Steve Coll, *Private Empire: ExxonMobil and American Power* (New York: Penguin Press, 2012).

21. Steve Graham, "Svante Arrhenius," National Aeronautics and Space Administration Earth Observatory, January 18, 2000, https://earthobservatory.nasa.gov/features/Arrhenius

22. Akshat Rathi, "A 1912 News Article Ominously Forecasted the Catastrophic Effects of Fossil Fuels on Climate Change," *Quartz*, October 24, 2016, https://qz.com/817354/scientists-have-been-forecasting-that-burning-fossil-fuels-will-cause-climate-change-as-early-as-1882/

23. Rob Monroe, "The Keeling Curve," Scripps Institution of Oceanography, April 3, 2013, https://scripps.ucsd.edu/programs/keelingcurve/2013/04/03/the-history-of-the-keeling-curve/

24. National Oceanic and Atmospheric Administration, "Monthly Average Mauna Loa $CO_2$," Global Monitoring Lab, https://www.esrl.noaa.gov/gmd/ccgg/trends/

25. National Oceanic and Atmospheric Administration (NOAA), "Global $CH_4$ Monthly Means," Global Monitoring Laboratory, https://www.esrl.noaa.gov/gmd/ccgg/trends_ch4/. Compared to 1,626.1 parts per million in July 1983, as reported by NOAA Earth System Research Laboratories (ESRL) data.

26. Robert D. Hershey Jr., "Experts Predict California Oil Field Will Be Biggest Find since 1968," *New York Times*, November 15, 1982, https://timesmachine.nytimes.com/timesmachine/1982/11/15/032610.html?pageNumber=1; and as recently as 2018, the American Association of Petroleum Geologists (AAPG) published a paper about the nearly 1 billion barrels of oil in California's Santa Maria Basin, the part of the extensive Monterey formation that contains the Point Arguello oil field; see James Ballard, "Undeveloped Petroleum Potential of the Offshore Santa Maria Basin, California," AAPG, April 23, 2018, http://www.searchanddiscovery.com/pdfz/documents/2018/11067ballard/ndx_ballard.pdf.html

27. In addition, 3D seismic mapping has led to the discovery of many more oil and gas reservoirs.

28. Adam Brandt et al., "Climate-Wise Choices in a World of Oil Abundance," *Environmental Research Letters*, April 5, 2018, https://iopscience.iop.org/article/10.1088/1748-9326/aaae76

29. As recently as 2010, scholars were writing about the looming global energy crisis and describing oil's nearly singular function in terms of miles driven. See Hewitt Crane, Edwin Kinderman, and Ripudaman Malhotra, *A Cubic Mile of Oil* (New York: Oxford University Press, 2010).

30. "Global Oils," Carnegie Endowment for International Peace, 2012, https://carnegieendowment.org/2012/12/18/global-oils/faw9

31. Bruce J. Bilodeau, Guido A. Haug, and Stephen P. Thurston, "Oil and Gas Development in West Coast 1985," *American Association of Petroleum Geologists Bulletin* 90, no. 10 (October 1986), https://pubs.geoscienceworld.org/aapgbull/article-abstract/70/10/1303/38108/Oil-and-Gas-Developments-in-West-Coast-in-1985?redirectedFrom=fulltext

32. Refinery retrofits were significant to handle Point Arguello oil. Estimated capital costs were $5 billion with increased operating costs of nearly $1 billion per year, in

today's dollars. See https://play.google.com/store/books/details?id=HXI5AQA
AMAAJ&rdid=book-HXI5AQAAMAAJ&rdot=1

33. Chevron, USA, "Development and Production Plan: Platform Hidalgo and
Associated Pipelines OCS Lease P 0450 (Supplement to Point Arguello OPP),"
May 1984, https://www.boem.gov/sites/default/files/about-boem/BOEM-
Regions/Pacific-Region/DPPs/5C2---1984-05-Platform-Hidalgo-and-Pipelines-
Development---Production-Plan-Lease-OCS-P-0450.pdf

34. Methane was likely captured in the fugitive emission controls that were required
to be installed (as long as they were designed to contain this smallest hydrocarbon). However, the regulators might have placed more stringent emission limits if
methane was expressly counted and reported. Including methane would also have
informed present-day climate accounting and mitigation strategies.

35. Compare the OCI+ GHG estimates for producing, processing, and shipping a barrel of Saudi Ghawar conventional oil (59 kg $CO_2$e/BOE) to a barrel of California
South Belridge depleted oil (230 kg $CO_2$e/BOE) or Kuwait Ratawi medium oil
(161 kg $CO_2$e/BOE). Although it is not modeled, Point Arguello oil GHGs are
expected to lie somewhere in between these two crudes. See https://dxgordon.
github.io/OCIPlus/#analysis?xSelect=ghgTotal&ySelect=industryGHG&group
Select=Oil&oiltypeSelect=Depleted%20Oil,Light%20Oil,Medium%20Oil

36. Chevron (as operator for its partners, Phillips Petroleum, Union Pacific Resources,
and Impkemix) discovered the Point Arguello oil field in 1981. Additional companies have bought and sold interests in the Point Arguello project since its discovery: Texaco, Pennzoil, Unocal, Getty Oil, Whiting Petroleum, Sun Exploration
and Production, Oryx Energy, Koch Industries, Oxbow Energy, Devon Energy,
Arguello Production, Kerr-McGee, Harvest Energy, Delta Petroleum, Champlin
Petroleum, Simmons Santa Barbara, Largo, Four Corners Pipeline, Venoco, Plains
Exploration and Production, and Freeport-McMoRan. Note that many of these
companies have since merged or gone out of business. Various sources identify
the myriad past Point Arguello partners, including https://www.latimes.com/
archives/la-xpm-1990-01-27-fi-694-story.html; https://www.nytimes.com/1999/
07/06/business/chevron-sells-26-stake-in-oilfield.html; https://www.sec.gov/
Archives/edgar/data/821483/000126327903000010/ex1022.txt

37. Seven international oil companies dominated the global petroleum industry from
the 1940s to the 1970s, including BP (originally the Anglo-Iranian Oil Company
and later the British Petroleum Company), Shell, Chevron (originally Standard
Oil of California or SoCal), Gulf (merged into Chevron), Texaco (merged into
Chevron), Exxon Mobil (originally Standard Oil of New Jersey or Esso), and
Socony (Standard Oil Company of New York that became Mobil and then merged
into Exxon), discussed in more detail in Daniel Sperling and Deborah Gordon,
*Two Billion Cars* (New York: Oxford University Press, 2009). Also see "The Secret

of the Seven Sisters," *Aljazeera*, April 28, 2013, https://www.aljazeera.com/pro-grammes/specialseries/2013/04/201344105231487582.html

38. "Are National Oil Companies the New International Oil Companies?," *KPMG*, October 2018, https://assets.kpmg/content/dam/kpmg/xx/pdf/2018/08/are-national-oil-companies-the-new-international-oil-companies.pdf

39. "California Oil Refinery History, " California Energy Commission, https://www.energy.ca.gov/data-reports/energy-almanac/californias-petroleum-market/californias-oil-refineries/california-oil

40. The agencies involved included US Minerals Management Service (lead), Santa Barbara County Resource Management Department (lead), California Lands Commission, California Coastal Commission, and California Secretary of Environmental Affairs. A cadre of local agencies also were involved under the banner of Santa Barbara County.

41. In 1979, Chevron paid $333.6 million for the right to develop and drill the Point Arguello oil field. This was the most expensive oil lease in US history at the time. See Thomas C. Hayes, "Alaska Spill Jolts Chevron, Too," *New York Times*, October 26, 1989, https://timesmachine.nytimes.com/timesmachine/1989/10/26/461089.html?pageNumber=81

42. Note: Since the platforms were in federal waters, which is over 3 nautical miles offshore in the Outer Continental Shelf (OCS), the proceeds were split between state and federal coffers. Ongoing legal fights challenged the percentages given to each government entity. See William F. Northrop, "Report to the Legislature on Proposed Oil and Gas Lease Sale Program, Pt. Conception – Pt. Arguello, Santa Barbara County," State Lands Commission, December 1981, https://www.slc.ca.gov/wp-content/uploads/2018/11/1981-ProposedOilGas.pdf

43. Note that countries, like Russia, that generate vast sums of government revenue from the sale of oil and gas do not disclose actual sums. See Alexander Malden and Joseph Williams, "Generating Government Revenue from the Sale of Oil and Gas: New Data and the Case for Improved Commodity Trading Transparency," Natural Resources Governance Institute, January 2018, https://resourcegovernance.org/sites/default/files/documents/generating-government-revenue-from-sale-of-oils-and-gas_0.pdf

44. While powerful oil and gas interests can influence political outcomes or induce corruption, this does not imply that petroleum products do not advantage everyday people. This is a conundrum that civil society actors continue to wrestle with. See Alexandra Gillies, *Crude Intentions: How Oil Corruption Contaminates the World* (New York: Oxford University Press, 2020).

45. William Hafker (an ExxonMobil retiree) was working for an industry competitor in Santa Barbara in those formative years while I was permitting Point Arguello. He was instrumental in helping us formulate this climate plan. See Deborah Gordon and Stephen D. Ziman, "Petroleum Companies Need a Credible Climate

Plan," Carnegie Endowment for International Peace, November 2018, https://carnegieendowment.org/files/Gordon_Petro_Companies_Need_Climate_Plan_Nov2018.pdf

46. Others argue that society needs better information and transparency to avoid future unnecessary price hikes and crashes. See Blake Clayton, *Market Madness: A Century of Oil Panics, Crises, and Crashes* (New York: Oxford University Press, 2015).

47. For more information on science-based targets see https://sciencebasedtargets.org/

48. Deborah Gordon and Madhav Acharya, "Oil Shake Up: Refining Transitions in a Low-Carbon Economy," Carnegie Endowment for International Peace, April 2018, https://carnegieendowment.org/files/Gordon_DrivingChange_Article_April2018_final.pdf

49. For methodology and sample oil and gas resources modeled, see Oil Climate Index + Gas Preview, Beta Web Tool Under Development, https://dxgordon.github.io/OCIPlus/

50. International Energy Agency, "Global Primary Energy, Electricity Generation, Final Consumption and $CO_2$ Emissions by Fuel, 2018," January 2020, https://www.iea.org/data-and-statistics/charts/global-primary-energy-electricity-generation-final-consumption-and-co2-emissions-by-fuel-2018

51. International Energy Agency, *World Energy Outlook 2018*, Table 1.5, https://www.iea.org/weo2018/

52. Frank Bruni, "She Predicted the Coronavirus. What Does She Foresee Next?," *New York Times*, May 2, 2020, https://www.nytimes.com/2020/05/02/opinion/sunday/coronavirus-prediction-laurie-garrett.html?referringSource=articleShare;; Laurie Garrett, *The Coming Plague: Newly Emerging Diseases in a World Out of Balance* (New York: Macmillan, 1994), https://www.lauriegarrett.com/the-coming-plague

53. Deborah Gordon, "Not All Oil Is Equal. As Economies Recover, Which Oils Should Stay in the Ground?," *Brink*, June 21, 2020, https://www.brinknews.com/not-all-oil-is-equal-which-types-should-oil-companies-turn-back-on-as-the-economy-recovers-paris-agreement-climate-change-post-covid/

54. See, for example, World Health Organization, "Climate Change and Infectious Disease," July 2003, https://www.who.int/globalchange/climate/en/chapter6.pdf;; Julie Kirkwood, "Climate Change Is Shifting the Course of Infectious Disease," *Clinical Laboratory News*, August 1, 2018, https://www.aacc.org/publications/cln/articles/2018/august/climate-change-is-shifting-the-course-of-infectious-diseases; Cyril Caminade, K. Marie McIntyre, and Anne E. Jones, "Impact of Recent and Future Climate Change on Vector-Borne Diseases," *Annals of the New York Academy of Sciences* 1436, no. 1 (2019): 157–73, https://www.ncbi.nlm.nih.gov/pmc/articles/PMC6378404/

CHAPTER 1

1. Council on Foreign Relations, "Oil Dependence and U.S. Foreign Policy, 1850–2017," https://www.cfr.org/timeline/oil-dependence-and-us-foreign-policy

2. Bjorn Lomborg, *The Skeptical Environmentalist: Measuring the Real State of the World* (Cambridge: Cambridge University Press), 2001.

3. Jimmy Carter, Presidential Debate, C-SPAN 2, 1976, http://missliberty.com/jimmy-carter-the-world-will-run-out-of-oil-by-2011/

4. Graham Jones, "World Oil and Gas 'Running Out,'" *CNN*, October 2, 2003, https://www.cnn.com/2003/WORLD/europe/10/02/global.warming/

5. US Energy Information Administration, "Does the World Have Enough Oil to Meet Our Future Needs?," https://www.eia.gov/tools/faqs/faq.php?id=38&t=6

6. Camila Ruz, "The Six Natural Resources Most Drained by Our 7 Billion People," *The Guardian*, October 3l, 2011, https://www.theguardian.com/environment/blog/2011/oct/31/six-natural-resources-population

7. "Are We Running Out of Oil?," Penn State, EME801, https://www.e-education.psu.edu/eme801/node/486

8. David Deming, "Are We Running Out of Oil?," National Center for Policy Analysis, Policy Backgrounder No. 159, January 29, 2003, ISBN #1-56808-123-5, http://www.ncpathinktank.org/pdfs/bg159.pdf

9. Alex Farrell and Adam Brandt, "Risks of the Oil Transition," *Environmental Research Letters*, October 30, 2006, https://nature.berkeley.edu/er100/readings/Farrell_2006_Risks.pdf

10. Adam Brandt et al., "Climate-Wise Choices in a World of Oil Abundance," *Environmental Research Letters*, April 5, 2018, https://iopscience.iop.org/article/10.1088/1748-9326/aaae76

11. National Aeronautics and Space Administration, "Cassini Reveals Surprises with Titan's Lakes," April 15, 2019, https://solarsystem.nasa.gov/news/904/cassini-reveals-surprises-with-titans-lakes/

12. M. K. Hubbert, "Nuclear Energy and the Fossil Fuels," Presented at American Petroleum Institute, Southern Division, March 7–9, 1956, https://www.resilience.org/stories/2006-03-08/nuclear-energy-and-fossil-fuels/. Also see reproduced Hubbert documents by the US government, https://www.osti.gov/servlets/purl/4655197

13. Ian Chapman, "The End of Peak Oil? Why This Topic Is Still Relevant Despite Recent Denials," Energy Policy, 2013, doi.org/10.1016, https://www.ourenergypolicy.org/wp-content/uploads/2015/06/The-End-of-Peak-Oil.pdf

14. Hubbert, "Nuclear Energy and the Fossil Fuels."

15. US Energy Information Administration, "STEO Current/Previous Forecast Comparisons: U.S. Energy Supply and Demand Survey," September 9, 2020, https://www.eia.gov/outlooks/steo/pdf/compare.pdf. Note that US crude oil and

dry natural gas production stood at 12.78 million barrels per day and 95.97 billion cubic feet per day, respectively, in Q4 2019, a reported all-time high. For updated data on an annual, quarterly, or monthly basis, see https://www.eia.gov/outlooks/steo/data/browser/#/?v=3&f=A&s=0&start=2016&end=2021&id=&linechart=COPRPUS&maptype=0&ctype=linechart

16. US Energy Information Administration, "International Petroleum and Other Liquids Production," Short-Term Energy Outlook Data Browser, https://www.eia.gov/outlooks/steo/data/browser/#/?v=6&f=Q&s=0&start=201601&end=202104&ctype=linechart&maptype=0&linechart=PAPR_OECD; "Global Energy Statistical Yearbook 2020," Enerdata, https://yearbook.enerdata.net/natural-gas/world-natural-gas-production-statistics.html. Note that total world petroleum production stood at 101.73 million barrels per day in Q4 2019 and world natural gas production stood at just over 4,000 billion cubic meters in 2019, both reported all-time highs. For historical data and current updates, see: https://www.eia.gov/international/data/world

17. DNG-VL, "Energy Transition Outlook 2018: Oil and Gas," 2018, https://eto.dnvgl.com/2018/#Timeline

18. Adam Brandt et al., "Peak Oil Demand: The Role of Fuel Efficiency and Alternative Fuels in a Global Oil Production Decline," *Environmental Science and Technology*, May 22, 2013, https://pubs.acs.org/doi/abs/10.1021/es401419t

19. US Energy Information Administration, "Short-Term Energy Outlook," May 11, 2021, https://www.eia.gov/outlooks/steo/report/global_oil.php

20. Chevron, "Climate Change Resilience: A Framework for Decision Making," https://www.chevron.com/-/media/shared-media/documents/climate-change-resilience.pdf

21. Colin Campbell and Jean Laherrere, "The End of Cheap Oil," *Scientific American*, March 1998, https://nature.berkeley.edu/er100/readings/Campbell_1998.pdf

22. Society of Petroleum Engineers, "Guidelines for Application of the Petroleum Resources Management System," November 2011, https://www.spe.org/industry/docs/PRMS_Guidelines_Nov2011.pdf; United Nations, "United Nations Framework Classification for Fossil Energy and Mineral Resources," http://www.unece.org/fileadmin/DAM/ie/se/pdfs/UNFC/UNFCemr.pdf

23. US Energy Information Administration, "Oil and Natural Gas Resource Categories Reflect Varying Degrees of Certainty," July 17, 2014, https://www.eia.gov/today-inenergy/detail.php?id=17151

24. Energy and Environment Study Institute, "Fact Sheet: Shale Gas & Oil Terminology Explained: Reserves and Estimates," November 2011, https://www.eesi.org/files/fracking_estimates_111611.pdf

25. Will Kenton, "Recoverable Reserves," *Investopedia*, April 17, 2018, https://www.investopedia.com/terms/r/recoverabel-reserve.asp

26. US Energy Information Administration, "U.S. Oil and Natural Gas Proved Reserves Declined in 2015 Because of Lower Prices," December 15, 2016, https://www.eia.gov/todayinenergy/detail.php?id=29172

27. See, for example, PWC, "Financial Reporting in the Oil and Gas Industry: International Financial Reporting Standards," July 19, 2017, https://www.pwc.com/gx/en/services/audit-assurance/assets/pwc-financial-reporting-in-the-oil-and-gas-industry-2017.pdf

28. Kate Kelly and Stanley Reed, "How Aramco's Huge I.P.O. Fell Short of Saudi Prince's Wish," *New York Times*, December 11, 2019, https://www.nytimes.com/2019/12/06/business/energy-environment/saudi-aramco-ipo.html#

29. Michael Coren, "Covid-19 Is Closing Canada's Carbon-Intensive Oil Sands for Business," *Quartz*, April 29, 2020, https://qz.com/1846830/covid-19-is-closing-canadas-carbon-intensive-oil-sands/

30. "Information about the Monterey Shale," Natural Gas Intelligence, 2015, https://www.naturalgasintel.com/information-about-the-monterey-shale/

31. "UPDATE 2 – U.S. EIA Cuts Recoverable Monterey Shale Oil Estimate by 95 Percent," *Reuters*, May 21, 2014, https://www.reuters.com/article/eia-monterey-shale/update-2-u-s-eia-cuts-recoverable-monterey-shale-oil-estimate-by-96-pct-idUSL1N0O713N20140521

32. Deborah Gordon, "The World's Growing Oil Resources," April 17, 2013, https://carnegieendowment.org/2013/04/17/world-s-growing-oil-resources-pub-51547

33. BP, "Statistical Review of World Energy," https://www.bp.com/en/global/corporate/energy-economics/statistical-review-of-world-energy.html

34. Adam R. Brandt et al., "Climate-wise Choices in a World of Oil Abundance," *Environmental Research Letters*, April 5, 2018, https://iopscience.iop.org/article/10.1088/1748-9326/aaae76/pdf

35. International Energy Agency, Chapter 8, "Outlook for Natural Gas," in *World Energy Outlook 2017*, 2018, 345.

36. This reserve estimate is far higher when counting all unconventional gas resources. See Table 2.1.

37. For details, see Tables 2.1 and 2.2. Note that consumption is approximately equal to production in a given year. Oil volumes do not consider biofuels, coal-to-liquids, or gas-to-liquids.

38. National Aeronautics and Space Administration, "Missions Cassini," https://solarsystem.nasa.gov/missions/cassini/overview/

39. National Aeronautics and Space Administration, "Titan," https://solarsystem.nasa.gov/moons/saturn-moons/titan/in-depth/

40. Nola Taylor Reed, "Titan: Facts about Saturn's Largest Moon," Space.com, March 27, 2018, https://www.space.com/15257-titan-saturn-largest-moon-facts-discovery-sdcmp.html#

41. Prasanta Mukhopadhyay (Muki), "Planet Mars: Prospects of Biogenic or Thermogenic Oil and Gas from Deeper Sources," AAPG, 2014, http://www.searchanddiscovery.com/documents/2014/70165muki/ndx_muki.pdf

42. "The Asteroid Trillionaires," PhysicsWorld, June 11, 2018, https://physicsworld.com/a/the-asteroid-trillionaires/

43. Louise Lerner, "7 Things You May Not Know about Catalysis," Argonne National Laboratory, December 14, 2011, https://www.anl.gov/article/7-things-you-may-not-know-about-catalysis

44. "Why the Next Generation of Rockets Will Be Powered by Methane," *Australia Science Channel*, September 3, 2019, https://australiascience.tv/why-the-next-generation-of-rockets-will-be-powered-by-methane/

45. For a user-friendly, rotating global map of conventional and continuous (unconventional shale) oil and gas resources, see US Geological Survey, "World Petroleum Assessment," https://certmapper.cr.usgs.gov/data/apps/world-energy/?resource=conventional

46. US Geological Survey, "World Oil and Gas Resource Assessments," https://www.usgs.gov/centers/cersc/science/world-oil-and-gas-resource-assessments?qt-science_center_objects=0#qt-science_center_objects

47. US Geological Survey, "U.S. Assessment of Undiscovered Oil and Gas," https://www.usgs.gov/media/images/map-us-assessments-undiscovered-oil-and-gas-resources

48. John Kingston, "The Undiscovered Oil and Gas of Antarctica," US Geological Survey, 91-597, 1001, https://pubs.usgs.gov/of/1991/0597/report.pdf

49. The governments of Argentina, Australia, Belgium, Chile, the French Republic, Japan, New Zealand, Norway, the Union of South Africa, the Union of Soviet Socialist Republics, the United Kingdom of Great Britain and Northern Ireland, and the United States are party to the Antarctic Treaty, which was entered into force on June 23, 1961. See "Antarctic Treaty," US Department of State, https://2009-2017.state.gov/t/avc/trty/193967.htm; also see Antarctic and Southern Ocean Coalition, "The Antarctic Oil Myth," https://www.asoc.org/component/content/article/9-blog/1184-the-antarctic-oil-myth

50. Ibid.

51. T. R. Klett et al., "Glossary," US Geological Survey, Digital Data Series DDS–69–D, May 2005, https://pubs.usgs.gov/dds/dds-069/dds-069-d/REPORTS/69_D_CH_25.pdf

52. US Geological Survey, "Definition of Terms," https://pubs.usgs.gov/ha/ha747/pdf/definition.pdf

53. T. R. Klett et al., "Glossary."

54. Gordon M. Kaufman, "Statistical Issues in the Assessment of Undiscovered Oil and Gas Resources," Sloan School of Management, Massachusetts Institute of Technology, MIT-CEEPR 92-010WP, September 1992, https://pdfs.semanticscholar.org/e1bc/e2cbf0bbed00912f332d4737308825ca0ccc.pdf

55. Note that field counts may consider newly discovered oil and gas accumulations as part of extended discovered fields or as new fields. See Emil Attanasi and Philip Freeman, "Statistics of Petroleum Exploration in the World Outside the United States and Canada Through 2015," US Geological Survey, Circular 1450, 2019, https://pubs.usgs.gov/circ/1450/circ1450.pdf

56. Li Guoyu, *World Atlas of Oil and Gas Basins* (Oxford: Wiley-Blackwell, 2011).

57. M. S. Masnadi et al., "Global Carbon Intensity of Crude Oil Production," *Science*, 361, no. 6505 (August 31, 2018), https://science.sciencemag.org/content/361/6405/851.summary

58. For background data, see US Energy Information Administration, "Top 100 U.S. Oil and Gas Fields," March 2015, https://www.eia.gov/naturalgas/crudeoil-reserves/top100/pdf/top100.pdf, and Enerdata, "Natural Gas Production, 2019," Global Energy Statistical Yearbook 2020, https://yearbook.enerdata.net/natural-gas/world-natural-gas-production-statistics.html

59. US Energy Information Administration, "U.S. Oil and Gas Wells by Production Rate," October 28, 2018, https://www.eia.gov/petroleum/wells/

60. See, for example, Velda Addison, "Exxon Mobil Grows Guyana Oil Find Estimates to More Than 8 Bboe," Hart Energy, January 27, 2020; *Oil & Gas Journal*, "Gazprom to Develop Giant Yamal Gas Field," July 17, 2018; E. Allison and B. Mandler, "Oil and Gas in the U.S. Arctic," American Geosciences Institute, 2018.

61. IFP Energies Nouvelles, "New Oil and Gas Discoveries in 2018," July 30, 2019, https://www.ifpenergiesnouvelles.com/article/new-oil-and-gas-discoveries-2018

62. Ahmed ElSakka, Ghareb Hamada, Eswaran Padmanabhan, and Ahmed Salim, "South East Asia Contains Abundant, Untapped Shale Reservoirs," *Oil & Gas Journal*, March 5, 2018.

63. Benjamin Roussey, "Mega Oil and Gas Discovery in the Permian Basin Doubles America's Energy Reserves," March 12, 2019, https://energycentral.com/c/og/mega-oil-and-gas-discovery-permian-basin-doubles-america%E2%80%99s-energy-reserves

64. In 2019 alone, the US Geological Survey assessed oil and gas basins in Greece, Albania, Saudi Arabia, Libya, Tunisia, Malta, Italy, Algeria, Australia, China, North Africa, and Alberta (Canada). Another twelve were studied in 2018. See https://www.usgs.gov/centers/cersc/science/world-oil-and-gas-resource-assessments?qt-science_center_objects=3#qt-science_center_objects; US Geological Survey, "World Petroleum Assessment," https://certmapper.cr.usgs.gov/data/apps/world-energy/

65. US Energy Information Administration, "World Petroleum and Other Liquids Consumption," September 2020, https://www.eia.gov/outlooks/steo/tables/pdf/3dtab.pdf

66. International Energy Agency, "Natural Gas Information: Overview," July 2020, https://www.iea.org/reports/natural-gas-information-overview

67. Since 2020 oil and gas supply and demand were temporarily altered during the COVID-19 pandemic and are not representative of past or projected future market conditions, 2018 and 2019 statistics are being used in this book.

68. Muyu Xu, Stephanie Kelly, and Yuka Obayashi, "China Drives Global Oil Demand Recover Out of Coronavirus Collapse," *Reuters*, June 3, 2020, https://www.reuters.com/article/us-global-oil-demand-analysis/china-drives-global-oil-demand-recovery-out-of-coronavirus-collapse-idUSKBN23A0XF

69. International Energy Agency, "Oil 2018," 2018, https://www.iea.org/oil2018/#section-1-1

70. Deborah Gordon and Madhav Acharya, "Oil Shake Up: Refining Transitions in a Low Carbon Economy," April 3, 2018, https://carnegieendowment.org/2018/04/03/oil-shake-up-refining-transitions-in-low-carbon-economy-pub-75954

71. American Petroleum Institute, "Why Seismic Surveys Are Needed," https://www.api.org/oil-and-natural-gas/wells-to-consumer/exploration-and-production/offshore/seismic-surveys-needed-in-the-atlantic; and Bureau of Ocean Energy Management, "High-Resolution Geophysical Survey Application,"https://www.boem.gov/High-Resolution-Geophysical-Survey-Application/

72. Schlumberger, "Upcoming Bid Rounds," http://www.multiclient.slb.com/latest-projects/licensing-rounds.aspx

73. International waters are 12 nautical miles (13.8 miles) beyond a nation's coastal low-water mark, according to the Law of the Sea. See United Nations Convention on the Law of the Sea, https://www.un.org/depts/los/convention_agreements/texts/unclos/unclos_e.pdf

74. Bureau of Ocean Energy Management, "Leasing," https://www.boem.gov/Leasing/

75. Note that "Drill, baby, drill" was first used at the 2008 Republican National Convention as a way to signal American energy independence and private land-owner control over their mineral rights. See https://thehill.com/policy/energy-environment/246453-drill-baby-drill-origin-story-revealed-in-tampa

76. US Department of Energy, "Quadrennial Technology Review 2015: Process Heating," https://www.energy.gov/sites/prod/files/2016/06/f32/QTR2015-6I-Process-Heating.pdf

77. *Oil & Gas Journal*, "Worldwide Gas Processing Survey 2015," https://www.ogj.com/ogj-survey-downloads.html

78. Ibid.

79. Deborah Gordon et al., "Know Your Oil: Creating a Global Oil-Climate Index," Carnegie Endowment for International Peace, 2015, https://carnegieendowment.org/2015/03/11/know-your-oil-creating-global-oil-climate-index-pub-59285

80. *Oil & Gas Journal*, "2018 Worldwide Refining Survey," January 1, 2019, https://www.ogj.com/ogj-survey-downloads.html

81. Ibid.

82. Ibid.

83. For the world's smallest refinery see ibid. For the world's largest refinery that was recently expanded, see ListNBest, "11 World's Largest Oil Refineries by Processing Capacity," https://www.listnbest.com/11-worlds-largest-oil-refineries-processing-capacity/

84. Liang Jing et al., "Global Carbon Intensity of Crude Oil Refining and Mitigation Potential," *National Climate Change* 10 (2020): 526–32, https://doi.org/10.1038/s41558-020-0775-3, https://rdcu.be/b4zHw

85. Ibid.

86. Ibid.

87. International Energy Agency, "Oil 2018."

88. E. Russell Braziel, *The Domino Effect* (Madison, WI: NTA Press, 2016).

89. Steve Pryor, "A Partial List of the over 6,000 Products Made from One Barrel of Oil (after Creating 19 Gallons of Gasoline)," July 26, 2016, https://www.linkedin.com/pulse/partial-list-over-6000-products-made-from-one-barrel-oil-steve-pryor/

90. Devyn Collado, "Making the Case: Drilling Down on Refining for a Climate Solution," *Medium*, August 13, 2020, https://medium.com/@dcollado9/making-the-case-drilling-down-on-refining-for-a-climate-solution-60d843bff2d4

91. Gordon et al., "Oil Shake Up."

92. For a 2017 Sankey diagram of oil and gas flows see International Energy Agency, https://www.iea.org/sankey/#?c=IEA%20Total&s=Balance

93. Dan Sperling and Deborah Gordon, *Two Billion Cars* (New York: Oxford University Press, 2008).

94. International Energy Agency, "Oil 2018."

95. As a point of reference, in 1978, the US electric power sector consumed an annual average of 1.75 mbpd oil. Percent reduction based on author's calculations using EIA historic data. See "Total Energy, Petroleum Consumption Estimates: Electric Power Sector," https://www.eia.gov/totalenergy/data/annual/

96. Liquid renewable fuels, such as synthetic diesel and kerosene, can be manufactured from reconfiguring biofuels or combining hydrogen with captured carbon, using renewable energy. Solar fuels use the sun for conversion energy. Lewis Research Group, "Solar Fuels," http://nsl.caltech.edu/home/solar-fuels/

97. The Sulfur Institute, "Sulphur Uses," https://www.sulphurinstitute.org/about-sulphur/glossary/#uses

98. US National Park Service, "Cabin Branch Pyrite Mine (1889–1920)," https://www.nps.gov/prwi/learn/historyculture/cabin-branch-mine.htm

99. Akshat Rathi, "To Hit Climate Goals, Bill Gates and His Billionaire Friends Are Betting on Energy Storage," *Quartz*, June 12, 2018, https://qz.com/1302711/to-hit-climate-goals-bill-gates-and-his-billionaire-friends-are-betting-on-energy-storage/

100. US Energy Information Administration, "What Drives Crude Oil Prices?," https://www.eia.gov/finance/markets/crudeoil/

101. Lutz Killian, "Oil Price Volatility: Origins and Effects," World Trade Report 2010, World Trade Organization, https://www.wto.org/english/res_e/publications_e/wtr10_forum_e/wtr10_kilian_e.htm

102. For information on past pandemics, see US Centers for Disease Control and Prevention, "Past Pandemics," August 2018, https://www.cdc.gov/flu/pandemic-resources/basics/past-pandemics.html

103. For a discussion on the barriers regarding energy innovation, see Richard Lester, "Energy Innovation," in *Production in the Innovation Economy*, ed. Richard Locke and Rachel Wellhausen (Cambridge, MA: MIT Press, 2014).

104. US Energy Information Administration, "Short-Term Energy Outlook," June 11, 2019, https://www.eia.gov/outlooks/steo/

105. Free on Board (FOB) is the price for loading and shipping the commodity but does not include the cost of delivery to its final destination. Instead, an FOB sale transfers the title to the buyer once the commodity is in transit. FOB pricing facilitates the export of large volumes that have multiple destinations en route. For more information see McKinsey, "FOB," *Energy Insights*, https://www.mckinseyenergyinsights.com/resources/refinery-reference-desk/fob/

106. Buyers and sellers who establish contracts and settle prices include producers, refiners, traders, and large distributors and consumers (such as large fuel retailers and wholesalers). Price assessment agencies (such as Platts, Argus, and others) act as third-party firms that publish spot prices. See McKinsey, "Spot Market," *Energy Insights*, https://www.mckinseyenergyinsights.com/resources/refinery-reference-desk/spot-market/

107. Francis Bator, "The Anatomy of Market Failure," *Quarterly Journal of Economics* 72 (August 1958), https://academic.oup.com/qje/article-abstract/72/3/351/1859737?redirectedFrom=fulltext

108. George Akerlof's 1970 paper, "The Market for Lemons," laid the ground for information economics. See "Secrets and agents," *The Economist*, June 23, 2016, https://www.economist.com/schools-brief/2016/07/23/secrets-and-agents; citizenscience.gov

109. K. J. Arrow, "Inter-Generational Equity and the Rate of Discount in Long-Term Social Investment," in *Contemporary Economic Issues*. International Economic Association Series, ed. M. R. Sertel (London: Palgrave Macmillan, 1999), https://doi.org/10.1007/978-1-349-14540-9_5; Louis Sohn and Edith Brown Weiss, "Intergenerational Equity in International Law," *Proceedings of the Annual Meeting of the American Society of International Law* 81 (April 8–11, 1987), https://www.jstor.org/stable/25658355

110. National Petroleum Council, "Unconventional Oil," Working Document of the NPC North American Resource Development Study, Paper #1–6, September 15,

2011, www.npc.org/Prudent_Development-Topic_Papers/1-6_Unconventional_
Oil_Paper.pdf

111. Alison Redford, "We Rise Together or We Fall Together," *Policy Options Politiques*,
February 1, 2012, https://policyoptions.irpp.org/magazines/sustainable-energy/
we-rise-together-or-we-fall-together/

112. Justin Worland, "The Reason Fossil Fuel Companies Are Finally Reckoning with
Climate Change," *Time Magazine*, January 16, 2020, https://time.com/5766188/
shell-oil-companies-fossil-fuels-climate-change/

113. Larry Fink, BlackRock, https://www.blackrock.com/corporate/investor-
relations/larry-fink-ceo-letter?cid=ppc:CEOLetter:PMS:US:NA

114. "Saudi Arabia Urges OPEC+ to Reduce Oil Production," Arabian Business
Industries, May 13, 2020, https://www.arabianbusiness.com/energy/446595-
saudi-arabia-urges-opec-to-reduce-oil-production

115. Ron Bousso, "For Big Oil, Reserve Size Matters Less Than Ever," *Reuters*, May 16,
2018, https://www.reuters.com/article/us-oilmajors-reserves/for-big-oil-reserve-
size-matters-less-than-ever-idUSKCN1IH1I2

### CHAPTER 2

1. The first person to commercialize kerosene—Abraham Pineo Gesner—moved to
Halifax in 1848. Gesner extracted "albertite" (the name he gave to tarry oil) in
1836 and successfully "cracked" these hydrocarbons to form "keroselain," railway
grease, paving asphalt, mineral caulk, ash for fertilizer, and solid coke. He renamed
this product "kerosene," which produced a brighter and cleaner flame than burning coal gas. Gesner's kerosene was already in use in a Halifax apartment and Nova
Scotia's lighthouse when it was used in Washington, DC's streetlights. See David
C. Cooper, ed., *Doctors of Another Calling: Physicians Who Are Known Best in
Fields Other Than Medicine* (Lanham, MD: University of Delaware Press, 2014).

2. Umar Ali, "The History of the Oil and Gas Industry from 347 AD to Today,"
March 7, 2019, https://www.offshore-technology.com/comment/history-oil-gas/

3. Rock oil came from the Marcellus play, one of the largest unconventional resource
basins in operation today. See US Energy Information Administration, "Marcellus
Shale Play," 2017, https://www.eia.gov/maps/pdf/MarcellusPlayUpdate_Jan2017.
pdf

4. Daniel Yergin, *The Prize: The Epic Quest for Oil, Money & Power* (New York: Free
Press, 1991).

5. Deborah Gordon and Samuel Wojcicki, "Drilling Down on Oil: The Case
of California's Complex Midway Sunset Field," Carnegie Endowment for
International Peace, Washington, DC, March 15, 2017, https://carnegieen-
dowment.org/2017/03/15/drilling-down-on-oil-case-of-california-s-complex-
midway-sunset-field-pub-68210

6.6. A half-century would pass before the world's single largest conventional oil field, Saudi Ghawar, began production in 1951. For more information on the industry and its evolution, see Deborah Gordon and Daniel Sperling, "Big Oil Can't Get Beyond Petroleum," *Washington Post*, June 20, 2010, http://www.washingtonpost.com/wp-dyn/content/article/2010/06/11/AR2010061103256.html

7. US Geological Survey, "The Sunset-Midway Oil Field, California," 1919, http://pubs.usgs.gov/pp/0117/report.pdf

8. Oil Climate Index + Gas Preview Web Tool (note: sort supply chain by Upstream+Midstream [indirect] emissions and hover over top and bottom bars); International Energy Agency, *World Energy Outlook 2018*, Chapter 11, November 13, 2018; for International Energy Agency *World Energy Outlooks* (previous years), see https://webstore.iea.org/search?q=World+Energy+Outlook&adv=false&cid=0&isc=false&pf=&pt=&sid=false

9. For figures and sources see Tables 2.1 and 2.2.

10. Ibid.

11. National Geographic Resource Library, "Petroleum," https://www.nationalgeographic.org/encyclopedia/petroleum/

12. Amit Kumar et al., "Lithium Recovery from Oil and Gas Produced Water: A Need for a Growing Energy Industry," *ACS Energy Letters* 4, no. 6 (2019): 1471–74, https://pubs.acs.org/doi/full/10.1021/acsenergylett.9b00779

13. Deborah Gordon et al., "Know Your Oil," Carnegie Endowment for International Peace, Washington, DC, March 2015, https://carnegieendowment.org/files/know_your_oil.pdf

14. "Gases – Densities," Engineering ToolBox, https://www.engineeringtoolbox.com/gas-density-d_158.html

15. US Energy Information Association, "Glossary," https://www.eia.gov/tools/glossary/index.php?id=C#conv_oil_nat_gas_prod and https://www.eia.gov/tools/glossary/index.php?id=Unconventional%20oil%20and%20natural%20gas%20production

16. For example, see OCI+ analysis of conventional light and medium oils and wet and dry gases that plots total emissions versus consumer GHG responsibilities: https://dxgordon.github.io/OCIPlus/#analysis?xSelect=consumerGHG&ySelect=ghgTotal&oiltypeSelect=Light%20Oil,Dry%20Gas,Wet%20Gas

17. For cumulative conventional oil production, see Adam Brandt et al., "Climate-Wise Choices in a World of Oil Abundance," *Environmental Research Letters*, April 2018, https://iopscience.iop.org/article/10.1088/1748-9326/aaae76 (from 1859 to 2015) and International Energy Agency, *World Energy Outlook 2017* (Table 4.1) and *2018* (Table 3.1) (for 2016–2019). For cumulative conventional gas production see Our World in Data, "Energy," https://ourworldindata.org/energy and IEA; *World Energy Outlook 2017* (Table 4.1) (for 2000–2019). Note that barrels of oil can be added directly to BOE gas.

18. International Energy Agency, *World Energy Outlook 2017,* https://www.iea.org/reports/world-energy-outlook-2017

19. International Energy Agency, *World Energy Outlook 2017*, Special Focus on Natural Gas, 2018, https://www.iea.org/reports/outlook-for-natural-gas

20. Water Quality Improvement Act of 1970 (P.L. 91-224).

21. US Congressional Budget Office, "The World Oil Market in the 1980s: Implications for the United States," May 1980, https://www.cbo.gov/sites/default/files/96th-congress-1979-1980/reports/80doc17.pdf

22. H. Rept. 96-1016, Part II (June 20, 1980), p. 6.

23. Jonathan L. Ramseur, "Oil Sands and the Oil Spill Liability Trust Fund: The Definition of 'Oil' and Related Issues for Congress," Congressional Research Service, February 15, 2017, https://fas.org/sgp/crs/misc/R43128.pdf

24. US Energy Information Administration, "Horizontally Drilled Wells Dominate U.S. Tight Formation Production," June 6, 2019, https://www.eia.gov/todayinenergy/detail.php?id=39752; US Energy Information Administration, "Hydraulically Fractured Horizontal Wells Account for Most New Oil and Natural Gas Wells," January 30, 2018, https://www.eia.gov/todayinenergy/detail.php?id=34732

25. *Oil Sands Magazine*, "Total Bitumen Production from Mining & In-Situ," https://www.oilsandsmagazine.com/projects/bitumen-production#Chart

26. Deborah Gordon et al., "Know Your Oil: Creating a Global Oil-Climate Index," Carnegie Endowment for International Peace, March 11, 2015, https://carnegieendowment.org/2015/03/11/know-your-oil-creating-global-oil-climate-index-pub-59285

27. Deborah Gordon and Chris Malins, "Uncovering Oil's Unknowns," Carnegie Endowment for International Peace, June 19, 2013, https://carnegieendowment.org/files/Debbie-UnknownOil-article-final.pdf

28. International Energy Agency, *World Energy Outlook 2017* (Table 4.1).

29. Ibid.

30. Continuous-type oil and gas deposits are present in fine-grain shale, carbonate, limestone, and other rocks. See Michael Ratner and Mary Tiemann, "An Overview of Unconventional Oil and Gas: Resources and Federal Oversight," US Congressional Research Service, 7-5700, R43148, April 22, 2015, https://fas.org/sgp/crs/misc/R43148.pdf

31. American Oil and Gas Historical Project (AOGHS), "Shooters – A 'Fracking' History," https://aoghs.org/technology/hydraulic-fracturing/

32. Mark Green, "Happy Birthday, Fracking!," American Petroleum Institute, March 17, 2014, https://www.api.org/news-policy-and-issues/blog/2014/03/17/march-17-happy-birthday-fracking

33. Gregory Zuckerman, "Breakthrough: The Accidental Discovery That Revolutionized American Energy," *The Atlantic*, November 6, 2013, https://www.theatlantic.com/business/archive/2013/11/breakthrough-the-accidental-discovery-that-revolutionized-american-energy/281193/

34. US Energy Information Administration, "Trends in U.S. Oil and Natural Gas Upstream Costs," March 2016, https://www.eia.gov/analysis/studies/drilling/pdf/upstream.pdf

35. Deborah Gordon and Frances Reuland, "Mapping, Measuring, and Managing Methane: The Critical Role of a Potent Climate Pollutant," Watson Institute for International and Public Affairs, Brown University, November 2019, https://watson.brown.edu/files/watson/imce/news/ResearchBriefs/2019/Methane%20Report-6%20November%202019.pdf

36. The earth's crust extends twenty-one miles below the surface; see https://www.extremetech.com/extreme/154357-earths-core-is-much-hotter-than-previously-thought-hotter-than-the-surface-of-the-sun#. At present, the world's deepest commercial oil and gas well (Russia Chayvo) is nearly 8 miles below the surface; see https://www.oilandgasiq.com/drilling-and-development/articles/z-44-chayvo-well-the-deepest-oil-extraction#

37. For estimates using OCI 2.0, see https://oci.carnegieendowment.org/#compare/russia-chayvo/angola-takula. Note: Hover over upstream graph to see GHG contribution from drilling.

38. Ibid.

39. Gongcheng Zhang et al., "Giant Discoveries of Oil and Gas Fields in Global Deepwaters in the Past 40 Years and the Prospect of Exploration," *Journal of Natural Gas Geoscience* 4, no. 1 (February 2019), https://www.sciencedirect.com/science/article/pii/S2468256X19300033

40. "Ultra-Deep Gas Production Set to Triple by 2025," Offshore Technology, June 5, 2018, https://www.offshore-technology.com/comment/ultra-deepwater-gas-production-set-triple-2025/

41. Ibid.

42. "Global Deepwater Production Hits 10 Million BOE/D," *Journal of Petroleum Technology*, December 30, 2019, https://pubs.spe.org/en/jpt/jpt-article-detail/?art=6422

43. Two additional nations, Finland and Sweden, border the Arctic Circle but do not have access to the Arctic seafloor. The Arctic basins include Amerasian Basin, Arctic Alaska Basin, East Barents Basin, East Greenland Rift Basin, West Greenland–East Canada Basin, West Siberian Basin, and the Yenisey-Khatanga Basin. See Hobart M. King, "Oil and Gas Resources of the Arctic," Geology.com, https://geology.com/articles/arctic-oil-and-gas/

44. National Aeronautics and Space Administration, "The Study of Earth as an Integrated System," https://climate.nasa.gov/nasa_science/science/

45. "Offshore Arctic Oil & Gas: Market Report to 2018," Infield Energy Analysts, 2012, https://www.infield.com/brochures/offshore-arctic-oil-gas-market-forecast-report.pdf

46. Atle Staalesen, "Norway Ramps Up Oil and Gas Production in Arctic Despite Looming Oil Crisis," Eye on the Arctic, January 10, 2019, https://www.rcinet.ca/

eye-on-the-arctic/2019/01/10/norway-oil-offshore-barents-sea-climate-change/; "The Future of Russian Oil Production in the Short, Medium, and Long Term," Oxford Institute for Energy Studies, September 2019, https://www.oxforden-ergy.org/wpcms/wp-content/uploads/2019/09/The-Future-of-Russian-Oil-Production-in-the-Short-Medium-and-Long-Term-Insight-57.pdf

47. Atle Staalesen, "Why Arctic Oil Development Could Soon Be in Crisis," *Independent Barents Observer*, March 26, 2020, https://www.arctictoday.com/why-arctic-oil-development-could-soon-be-in-crisis/

48. Atle Staalesen, "Putin Signs Arctic Master Plan," *Barents Observer*, March 6, 2020, https://thebarentsobserver.com/en/arctic/2020/03/putin-signs-arctic-master-plan

49. US National Oceanic and Atmospheric Administration, "What Is the Gulf Stream?," https://scijinks.gov/gulf-stream/; Moises Velasquez-Manoff and Jeremy White, "In the Atlantic Ocean, Subtle Shifts Hint at Dramatic Dangers," New York Times, March 2, 2021, https://www.nytimes.com/interactive/2021/03/02/climate/atlantic-ocean-climate-change.html

50. Rigzone, "What Is Tight Gas, and How Is It Produced?," https://www.rigzone.com/training/insight.asp?insight_id=346&c_id=

51. "Coal," Geology.com, https://geology.com/rocks/coal.shtml

52. American Geosciences Institute, "What Is Coalbed Methane?," https://www.americangeosciences.org/critical-issues/faq/what-coalbed-methane

53. US Environmental Protection Agency, "Coalbed Methane Outreach Program," https://www.epa.gov/cmop/frequent-questions; Cheryl Katz, "In Push to Find Methane Leaks, Satellites Gear Up for the Hunt," *Yale Environment 360*, June 15, 2021, https://e360.yale.edu/features/in-push-to-find-methane-leaks-satellites-gear-up-for-the-hunt

54. International Energy Agency, *World Energy Outlook 2018*, Table 4.1.

55. International Energy Agency, *World Energy Outlook 2017*, Table 8.2.

56. Valerio Paolini et al., "Environment Impact of Biogas: A Short Review of Current Knowledge," *Journal of Environmental Science and Health*, April 13, 2018, https://www.tandfonline.com/doi/full/10.1080/10934529.2018.1459076

57. US Energy Information Administration, "Biomass Explained," updated November 12, 2019, https://www.eia.gov/energyexplained/biomass/landfill-gas-and-biogas.php

58. Manjola Banj et al., "Support for Biogas in the EU Electricity Sector – A Comparative Analysis," *Biomass and Bioenergy* 128 (September 2019), https://www.sciencedirect.com/science/article/pii/S0961953419302624?via%3Dihub

59. L. Yang and X. Ge, "Chapter Three: Biogas and Syngas Upgrading," in *Advances in Bioenergy*, vol. 1, Edited by Yebo Li, Xumeng Ge, Elsevier, 125-186, (2016), https://doi.org/10.1016/bs.aibe.2016.09.003

60. Clifford Krauss, "Big Oil Bets on Carbon Removal," *New York Times*, April 8, 2019, https://www.nytimes.com/2019/04/07/business/energy-environment/climate-change-carbon-engineering.html

61. Note that this article incorrectly refers to acid gas as "sour" gas. W. F. J. Burgers et al., "Worldwide Development Potential for Sour Gas," *Energy Procedia* 4 (2011), https://www.sciencedirect.com/science/article/pii/S1876610211003018

62. Note that there are pitfalls developing acid gas. For example, it is unclear whether Chevron will ultimately be able to successfully operate Gorgon's acid gas field despite permit requirements to deploy CCS in Australia. See Lisa Cox, "Western Australia LNG Plant Faces Calls to Shut Down Until Faulty Carbon Capture System Is Fixed, *The Guardian*, January 14, 2021, https://www.theguardian.com/environment/2021/jan/15/western-australia-lng-plant-faces-calls-to-shut-down-until-faulty-carbon-capture-system-is-fixed

63. Oil Climate Index + Gas Preview Web Tool, https://dxgordon.github.io/OCIPlus/#supply-chain?model=00000000&carbonToggle=off&carbonTax=20.00&stepSelect=ghgTotal&sortSelect=true&oiltypeSelect=CO2%20EOR%20Oil,Sour%20Gas,Dry%20Gas

64. James G. Speight, "Gas in Geopressurized Zones," in *Natural Gas,* 2nd ed., Science Direct, 59-98, (2019), https://www.sciencedirect.com/topics/earth-and-planetary-sciences/geopressure

65. US Department of Energy, "Methane Hydrate," https://www.energy.gov/fe/science-innovation/oil-gas-research/methane-hydrate

66. Ibid.; Carolyn D. Ruppel, "Methane Hydrates and Contemporary Climate Change," *Nature Education Knowledge* 3, no. 10 (2011), https://www.nature.com/scitable/knowledge/library/methane-hydrates-and-contemporary-climate-change-24314790/

67. US Geological Survey, "Gas Hydrate Breakdown Unlikely to Cause Massive Greenhouse Gas Release," February 9, 2017, https://www.usgs.gov/news/gas-hydrate-breakdown-unlikely-cause-massive-greenhouse-gas-release

68. US Department of Energy, "Methane Hydrate, Science and Technology: A 2017 Update," July 2017, https://netl.doe.gov/sites/default/files/netl-file/2017-Methane-Hydrate-Primer%5B1%5D.pdf

69. Y. C. Beaudoin, S. R. Dallimore, and R. Boswell, eds., *Frozen Heat: A UNEP Global Outlook on Methane Gas Hydrates*, vol. 2. United Nations Environment Programme, GRID-Arendal, 2014. https://sustainabledevelopment.un.org/content/documents/1993GasHydrates_Vol2_screen.pdf; Evgenios Zogopoulos, "Fire from Ice: A Case Study of Methane Hydrates in the Eastern Mediterranean," *Energy Industry Review*, August 13, 2019, https://energyindustryreview.com/analysis/fire-from-ice-a-case-study-of-methane-hydrates-in-the-eastern-mediterranean/

70. For current development prospects, see, for example, US Geological Survey, "Test Well Confirms Two Gas Hydrate Reservoirs in Alaska North Slope," April 26, 2019, https://www.usgs.gov/news/test-well-confirms-two-gas-hydrate-reservoirs-alaska-north-slope; DNV GL, "Oil and Gas Forecast to 2050," https://www.ourenergypolicy.org/wp-content/uploads/2017/09/DNV-GL_Energy-Transition-Outlook-2017_oil-gas_lowres-single_3108_3.pdf; US Department of Energy, "Data from Innovative Methane Hydrate Test on

Alaska's North Slope," March 11, 2013, https://www.energy.gov/fe/articles/data-innovative-methane-hydrate-test-alaskas-north-slope-now

71. Daniel Kressey, "Giant Dome Fails to Fix Deepwater Horizon Oil Disaster," *Nature News Blog*, May 10, 2010, http://blogs.nature.com/news/2010/05/_giant_dome_fails_to_fix_deepw.html

72. Nancy Bazilchuk, "Giant Gas Craters Discovered at the Bottom of the Barents Sea," *ScienceNordic*, October 5, 2018, http://sciencenordic.com/giant-gas-craters-discovered-bottom-barents-sea

73. Nicola Jones, "The World Eyes Another Unconventional Source of Fossil Fuels," *Yale Environment 360*, August 21, 2017, https://e360.yale.edu/features/the-world-eyes-yet-another-unconventional-source-of-fossil-fuels-methane-hydrates

74. International Energy Agency, *World Energy Outlook 2018*, Table 3.1.

75. International Energy Agency, *World Energy Outlook 2019*, November 2019, https://www.iea.org/data-and-statistics/charts/us-shale-oil-production-in-the-stated-policies-scenario-2005-2030

76. International Energy Agency, *World Energy Outlook 2018*.

77. Cally Carswell, "Unique Oil Spill in East China Sea Frustrates Scientists," *Nature* 554 (January 24, 2018): 17–18, https://www.nature.com/articles/d41586-018-00976-9; Alexis Madrigal, "The Great Crude Oil Fireball Test," *The Atlantic*, February 5, 2018, https://www.theatlantic.com/technology/archive/2018/02/the-great-crude-oil-fireball-test/552029/; and Scandia National Laboratory, "Crude Oil Properties Relevant to Handling and Fire Safety in Transport," June 19, 2015, https://energy.sandia.gov/crude-oil-properties-relevant-to-handling-and-fire-safety-in-transport/

78. Schlumberger, "Oilfield Glossary," https://www.glossary.oilfield.slb.com/en/Terms/c/condensate.aspx

79. US Congressional Research Service, "Natural Gas Liquids: The Unknown Hydrocarbons," October 26, 2018, https://www.everycrsreport.com/reports/R45398.html#_Ref528046797

80. John Manfreda, "The Real History of Fracking," Oilprice.com, April 13, 2015, https://oilprice.com/Energy/Crude-Oil/The-Real-History-Of-Fracking.html

81. Ibid.; American Oil and Gas Historical Project (AOGHS), "Shooters – A Fracking History," https://aoghs.org/technology/hydraulic-fracturing/

82. US Energy Information Administration, "EIA Adds New Play Production Data to Shale Gas and Tight Oil Reports," February 15, 2019, https://www.eia.gov/today-inenergy/detail.php?id=38372#

83. Abbas Ghandi et al., "Energy Intensity and Greenhouse Gas Emissions from Crude Oil Production in the Eagle Ford Region: Input Data and Analysis Methods," Prepared for Argonne National Laboratory, September 2015, https://www.researchgate.net/publication/303592051_Energy_Intensity_and_Greenhouse_Gas_Emissions_from_Crude_Oil_Production_in_the_Eagle_Ford_Region_Input_Data_and_Analysis_Methods

84. Andrew Hecht, "Crude Oil vs. Natural Gas," The Balance, January 6, 2020, https://www.thebalance.com/crude-oil-versus-natural-gas-competing-energy-markets-808876

85. Deborah Gordon. "The Carbon Contained in Global Oils," Carnegie Endowment for International Peace, Washington, DC, December 2012, https://carnegieendowment.org/files/global_oils.pdf

86. Manik Talwani, "The Orinoco Heavy Oil Belt in Venezuela (or Heavy Oil to the Rescue?)," James A. Baker III Institute for Public Policy, Rice University, September 2002, https://www.bakerinstitute.org/media/files/Research/8bb18b4e/the-orinoco-heavy-oil-belt-in-venezuela-or-heavy-oil-to-the-rescue.pdf

87. "From Diluted Bitumen to Synthetic Crude: Upgrading Explained," *Oil Sands Magazine*, https://www.oilsandsmagazine.com/technical/bitumen-upgrading

88. In addition to Venezuela and Canada, China, Indonesia, Russia, Brazil, Alaska, California, Utah, Wyoming, and Colorado contain isolated deposits. See Rigzone, "Where Is Heavy Oil Found?," https://www.rigzone.com/training/heavyoil/insight.asp?i_id=194

89. William Walter Kay, "Valuing Venezuela's Orinoco Belt," GlobalResearch, February 2, 2019, https://www.globalresearch.ca/valuing-venezuelas-orinoco-oil-belt/5667266

90. "Oil Sands History and Milestones," https://www.canadasoilsands.ca/en/what-are-the-oil-sands/oil-sands-history-and-milestones

91. "In Situ Bitumen Extraction," *Oil Sands Magazine*, https://www.oilsandsmagazine.com/technical/in-situ

92. Chevron, BP, Statoil (Equinor), and Total have each been instrumental in working with PdVSA to develop their heavy oil over the decades. Also see Matt Ferchen, "Crude Complications: Venezuela, China, and the United States," Carnegie Endowment for International Peace, October 23, 2014, https://carnegietsinghua.org/2014/10/23/crude-complications-venezuela-china-and-united-states/hsk9

93. "COSIA Is Mobilizing the World's Best Minds to Improve Environmental Performance," https://www.cosia.ca/about; Government of Canada, "University of Calgary: Global Research Initiative in Sustainable Low Carbon Unconventional Resources," http://www.cfref-apogee.gc.ca/results-resultats/abstracts-resumes/competition_2/university_of_Calgary-eng.aspx

94. See, for example, "Scientists Extract H2 Gas from Oil and Bitumen, Giving Potential Pollution-Free Energy," American Association for the Advancement of Science, August 19, 2019, https://www.eurekalert.org/pub_releases/2019-08/gc-seh081819.php; Green Car Congress, August 20, 2019, https://www.greencarcongress.com/2019/08/20190820-proton.html; Rohallah Hashemi et al., "Nanoparticle Technology for Heavy Oil In-Situ Upgrading and Recovery Enhancement: Opportunities and Challenges," *Applied Energy* 133 (November 15, 2014), https://www.sciencedirect.com/science/article/abs/pii/

S030626191400752; and Ian Head and Neil Gray, "Microbial Biotechnology 2020: Microbiology of Fossil Fuel Resources, *Microbial Technology* 9, no. 5 (September 2016), https://www.ncbi.nlm.nih.gov/pmc/articles/PMC4993181/

95. Oil & Gas iQ, "Z-44 Chavo Well: The Deepest Oil Extraction," https://www.oilandgasiq.com/drilling-and-development/articles/z-44-chayvo-well-the-deepest-oil-extraction

96. Zhang et al., "Giant Discoveries of Oil and Gas Fields in Global Deepwaters in the Past 40 Years and the Prospect of Exploration."

97. Rigzone, "How Do FPSOs Work?," https://www.rigzone.com/training/insight.asp?insight_id=299&c_id=

98. "GHGSat Announces Research Project to Demonstrate Satellite-Based Measurement of Methane Emissions from Offshore Sources," SpaceRef, July 7, 2021, http://spaceref.com/news/viewpr.html?pid=57719

99. Gas is produced along with Arctic oil, condensates, and NGLs, but much of this gas is reinjected to increase oil production. For national production figures, see US Energy Information Administration, https://www.eia.gov/dnav/pet/hist/LeafHandler.ashx?n=pet&s=manfpak2&f=a (U.S. Alaska North Slope); "The Future of Russian Oil Production in the Short, Medium, and Long Term," Oxford Institute for Energy Studies, September 2019,https://www.oxfordenergy.org/wpcms/wp-content/uploads/2019/09/The-Future-of-Russian-Oil-Production-in-the-Short-Medium-and-Long-Term-Insight-57.pdf (Russia Arctic); and Atle Staalesen, "Norway Ramps Up Oil and Gas Production in Arctic Despite Looming Oil Crisis," Eye on the Arctic, January 10, 2019, https://www.rcinet.ca/eye-on-the-arctic/2019/01/10/norway-oil-offshore-barents-sea-climate-change/ (Norway Arctic).

100. Zhang et al., "Giant Discoveries of Oil and Gas Fields in Global Deepwaters in the Past 40 Years and the Prospect of Exploration."

101. Wendy Koch, "3 Reasons Why Shell Halted Drilling in the Arctic," *National Geographic*, September 28, 2015, https://www.nationalgeographic.com/news/energy/2015/09/150928-3-reasons-shell-halted-drilling-in-the-arctic/

102. In 2020, BP sold its Alaska assets. See Rachel Adams-Heard, "What Happens When an Oil Giant Walks Away," *Bloomberg Green*, April 15, 2021, https://www.bloomberg.com/graphics/2021-tracking-carbon-emissions-BP-hilcorp/

103. Julian Turner, "The Cold Thaw: Inside Russia's $300bn Arctic Oil and Gas Investment," Offshore Technology, May 5, 2020, https://www.offshore-technology.com/features/the-cold-thaw-inside-russias-300bn-arctic-oil-and-gas-investment/

104. Yereth Rosen, "Alaska Officials Probing BP Oil, Gas Wells at Prudhoe Bay after Spill," *Reuters*, January 14, 2019, https://www.reuters.com/article/us-bp-alaska-wells/alaska-officials-probing-bp-oil-gas-wells-at-prudhoe-bay-after-spill-idUSKCN1P909H

105. US Department of Energy, "US Energy Sector Vulnerabilities to Climate Change and Extreme Weather," July 2013, https://www.energy.gov/sites/prod/files/2013/07/f2/20130710-Energy-Sector-Vulnerabilities-Report.pdf

106. Primary recovery uses the reservoir's natural pressure to lift oil and gas to the surface and recover about 10 percent of the original oil in place (OOIP). Secondary techniques are used to tap an additional 10 to 30 percent of OOIP by injecting water or gas to displace oil and driving it to the surface. See US Department of Energy, Office of Fossil Energy, "Enhanced Recovery," https://www.energy.gov/fe/science-innovation/oil-gas-research/enhanced-oil-recovery

107. Christophe McGlade, G. Sondak, and Mei Han, "Commentary: Whatever Happened to Enhanced Oil Recovery?," November 28, 2018, https://www.iea.org/newsroom/news/2018/november/whatever-happened-to-enhanced-oil-recovery.html

108. GlassPoint, "GlassPoint and Occidental of Oman Sign Agreement to Cooperate on Project to Facilitate Oil Production in Oman," https://www.glasspoint.com/glasspoint-and-occidental-of-oman-sign-agreement-to-cooperate-on-project-to-facilitate-oil-production-in-oman/. For more information on this particular case study on solar steam, see Julia Benz, "Leading Emission Reduction Opportunities and Renewable Technologies for Oil and Gas Production," Senior thesis, Brown University, March 26, 2021.

109. PetroWiki, "In-Situ Combustion," https://petrowiki.org/In-situ_combustion

110. Chegenizadeh Negin, Saeedi Ali, and Quan Xie, "Application of Nanotechnology for Enhancing Oil Recovery – A Review," *Petroleum* 2, no. 4 (December 2016), https://doi.org/10.1016/j.petlm.2016.10.002

111. Christopher Matthews, "Silicon Valley to Big Oil: We Can Manage Your Data Better Than You," *Wall Street Journal*, July 24, 2018, https://www.wsj.com/articles/silicon-valley-courts-a-wary-oil-patch-1532424600

112. Oil shale (kerogen) should not be confused with shale oil—an entirely unrelated hydrocarbon produced by hydraulic fracturing and horizontal drilling, as discussed previously.

113. Retorting involves mining and crushing the rock in a cylindrical vessel and heating it for prolonged periods at 500°C in the absence of oxygen to thermally break down the kerogen into lighter hydrocarbons that can then be manufactured into a limited slate of synthetic gasoline, diesel, and jet fuel. The process also involves removing high sulfur loads, dealing with heavy metals and toxic gases, and consuming significant water. See Anthony Andrews, "Oil Shale: History, Incentives, and Policy," US Congressional Research Service, RL 33359, April 13, 2006, https://fas.org/sgp/crs/misc/RL33359.pdf

114. For a comparison of oil shale lifecycle GHGs to conventional oil, see figure 1: Adam Brandt, "Greenhouse Gas Emissions from Liquid Fuels Produced from Estonian Oil Shale," Prepared for European Commission, Joint Research

Center, November 19, 2010, https://circabc.europa.eu/sd/a/9ab55170-dc88-4dcb-b2d6-e7e7ba59d8c3/Brandt_Estonian_Oil_Shale_Final.pdf

115. Oil-deficient regions with oil shale include Jordan, Israel, Egypt, Morocco, Estonia, Sweden, Poland, Slovakia, Turkey, Thailand, Korea, Western China, and Utah. Oil-bearing regions with oil shale include Australia, Alaska, Siberia, Scotland, Canada, Colorado, Wyoming, Tennessee, and Brazil. For more information see "Geology and Resources of Some World Oil-Shale Deposits," Geology.com, https://geology.com/usgs/oil-shale/

116. Jennifer A. Dlouhy, "Shell Quits Colorado Oil Shale Effort," *Houston Chronicle*, September 25, 2013, https://www.houstonchronicle.com/business/energy/article/Shell-quits-Colorado-oil-shale-effort-4844591.php

117. Richard C. Selley and Stephen A. Sonnenberg, "Nonconventional Petroleum Resources," in *Elements of Petroleum Geology*, 3rd ed., Science Direct, 427-482 (2015), https://doi.org/10.1016/B978-0-12-386031-6.00009-6

118. Daniel Strohl, "Fact Check: Henry Ford Didn't Design the Model T as a Multi-Fuel Vehicle," *Hemmings Daily*, April 23, 2017, https://www.hemmings.com/blog/2017/04/23/fact-check-henry-ford-didnt-design-the-model-t-as-a-multi-fuel-vehicle/

119. Beyond the United States and Brazil, the European Union, China, India, Canada, Thailand, and Argentina accounted for 14 percent of total ethanol production. Statista, "Fuel Ethanol Production Worldwide in 2019, by Country," https://www.statista.com/statistics/281606/ethanol-production-in-selected-countries/

120. In addition to these leading countries, Argentina, Thailand, and Malaysia are also major biodiesel producers. Statista, "Leading Biodiesel Producers Worldwide in 2017, by Country," https://www.statista.com/statistics/271472/biodiesel-production-in-selected-countries/

121. Renewable diesel is more similar to conventional diesel than biodiesel and can be transported in existing pipelines and blended at high ratios into petroleum-based diesel. See Amy Kalt, "Green Grow the (Refineries) – Low-Carbon Programs. Spur More Renewable Diesel," RBN Energy, July 6, 2020, https://rbnenergy.com/green-grow-the-refineries-low-carbon-programs-spur-more-renewable-diesel-refineries; Adam Klauber, Ned Harvey, Kathleen Wight, "The Low-Carbon Jet Fuel Market Is Cleared for Take-Off," Rocky Mountain Institute, October 22, 2020, https://rmi.org/the-low-carbon-jet-fuel-market-is-cleared-for-take-off/

122. Biofuel GHGs vary significantly depending on the organic feedstock and how it is produced and transformed. Improved emissions performance is ongoing. For an older resource see US Department of Energy, https://www.energy.gov/sites/prod/files/edg/media/BiofuelsMythVFact.pdf. For a newer source see Georgina Gustin, "As Trump Touts Ethanol, Scientists Question the Fuel's Climate Claims," Inside Climate News, June 14, 2019, https://insideclimatenews.org/news/14062019/trump-ethanol-climate-change-benefit-science-questions-emissions-data-usda

123. Biomass Research and Development Board, "Increasing Feedstock Production for Biofuels," 2019, https://afdc.energy.gov/files/pdfs/increasing_feedstock_revised.pdf; Brian Barth, "The Next Generation of Biofuels Could Come from These Five Crops," Smithsonian.com, October 3, 2017, https://www.smithsonianmag.com/innovation/next-generation-biofuels-could-come-from-these-five-crops-180965099/

124. Liquefied natural gas (LNG) is not the same as GTLs (discussed earlier) or NGLs (discussed previously). LNG is when natural gas (methane) is temporarily turned into a liquid for transport. At the final destination, regasification equipment then turns LNG back into a gas. See Penn State, "Oil: International Evolution," Chapter 16, https://www.e-education.psu.edu/egee120/node/411

125. For example, Shell's GTL process partially oxidizes natural gas into hydrogen and carbon monoxide (syngas), removes impurities, converts this into a waxy hydrocarbon intermediary using a cobalt catalyst, and cracks this intermediary into diesel, jet fuel, and lubricant oil. See Shell, "Gas-to-Liquids," https://www.shell.com/energy-and-innovation/natural-gas/gas-to-liquids.html; and National Energy Technology Laboratory, "Fischer-Tropsch Synthesis," Figure 1, https://www.netl.doe.gov/research/coal/energy-systems/gasification/gasifipedia/ftsynthesis

126. Shell, "Pearl GTL," https://www.shell.com/about-us/major-projects/pearl-gtl/pearl-gtl-an-overview.html; "Pearl GTL Project Is Operating Efficiently: Shell Qatar Chairman," *The Peninsula*, February 5, 2019, https://www.thepeninsulaqatar.com/article/05/02/2019/Pearl-GTL-project-is-operating-efficiently-Shell-Qatar-Chairman

127. Han Hao et al., "Energy Consumption and GHG Emissions of GTL Fuel by LCA," *Applied Energy* 87, no. 10 (October 2010), https://www.sciencedirect.com/science/article/abs/pii/S0306261910000930

128. Rachel Cernansky, "Is Coal to Liquid More Efficient Than Gasoline?," Planet Green, August 29, 2012, https://science.howstuffworks.com/environmental/energy/coal-liquid-efficient-gasoline1.htm

129. See "Innovation and the Environmental Performance of Oil and Gas Supply," in International Energy Agency, *World Energy Outlook 2018*, Chapter 11, November 13, 2018. Note, a gigatonne (Gt) equals a billion metric tonnes.

130. The International Energy Agency assumes a 100-year global warming potential (GWP) of 30x for methane, which undercounts $CO_2e$ in the near term. Using a 20-year GWP of 86x for methane significantly boosts $CO_2e$, as discussed in forthcoming chapters.

131. Zeke Hausfather, "How Much 'Carbon Budget' Is Left to limit Global Warming to 1.5C?," *Carbon Brief*, April 9, 2018, https://www.carbonbrief.org/analysis-how-much-carbon-budget-is-left-to-limit-global-warming-to-1-5c

132. SLAC National Accelerator Laboratory, "Study Shows a Much Cheaper Catalyst Can Generate Hydrogen in a Commercial Device," October 14, 2019, https://phys.org/news/2019-10-cheaper-catalyst-hydrogen-commercial-device.html

133. US Department of Energy, Office of Energy Efficiency & Renewable Energy, "5 Fast Facts about Hydrogen and Fuel Cells," October 4, 2017, https://www.energy.gov/eere/articles/5-fast-facts-about-hydrogen-and-fuel-cells

134. Noe van Hulst, "The Clean Hydrogen Future Has Already Begun," International Energy Agency, April 23, 2019, https://www.iea.org/commentaries/the-clean-hydrogen-future-has-already-begun

135. "Hydrogen from Renewable Power," IRENA, September 2018, https://www.irena.org/-/media/Files/IRENA/Agency/Publication/2018/Sep/IRENA_Hydrogen_from_renewable_power_2018.pdf

136. GasTerra, "Hydrogen and CCS: A Smart Combination," https://www.gasterra.nl/en/news/hydrogen-and-ccs-a-smart-combination

137. Ewa Krukowska and Alexandra Weber, "Hydrogen and Recycling to Get a Boost in EU Green-Industry Plan," Bloomberg Quint, January 28, 2020, https://www.bloombergquint.com/business/hydrogen-and-recycling-to-get-a-boost-in-eu-green-industry-plan

138. For details on hydrogen production pathways see US Department of Energy, Office of Energy Efficiency & Renewable Energy, "Alternative Fuels Data Center: Hydrogen Production and Distribution," https://afdc.energy.gov/fuels/hydrogen_production.html; and Marshall Miller, Arun Raju, and Partho Sorothi Roy, "The Development of Lifecycle Data for Hydrogen Fuel Production and Delivery," National Center for Sustainable Transportation and Institute of Transportation Studies, UC Davis, October 2017.

139. Bret Williams, "New Green Hydrogen Project Could Be the Next Step towards a Circular Economy," *Hydrogen Fuel News*, March 6, 2019, http://www.hydrogenfuelnews.com/new-green-hydrogen-project-could-be-the-next-step-towards-a-circular-economy/8537052/

140. For US trends see US Energy Information Administration, "Hydrogen for Refineries Is Increasingly Provided by Industrial Suppliers," January 20, 2016, https://www.eia.gov/todayinenergy/detail.php?id=24612; US Energy Information Administration, "U.S. Gulf Coast Refinery Demand for Hydrogen Increasingly Met by Merchant Suppliers," March 15, 2019, https://www.eia.gov/todayinenergy/detail.php?id=38712; and International Energy Agency, "The Future of Hydrogen," June 2019, https://www.iea.org/reports/the-future-of-hydrogen

141. Industrial examples include Shell Sky Scenario, https://www.shell.com/energy-and-innovation/the-energy-future/scenarios/shell-scenario-sky.html; Equinor, "Hydrogen," https://www.equinor.com/en/how-and-why/climate-change/hydrogen.html; "A Future with Hydrogen," Jacobs Engineering, January 14, 2019, https://www.jacobs.com/news/215/a-future-with-hydrogen

142. "After Many False Starts, Hydrogen Power Might Now Bear Fruit," *The Economist*, July 4, 2020, https://www.economist.com/science-and-technology/2020/07/04/after-many-false-starts-hydrogen-power-might-now-bear-fruit; Other national examples include Bianca Nogrady, "From German Trains to South Korean Buses, Hydrogen Fuel Is Back in the Energy Picture," November 6, 2018, https://ensia.com/features/hydrogen-fuel/; Raphael Schoentgen, "The Green Hydrogen Revolution Has Started, and It Won't Be Stopped," World Economic Forum, November 23, 2018, https://www.weforum.org/agenda/2018/11/the-green-hydrogen-revolution-has-started-and-won-t-be-stopped/

143. Oliver Holmes, "Space: How Far Have We Gone—and Where Are We Going?," *The Guardian*, November 19, 2018, https://www.theguardian.com/science/2018/nov/19/space-how-far-have-we-gone-and-where-are-we-going

144. Climate Champions, "Green Hydrogen Catapult," December 8, 2020, https://racetozero.unfccc.int/green-hydrogen-catapult/

145. Ian M. Head and Neil D. Grey, "Microbial Biotechnology 2020: Microbiology of Fossil Fuel Resources," *Microbial Biotechnology*, September 2016 9 (5) 626-634), doi:10.1111/1751-7915.12396; S. R. Larter, M. Strous, and Steven Bryant, "Extracting Energy from Petroleum Reservoirs at Large Scale without $CO_2$ Emissions – Is It Possible? Is the Attempt Desirable?," *Geophysical Research Abstracts* 18 (2016), http://adsabs.harvard.edu/abs/2016EGUGA..1817380L

## CHAPTER 3

1. For opening keynote see Deborah Gordon, "Confronting Complexity: Technological Oil Shifts and Climate Impacts," Carnegie Endowment for International Peace, January 9, 2013, https://carnegieendowment.org/2013/01/09/welcome-and-confronting-complexity-technological-oil-shifts-and-climate-impacts-pub-50584

2. Unconventional Oil Symposium, Carnegie Endowment for International Peace, January 9, 2013, https://carnegieendowment.org/2013/01/09/unconventional-oil-symposium-event-3888

3. According to the US Environmental Protection Agency, a standard barrel of oil emits 432 kilograms of $CO_2$ and a therm of gas (which equals 100,000 Btu) releases 0.53 kilograms of $CO_2$. See US Environmental Protection Agency, "Greenhouse Gas Equivalencies Calculator," https://www.epa.gov/energy/greenhouse-gases-equivalencies-calculator-calculations-and-references

4. This chapter uses adapted text from the report "Know Your Oil: Creating a Global Oil-Climate Index," written by the OCI+ research partners and published by the Carnegie Endowment for International Peace, 2015.

5. "Supply-side" emissions are the direct or indirect responsibility of the industry itself and are considered Scope 1 and Scope 2 GHGs. Scope 1 GHGs are emitted directly by industry in their operations, while Scope 2 GHGs are indirectly emitted from those who produce inputs that the industry procures for use in oil and gas operations (such as electricity and other purchases).

6. Scope 3 "end-use" GHG emissions are composed of all end uses of the entire slate of products the oil and gas industry sells into the marketplace, such as gasoline for cars and jet fuel for airplanes.

7. "Dr. Adam Brandt," Stanford Profiles, https://profiles.stanford.edu/adam-brandt

8. Assumes Midway Sunset API gravity of 12.1 degrees (heating value of 5.99 MMBtu per barrel) and uses EPA crude oil emission factors reported in https://www.epa. gov/sites/production/files/2015-12/documents/emission-factors_nov_2015.pdf

9. In the short term, GHGs are even higher, estimated at 807 kg $CO_2$e/bbl using twenty-year GWPs. See OCI+ Preview Web Tool, https://dxgordon.github.io/ OCIPlus/#oil/u.s.-california-midway-sunset?model=00100000&carbonToggle= off&carbonTax=20.00

10. Hassan El-Houjeiri, Adam Brandt, and James Duffy, "Open-Source LCA Tool for Estimating Greenhouse Gas Emissions from Crude Oil Production Using Field Characteristics," *Environmental Science and Technology*, May 1, 2013, doi:10.1021/ es304570m

11. OPGEE, Stanford University, https://eao.stanford.edu/research-areas/opgee

12. California Air Resources Board, "LCFS Crude Oil Lifecycle Assessment," https:// ww2.arb.ca.gov/resources/documents/lcfs-crude-oil-life-cycle-assessment

13. Daniel Sperling and Deborah Gordon, *Two Billion Cars: Driving Toward Sustainability* (New York: Oxford University Press, 2009).

14. Jonathan Koomey et al., "Getting Smart about Oil in a Warming World," Carnegie Endowment for International Peace, October 5, 2016, https://carnegieendow-ment.org/files/Gordon-Oil_in_a_warming_world1.pdf

15. Argonne National Laboratory, "GREET Model," https://greet.es.anl.gov/

16. Hao Cai et al., "Well-to-Wheel Greenhouse Gas Emissions of Canadian Oil Sands Products: Implications for U.S. Petroleum Fuels," *Environmental Science and Technology* 49, no. 13 (2015): 8219–27, https://doi.org/10.1021/acs.est.5b01255

17. Deborah Gordon, "The Carbon Contained in Global Oil," Carnegie Endowment for International Peace, December 2012, https://carnegieendowment.org/files/ global_oils.pdf

18. Koomey, "Getting Smart about Oil in a Warming World."

19. "Dr. Joule Bergerson," University of Calgary, https://www.ucalgary.ca/jbergers/

20. "Jonathan Koomey," https://www.koomey.com/about.html

21. The California Air Resources Board required OPGEE to be executed in Excel, which is why PRELIM and OPEM were developed in Excel as well.

22. OPGEE: The Oil Production Greenhouse Gas Emissions Estimator, https://eao. stanford.edu/research-areas/opgee

23. PRELIM: The Petroleum Refinery Life Cycle Inventory Model, https://www. ucalgary.ca/lcaost/prelim

24. OPEM Version 2.0, https://dxgordon.github.io/OCIPlus/#methodology

25. OPGEE funding sources include the California Air Resources Board, Stanford University School of Earth Sciences, US Department of Energy, Ford, Saudi Aramco, and various NGOs and philanthropic organizations. Brandt and his students have published over twenty-five peer-reviewed papers documenting the use of OPGEE.

26. Adam R. Brandt, Yuchi Sun, and Kourosh Vafi, "Uncertainty in Regional-Average Petroleum GHG Intensities: Countering Information Gaps with Targeted Data Gathering," *Environmental Science and Technology* 49, no. 1 (January 6, 2015): 679–86, doi:10.1021/es505376t.

27. Mohammad Masnadi et al., "Global Carbon Intensity of Crude Oil Production," *Science* 361, no. 6405 (August 31, 2018), https://science.sciencemag.org/content/361/6405/851.summary

28. Adam Brandt, et. al., OPGEE 3.0a Candidate Model," presentation to the California Air Resources Board, October 14, 2020, https://ww2.arb.ca.gov/sites/default/files/2020-10/101420presentation_stanford_opgee.pdf; See Github, "arbrandt/OPGEE," https://github.com/arbrandt/OPGEE

29. While oil companies consider refining as downstream operations, the OCI+ refers to refining and product transport as midstream industrial operations.

30. Liang Jing et al., "Global Carbon Intensity of Crude Oil Refining," *Nature Climate Change*, 2020, https://www.nature.com/articles/s41558-020-0775-3

31. Transport GHG calculations are built into the various OCI+ models, but emissions can be reallocated and regrouped together for purposes of real-world comparisons.

32. Although default crude oil transport between OCI+ fields and Houston is fixed in the BETA Preview version, this is being updated with regional patterns, and users can update this further if oil is shipped elsewhere rather than to Houston, which is the largest refining hub in the world.

33. Default gas transport hubs include Louisiana (North America); Rio (Latin America); Moscow (Russia); Hamburg (European Union); Riyadh (MENA); Lagos (Sub-Saharan Africa); Guangzhou (China); Shanghai (Central Asia); and Perth (Australia-Pacific). Pipeline and sea route distances were estimated for each field modeled in the OCI+.

34. For transport GHGs, hover over the downstream emissions bar for an oil in the OCI Web Tool. For example, California South Belridge is estimated to have transport emissions of 12 kg $CO_2$e per bbl, which is about 2 percent of its total GHGs of 690 kg $CO_2$e per bbl. See https://oci.carnegieendowment.org/#oil/u.s.-california-south-belridge. Note that no transport scenario run on oil resulted in transport GHGs over 5 percent of total emissions. This does not hold for LNG transport GHG emissions, however, which can account for a larger share (15 percent or more) of gas lifecycle emissions. See https://dxgordon.github.io/OCIPlus/#oil/australia-gorgon?model=00002000&carbonToggle=off&carbonTax=20.00

35. See the web tool https://sea-distances.org/ to compare ports of departure and arrival; Jeremy Bender, "These 8 Narrow Chokepoints Are Critical to the World's Oil Trade," *Business Insider*, April 1, 2015, https://www.businessinsider.com/worlds-eight-oil-chokepoints-2015-4

36. For more details on climate forcing measurements, see National Oceanic and Atmospheric Administration, "Climate Forcing," https://www.climate.gov/maps-data/primer/climate-forcing

37. See, for example, previous IPCC Assessment Reports (referred to as ARs) with their various GWP assignments: https://www.ipcc.ch/site/assets/uploads/2018/03/ipcc_far_wg_I_full_report.pdf; https://www.ipcc.ch/site/assets/uploads/2018/02/ar4-wg1-chapter2-1.pdf; https://www.ipcc.ch/site/assets/uploads/2018/02/WG1AR5_Chapter08_FINAL.pdf

38. Michelle Cain et al., "Improved Calculation of Warming-Equivalent Emissions for Short-Lived Climate Pollutants," *Nature Climate and Atmospheric Science*, 2019, https://www.researchgate.net/publication/335608530_Improved_calculation_of_warming-equivalent_emissions_for_short-lived_climate_pollutants

39. Highly integrated efforts to trace oil and gas pollution sources are on the horizon. See Al Gore, "We Can Solve the Climate Crisis by Tracing Pollution Back to Its Source," *Medium*, July 15, 2020, https://medium.com/@algore/we-can-solve-the-climate-crisis-by-tracing-pollution-back-to-its-sources-4f535f91a8dd

40. National Aeronautics and Space Administration, "NASA Carbon Monitoring System," https://carbon.nasa.gov/

41. National Aeronautics and Space Administration, Carbon Monitoring System, https://carbon.nasa.gov/objectives.html?; https://carbon.nasa.gov/app_initiative.html

42. Deborah Gordon and Frances Reuland, "Mapping Measuring, and Managing Methane: The Critical Role of a Potent Climate Pollutant," Watson Institute for International and Public Affairs, November 2019, https://watson.brown.edu/files/watson/imce/news/ResearchBriefs/2019/Methane%20Report-6%20November%202019.pdf

43. For example, see Spherical Analytics that is creating data platforms for Environmental, Social, and Governance (ESG) risk mitigation that could be accessible to NGOs and the public.

44. David Livingston, "A Blockchain-Based Energy Future," Aspen Institute and Carnegie Endowment for International Peace, August 31, 2017, https://carnegieendowment.org/2017/08/31/blockchain-based-energy-future-pub-72973

45. For example, see the Methane Intelligence Quotient (MiQ) concept put forth by the Rocky Mountain Institute and SYSTEMIQ, "Global Methane Solutions," https://www.globalmethanesolutions.org/; also see Taku Ide and T. J. Kirk, "Climate Intelligence: A Clear Path to Insights and Action for the Oil and Gas

Sector," Rocky Mountain Institute, September 2020, which can be found at https://rmi.org/cae

46. For details on upstream OPGEE data inputs see OCI+ Preview Web Tool Methodology, https://dxgordon.github.io/OCIPlus/#methodology

47. For details on midstream PRELIM data inputs see OCI+ Preview Web Tool Methodology, https://dxgordon.github.io/OCIPlus/#methodology

48. For more information on transport (shipping) emission factors see https://greet.es.anl.gov/. For combustion emission factors see US, Environmental Protection Agency, "Emission Factors for GHG Emission Inventories," last modified November 19, 2015,https://www.epa.gov/sites/production/files/2015-11/documents/emission-factors_nov_2015.pdf

49. Currently, the OCI+ assumes that lighter petroleum liquid products (petrochemical feedstocks, gasoline, diesel, jet fuel, NGLs, LPG) are transported via pipeline 1,500 miles (2,414 kilometers) and then distributed by heavy-duty tanker 236 miles (280 km) to metropolitan areas. Heavy liquid products (fuel oil, residual fuels) are assumed to travel by rail 1,525 miles (2,455 km) and then by tanker 280 miles (243 nm, 450 km), and by tanker truck 306 miles (493 km). Solid fuels (petroleum coke and sulfur) are assumed to travel 2,895 miles (2,515 nm, 4,660 km) by tanker and then by rail 860 miles (1,380 km). Gas is assumed to move via pipeline 1,275 miles (2,050 km) or transported as LNG in a tanker 6,000 miles (5,215 nm, 9,660 km).

50. For November 2019 CMS workshop agenda, see https://carbon.nasa.gov/meeting_2019/Presentations/CMS%20Applications%20Workshop%202019_Final%20Agenda.pdf

51. Deborah Gordon and Frances Reuland, "Mapping, Measuring, and Monitoring Methane: The Critical Role of a Potent Climate Pollutant," Watson Institute for International and Public Affairs, November 2019, https://watson.brown.edu/files/watson/imce/news/ResearchBriefs/2019/Methane%20Report-6%20November%202019.pdf

52. Eric Kort et al., "Four Corners: The Largest US Methane Anomaly Viewed from Space," *Geophysical Research Letters*, September 16, 2014, https://agupubs.onlinelibrary.wiley.com/doi/full/10.1002/2014GL061503

53. Daniel Jacob et al., "Satellite Observations of Atmospheric Methane and Their Value for Quantifying Methane Emissions," *Atmospheric Chemistry and Physics* 16 (November 18, 2016), https://www.atmos-chem-phys.net/16/14371/2016/

54. Stefan Schwietzke et al., "Upward Revision of Global Fossil Fuel Methane Emissions Based on Isotope Database," *Nature* 538 (October 6, 2016), https://www.nature.com/articles/nature19797

55. B. Franco et al., "Evaluating Ethane and Methane Emissions Associated with the Development of Oil and Natural Gas Extraction in North America,"

*Environmental Research Letters* 11, no. 4 (April 7, 2016), https://iopscience.iop.org/article/10.1088/1748-9326/11/4/044010

56. Natalie Kille, "Separation of Methane Emissions from Agricultural and Natural Gas Sources in the Colorado Front Range," *AGU Research Letter*, March 21, 2019, https://agupubs.onlinelibrary.wiley.com/doi/abs/10.1029/2019GL082132

57. European Space Agency, "Mapping Methane Emissions on a Global Scale," *SciTechDaily*, May 4, 2020, https://scitechdaily.com/mapping-methane-emissions-on-a-global-scale/

58. Personal communication with Dr. Daniel Jacob, Harvard University, May 7, 2020. See: Atmospheric Chemistry Modeling Group, Harvard University, http://acmg.seas.harvard.edu/presentations.html

59. Christopher Elvidge et al., "Methods for Global Survey of Natural Gas Flaring from Visible Infrared Imaging Radiometer Suite Data," *Energies* 9, no. 1 (December 25, 2015): 14, https://www.researchgate.net/publication/288496788_Methods_for_Global_Survey_of_Natural_Gas_Flaring_from_Visible_Infrared_Imaging_Radiometer_Suite_Data

60. Daniel Cusworth et al., "Multi-satellite Imaging of a Gas Well Blowout Provides New Insights for Methane Monitoring," *Geophysical Research Letters* 48, Issue 2 (January 28, 2021), https://agupubs.onlinelibrary.wiley.com/doi/full/10.1029/2020GL090864

61. Christopher Elvidge et al., "VIIRS Night-Time Lights," *International Journal of Remote Sensing* 38, no. 21 (June 26, 2017), https://www.tandfonline.com/doi/full/10.1080/01431161.2017.1342050

62. Christopher Elvidge et al., "Methods for Global Survey of Natural Gas Flaring from Visible Infrared Imaging Radiometer Suite Data," *Energies* 9 (2016): 14, doi:10.3390/en9010014. For more information on VIIRS see National Aeronautics and Space Administration, "Visible Infrared Imaging Radiometer Suite (VIIRS)," EarthData, https://earthdata.nasa.gov/earth-observation-data/near-real-time/download-nrt-data/viirs-nrt; National Oceanic and Atmospheric Administration, " Visible Infrared Imaging Radiometer Suite (VIIRS)," STAR Calibration Center, https://ncc.nesdis.noaa.gov/VIIRS/

63. Zubin Bamji, "Global Gas Flaring Inches Higher for the First Time in Five Years," *World Bank Blogs*, June 14, 2019, https://blogs.worldbank.org/opendata/global-gas-flaring-inches-higher-first-time-five-years; "Global Gas Flaring Tracker Report," Global Gas Flaring Reduction Partnership and World Bank, July 2020, https://pubdocs.worldbank.org/en/503141595343850009/WB-GGFR-Report-July2020.pdf

64. Ibid.

65. For an image of the OCI+ analysis plus VIIRS flaring map see https://dxgordon.github.io/OCIPlus/#map?yearSelect=2017&methane=off

66. National Aeronautics and Space Administration, Carbon Monitoring System, https://carbon.nasa.gov and https://carbon.nasa.gov/cgi-bin/search_projects.pl

67. "Estimation of Flare Gas Volumes from Satellite Data," http://pubdocs.worldbank.org/en/251461483541510567/ACS.pdf; Mikhail Zhizhin, "Five-Year Survey of the U.S. Natural Gas Flaring Observed from Space with VIIRS," National Oceanic and Atmospheric Administration and Cooperative Institute for Research in Environmental Sciences, https://www.esrl.noaa.gov/gmd/publications/annual_meetings/2017/slides/5-Zhizhin.pdf;

68. 30 CFR 250.1160. See Legal Information Institute, "When May I Vent or Flare Gas?," https://www.law.cornell.edu/cfr/text/30/250.1160

69. National Oceanic and Atmospheric Administration, "What Is Lidar?," https://oceanservice.noaa.gov/facts/lidar.html. For news on mobile methane sensing system designs for monitoring offshore oil and gas platforms, see Advanced Research Projects Agency–Energy, https://arpa-e.energy.gov/technologies/projects/mobile-methane-sensing-system

70. For an image of the OCI+ analysis plus the methane heat map see https://dxgordon.github.io/OCIPlus/#map?yearSelect=off&methane=on; for details on creating the methane map see Tia Scarpelli et al., "A Global, Gridded Inventory of Methane Emissions from Fuel Exploitation Based on National Reports to the UNFCCC," Harvard University, National Energy Technology Laboratory, and Jet Propulsion Laboratory, http://acmg.seas.harvard.edu/presentations/2018/TScarpelli_agu_final.pdf

71. Sudhanshu Pandey et al., "Methane Leakage from a Gas Well Blowout in Ohio Detected from Space," *Geophysical Research Abstract* 21 (2019), https://meetingorganizer.copernicus.org/EGU2019/EGU2019-18072-1.pdf

72. "Exxon's XTO Caps Leaking Ohio Gas Well, 20 Days after Blowout," Reuters, March 7, 2018, https://www.reuters.com/article/us-exxon-xto-natgas-ohio/exxons-xto-caps-leaking-ohio-gas-well-20-days-after-blowout-idUSKCN1GJ355

73. Carbon Mapper, a non-profit entity with a mission to guide the adoption of digital public goods that facilitate timely action to mitigate human impacts to Earth's climate and ecosystems, https://carbonmapper.org/

74. Daniel Jacob et al., "Satellite Observations of Atmospheric Methane and Their Value for Quantifying Methane Emissions," *Atmospheric Chemistry and Physics* 14, no. 22 (November 18, 2016), https://www.atmos-chem-phys.net/16/14371/2016/; Carbon Mapper, "Technology," https://carbonmapper.org/our-mission/technology/

75. Daniel Zavala-Araiza et al., "Reconciling Divergent Estimates of Oil and Gas Methane Emissions," *PNAS* 112, no. 51 (December 7, 2015), https://www.pnas.org/content/112/51/15597

76. Riley Duren et al., "California's Methane Super-Emitters," *Nature* 575 (November 6, 2019), https://www.nature.com/articles/s41586-019-1720-3?draft=collection

77. United Nations Environment Program, "Global Methane Assessment: Benefits and Costs of Mitigating Methane Emissions," May 6, 2021, https://www.unep.org/resources/report/global-methane-assessment-benefits-and-costs-mitigating-methane-emissions; United Nations Environment Program, "Air Pollution and Climate Change: Two Sides of the Same Coin," April 23, 2019, https://www.unenvironment.org/news-and-stories/story/air-pollution-and-climate-change-two-sides-same-coin

78. Debra Silverman, "Diesel Exhaust and Lung Cancer—Aftermath of Becoming an IARC Group 1 Carcinogen," *American Journal of Epidemiology* 187, no. 6 (June 2018), https://academic.oup.com/aje/article/187/6/1149/4924564

79. US Environmental Protection Agency, "Black Carbon Research," https://www.epa.gov/air-research/black-carbon-research

80. This report also has a discussion about the interactions between climate change and air pollution. See Deborah Gordon, "The Role of Transportation in Driving Climate Change," Carnegie Endowment for International Peace, No. 117, December 2010, https://carnegieendowment.org/files/transport_climate_disruption.pdf

81. Real-time Air Quality Index Map, https://aqicn.org/map/china/

82. Both China and India, along with other Asian countries, vie for the world's worst air quality. Current rankings can be found at IQAir, https://www.iqair.com/us/world-air-quality-ranking. According to the World Health Organization, in 2018, these were "The Most Polluted Cities in the World, Ranked," CBS News, https://www.cbsnews.com/pictures/the-most-polluted-cities-in-the-world-ranked/2/

83. Deborah Gordon et al., "Know Your Oil: Creating the Oil-Climate Index," Carnegie Endowment for International Peace, March 2015, https://carnegieendowment.org/files/know_your_oil.pdf

84. This range increases dramatically when anthropogenic $CO_2$ is injected and sequestered to produce oil in UAE, in which case upstream emissions are estimated to drop to −200 kg per BOE compared with Duri at +246 kg per BOE. See OCI+ Preview Beta Web Tool (2020), https://dxgordon.github.io/OCIPlus/#analysis?xSelect=upstream&ySelect=ghgTotal&oiltypeSelect=Heavy%20Oil,CO2%20EOR%20Oil.

85. This range increases to seventeen times when comparing Argentina Vaca Muerta to UK Brent using a twenty-year GWP. See Oil Climate Index + Gas Preview Beta Web Tool (2020), https://dxgordon.github.io/OCIPlus/#analysis?xSelect=ghgPerMJ&ySelect=upstream&oiltypeSelect=Shale%20Gas,Depleted%20Gas

86. This midstream refining GHG range compares California South Belridge to Nigeria Obagi. See Oil Climate Index + Gas Preview Web Tool under development (2020). For a global assessment see Jing et al., "Global Carbon Intensity of Crude Oil Refining," Figure 4.

87. Downstream transport GHG estimates are less certain owing to the lack of actual data regarding actual oil, gas, and petroleum product movements. Download the OCI+ Preview workbook at https://dxgordon.github.io/OCIPlus/#methodology

88. An operator's GHG emissions arise directly and indirectly from different sources (or scopes). Scope 1 emissions are emitted directly by the operator itself. Scope 2 emissions arise from energy and services that the operator purchases from another source. Scope 3 emissions arise from the use of the products an operator sells. Together, these Scope 1, 2, and 3 emissions compose the GHG inventory. See https://ghgprotocol.org/sites/default/files/Guidance_Handbook_2019_FINAL.pdf

89. OCI+ Beta Web Tool (2020), https://dxgordon.github.io/OCIPlus/#analysis?xSelect=ghgPerMJ&ySelect=downstream&groupSelect=Oil; https://dxgordon.github.io/OCIPlus/#analysis?xSelect=ghgPerMJ&ySelect=downstream&groupSelect=Gas

CHAPTER 4

1. For aerial views of petroleum coke, see Getty Images, "Aerial Views of the Oil Sands as Heavy Crude Trades Below U.S. Benchmark," https://www.gettyimages.com/detail/news-photo/the-syncrude-canada-ltd-petroleum-coke-fields-stand-in-this-news-photo/451281082

2. Please see the following information about this 2012 event at the Carnegie Endowment. Carnegie Endowment for International Peace, "Unconventional Oil: Illuminating the Global Paradigm Shift to New Petroleum Fuels," February 8, 2012, https://carnegieendowment.org/2012/02/08/unconventional-oil-illuminating-global-paradigm-shift-to-new-petroleum-fuels-event-3538

3. See Richard Lattanzio, "Canadian Oil Sands: Life-Cycle Assessments of Greenhouse Gas Emissions," Congressional Research Service," March 10, 2014, footnote 10, https://fas.org/sgp/crs/misc/R42537.pdf. This report cites the *Final Environmental Impact Statement* (Final EIS) for the Keystone XL Pipeline Project and several studies published after the release of the 2011 Final EIS, including Jacobs Consultancy, *EU Pathway Study: Life Cycle Assessment of Crude Oils in a European Context*, 2012; IHS Cambridge Energy Research Associates (CERA), *Oil Sands, Greenhouse Gases, and U.S. Oil Supply Getting the Numbers Right*—2012 Update; Adam Brandt, *Upstream GHG Emissions from Canadian Oil Sands as a Feedstock for European Refineries*, 2011; and Joule Bergerson et al., *Life Cycle Greenhouse Gas Emissions of Current Oil Sands Technologies: Surface Mining and In Situ Applications*, 2012. The Final Environmental Impact Statement retained a focus on the data and results from Alberta Energy Research Institute (AERI)/Jacobs 2009, AERI/TIAX 2009, National Energy Technology Laboratory (NETL) 2008, and NETL 2009.

4. Deborah Gordon, "The Carbon Contained in Global Oils," Carnegie Endowment for International Peace, December 2012, https://carnegieendowment.org/files/global_oils.pdf

5. Alberta Government, "Mines and Minerals Act," M-17, RSA, 2000, updated July 18, 2019, https://open.alberta.ca/publications/m17#summary

6. US Energy Information Administration, "U.S. Exports of Petroleum Coke," https://www.eia.gov/dnav/pet/hist/LeafHandler.ashx?n=PET&s=MCKEXUS1&f=M

7. Ibid.

8. C-SPAN, "President Carter's Fireside Chat on Energy," February 2, 1977, https://www.c-span.org/video/?153913-1/president-carters-fireside-chat-energy

9. Deborah Gordon, *Steering a New Course: Transportation, Energy, and the Environment* (Washington, DC: Island Press, 1991); Daniel Sperling and Deborah Gordon, *Two Billion Cars: Driving Toward Sustainability* (New York: Oxford University Press, 2009).

10. Lawrence Berkeley National Laboratory, "In Memoriam, Arthur H. Rosenfeld, Eminent Physicist, Inspirational Researcher, Energy Efficiency Maven," January 27, 2017, https://eta.lbl.gov/about-us/arthur-h-rosenfeld; and Julie Chao, "Art Rosenfeld, California's Father of Energy Efficiency, Dies at 90," University of California, January 27, 2017,https://www.universityofcalifornia.edu/news/art-rosenfeld-california-s-godfather-energy-efficiency-dies-90

11. Leo Levinson and Deborah Gordon, "DRIVE+: Promoting Cleaner and More Fuel Efficient Motor Vehicles through a Self-Financing System of State Sales Tax Incentives," *Journal of Policy Analysis and Management* 9, no. 3 (Summer 1990): 409–15, https://www.jstor.org/stable/3325286?origin=crossref&seq=1

12. Rocky Mountain Institute and NITI Aayog, "Valuing Society First: An Assessment of the Potential for a Feebate Policy in India," November 2017, https://niti.gov.in/writereaddata/files/document_publication/Valuing_Society_First_Feebates_Policy.pdf

13. United Nations, "United Nations (UN) Sustainable Development Goals," https://www.un.org/sustainabledevelopment/sustainable-development-goals.

14. See Figure SPM.3b in the following climate change report: Intergovernmental Panel on Climate Change, "Summary for Policymakers: Special Report: Global Warming of 1.5 °C," October 6, 2018, https://www.ipcc.ch/sr15/graphics/

15. US Energy Information Administration, "Short-term Energy Outlook," July 7, 2021, https://www.eia.gov/outlooks/steo/report/global_oil.php

16. International Energy Agency, "The Oil and Gas Industry in Energy Transitions," 2020, https://www.iea.org/reports/the-oil-and-gas-industry-in-energy-transitions.

17. International Energy Agency, "Changes in the Average Global Emissions Intensity of Oil and Natural Gas Operations in the Sustainable Development Scenario, 2018–2030," March 9, 2020, https://www.iea.org/data-and-statistics/

charts/changes-in-the-average-global-emissions-intensity-of-oil-and-natural-gas-operations-in-the-sustainable-development-scenario-2018-2030

18. Compares Kuwait Ratawi (high methane settings) to Nigeria Obagi (low methane settings), OCI+ Preview Web Tool, https://dxgordon.github.io/OCIPlus/#oil/kuwait-ratawi?model=20100002&carbonToggle=off&carbonTax=20.00 and https://dxgordon.github.io/OCIPlus/#oil/nigeria-obagi?model=10100001&carbonToggle=off&carbonTax=20.00

19. Kristin Oye Gjerde, "Draugen Gas – Flaring or Reinjection," Norwegian Petroleum Museum, https://draugen.industriminne.no/en/2018/04/27/draugen-gas-flaring-or-reinjection/

20. World Bank Flaring Initiative, http://pubdocs.worldbank.org/en/603281560185748682/pdf/Gas-flaring-volumes-Top-30-countries-2014-2018.pdf

21. Raphael Calel and Paasha Mahdavi, "Opinion: The Unintended Consequences of Antiflaring Policies—and Measures for Mitigation," *Proceedings of the National Academy of Sciences* 117, no. 23 (June 9, 2020), https://www.pnas.org/content/117/23/12503

22. Personal communication with Stephen Ziman, a former Chevron employee and current industry retiree, on July 26, 2020.

23. Chad Wocken et al., "End-Use Technology Study: An Assessment of Alternative Uses for Associated Gas," University of North Dakota Energy and Environmental Research Center, September 13, 2012, https://undeerc.org/bakken/pdfs/CW_Tech_Study_July12.pdf

24. Amy Dalrymple, "Study Finds Storing Natural Gas Underground Feasible," *Bismarck Tribune*, June 12, 2019, https://bismarcktribune.com/bakken/study-finds-storing-natural-gas-underground-feasible/article_929ff797-7e88-5d55-9d68-9b3cd42c8992.html

25. US Department of Energy, "Combined Heat and Power Technology Fact Sheet Series," DOE/EE-1329, July 2016, https://www.energy.gov/sites/prod/files/2016/09/f33/CHP-Microturbines_0.pdf

26. Compares Norway Ekofisk to Texas Spraberry, OCI+ Preview Web Tool, https://dxgordon.github.io/OCIPlus/#compare/norway-ekofisk/u.s.-texas-spraberry?model=00100000

27. Christophe McGlade et al., "What Ever Happened to Enhanced Oil Recovery?," US International Energy Agency, November 28, 2018, https://www.iea.org/newsroom/news/2018/november/whatever-happened-to-enhanced-oil-recovery.html

28. The world's largest EOR project using $CO_2$ is Brazil's Lula oil field operated by Petrobras. Others are under way in Canada, China, Saudi Arabia, the United Arab Emirates, Turkey, and Croatia. See McGlade et al., "What Ever Happened to Enhanced Oil Recovery?"

29. US Department of Energy National Energy Technology Laboratory, "Carbon Dioxide Enhance Oil Recovery," March 2010, https://www.netl.doe.gov/sites/default/files/netl-file/CO2_EOR_Primer.pdf

30. Compares United Arab Emirates to Wyoming LaBarge $CO_2$ EOR, using OCI+ Preview Web Tool, https://dxgordon.github.io/OCIPlus/#compare/uae-man-made-co2-eor-example/u.s.-wyoming-natural-co2-eor-example

31. The world's largest EOR project using $CO_2$ is Brazil's Lula oil field operated by Petrobras. Others are underway in Canada, China, Saudi Arabia, the United Arab Emirates, Turkey, and Croatia.

32. Christopher Smith, "ADNOC Expanding Carbon Dioxide, EOR Capabilities," *Oil & Gas Journal*, November 14, 2019, https://www.ogj.com/drilling-production/production-operations/article/14072017/adnoc-expanding-carbon-dioxide-eor-capacities

33. Michael Parker et al., "$CO_2$ Management at ExxonMobil's LaBarge Field, Wyoming, USA," *Energy Procedia* 4 (2011): 5455–70, https://www.sciencedirect.com/science/article/pii/S1876610211008101

34. "ExxonMobil Shute Creek Treating Facility," Subpart RR Monitoring, Reporting, and Verification Plan, submitted to US Environmental Protection Agency, February 2018, https://www.epa.gov/sites/production/files/2018-06/documents/shutecreekmrvplan.pdf

35. Clea Kolster et al., "$CO_2$ Enhanced Oil Recovery: A Catalyst for Gigatonne-Scale Carbon Capture and Storage Deployment?," *Energy and Environmental Science* 12 (2017), https://pubs.rsc.org/en/content/articlelanding/2017/ee/c7ee02102j#!divAbstract; and US International Energy Agency, "Carbon Capture, Utilization, and Storage: A Critical Tool in the Climate Energy Toolbox," https://www.iea.org/topics/carbon-capture-and-storage

36. Assumes estimated Permian water-to-oil ratio is 17:1 and oil production volume is 555,000 bpd. Olympic pools contain approximately 450,000 gallons water per ten-lane Olympic pool with regulation dimensions; see FINA Facility Rules: 2015–2017, https://fina.org/sites/default/files/finafacilities_rules.pdf. There are forty-two gallons per barrel; OCI+ Preview Web Tool, https://dxgordon.github.io/OCI+Plus/

37. For example, Pennsylvania's Marcellus (Gas) Formation also generates high volumes of produced water. In the United States, on average, ten barrels of water are produced for every barrel of crude and ninety-seven barrels of water for every MMcf of gas. See E. Allison and B. Mandler, "Water in the Oil and Gas Industry," American Geosciences Institute, 2018, https://www.americangeosciences.org/geoscience-currents/water-oil-and-gas-industry (footnote 3, Tables 4-5 and 4-6); Argonne National Laboratory, "Produced Water Volumes and Management Practices in the United States," 2009, https://www.circleofblue.org/wp-content/uploads/2010/09/ANL_EVS__R09_produced_water_volume_report_2437.pdf

38. John Kemp, "Water Is the Biggest Output of U.S. Oil and Gas Wells," *Reuters*, November 18, 2014, https://www.reuters.com/article/us-usa-shale-water-kemp-idUSKCN0J223P20141118

39. Richard Nemec, "Draft California Rules for Oil/Gas Steam, Water and Storage Released," Natural Gas Intelligence, January 22, 2016, https://www.naturalgasin-tel.com/articles/105088-draft-california-rules-for-oilgas-steam-water-and-storage-released

40. Compares California Wilmington at 125 percent water intensity to 10 percent water intensity, OCI 2.0, Archived Carnegie Endowment for International Peace, https://oci.carnegieendowment.org/#oil/u.s.-california-wilmington?opgee=run0 30&prelim=run01&showCoke=1 and https://oci.carnegieendowment.org/#oil/ u.s.-california-wilmington?opgee=run010&prelim=run01&showCoke=1

41. Deborah Gordon and Katherine Garner, "Texas's Oil and Water Tightrope," Carnegie Endowment for International Peace, March 11, 2014, https://carnegieen-dowment.org/2014/03/11/texas-s-oil-and-water-tightrope-pub-54879; Somini Sengupta and Weiyi Cai, "A Quarter of Humanity Faces Looming Water Crises," *New York Times*, August 6, 2019, https://www.nytimes.com/interactive/2019/08/ 06/climate/world-water-stress.html

42. Cooperative Institute for Research in Environmental Sciences at the University of Colorado Boulder, "Wastewater Injection and Induced Seismicity," September 22, 2016, https://cires.colorado.edu/news/wastewater-injection-and-induced-seismicity

43. US Geological Survey, "Does Fracking Cause Earthquakes?," https://www.usgs. gov/faqs/does-fracking-cause-earthquakes?qt-news_science_products=0#qt-news_science_products; University of Texas at Austin, "Where Water Goes after Fracking Is Tied to Earthquake Risk," *Science Daily*, November 1, 2018, https:// www.sciencedaily.com/releases/2018/11/181101133823.htm

44. US Department of Energy, "Produced Water R&D," https://www.energy.gov/fe/ science-innovation/oil-gas/shale-gas-rd/produced-water-rd

45. Amit Kumar et al., "Lithium Recovery from Oil and Gas Produced Water: A Need for a Growing Energy Industry," *ACS Energy Letters* 4, no. 6 (June 2019), https://www.researchgate.net/publication/333643378_Lithium_ Recovery_from_Oil_and_Gas_Produced_Water_A_Need_for_a_Growing_ Energy_Industry; see also "MGX Minerals and Eureka Resources Announce Joint Venture to Recover Lithium from Produced Water in Eastern United States," *PRNewsWire*, March 5, 2019, https://www.prnewswire.com/news-releases/mgx-minerals-and-eureka-resources-announce-joint-venture-to-recover-lithium-from-produced-water-in-eastern-united-states-300806407. html

46. Deborah Gordon and Madhav Acharya, "Oil Shake-Up: Refining Transitions in a Low- Carbon Economy," Carnegie Endowment for International Peace, April

3, 2018, https://carnegieendowment.org/2018/04/03/oil-shake-up-refining-transitions-in-low-carbon-economy-pub-75954

47. Kavan Motazedi, Jessica Abella, and Joule Bergerson, "Techno-Economic Evaluation of Technologies to Mitigate Greenhouse Gas Emissions at North American Refineries," *Environmental Science and Technology* 51, no. 3 (2017), https://pubs.acs.org/doi/abs/10.1021/acs.est.6b04606

48. Statista, "Global Refinery Capacity for Crude Oil from 1970 to 2019," https://www.statista.com/statistics/264333/global-refinery-capacity-for-crude-oil/

49. Compares Indonesia Duri baseline to 100 percent renewable hydrogen, OCI+ Preview Web Tool, https://dxgordon.github.io/OCIPlus/#oil/indonesia-duri?model=00100000&carbonToggle=off&carbonTax=20.00 and https://dxgordon.github.io/OCIPlus/#oil/indonesia-duri?model=00120000&carbonToggle=off&carbonTax=20.00

50. US Department of Energy, "Hydrogen Production: Natural Gas Reforming," https://www.energy.gov/eere/fuelcells/hydrogen-production-natural-gas-reforming

51. Michael Liebreich, "Beyond Three Thirds, the Road to Deep Decarbonization," *Bloomberg New Energy Finance*, March 13, 2018, https://about.bnef.com/blog/liebreich-beyond-three-thirds-road-deep-decarbonization/

52. Rotten Tomatoes, "*The Graduate* Quotes," https://www.rottentomatoes.com/m/graduate/quotes/

53. Compares Qatar Idd El Shargi baseline to low methane and zero natural gas combustion, OCI+ Preview Web Tool, https://dxgordon.github.io/OCIPlus/#oil/qatar-idd-el-shargi?model=01100001&carbonToggle=off&carbonTax=20.00

54. Water, air quality, and public health concerns are associated with petrochemical production. Petrochemical facilities release an array of organic compounds, including known carcinogens such as benzene, toluene, ethylbenzene, and xylene (BTEX) along with persistent organic pollutants (POPs) that do not break down over time. And petrochemical exposure can be dangerous and pervasive, occurring through inhalation, skin absorption, and ingestion. See Michael Corkery, "Deluged by Plastics but Bustling to Make More," *New York Times*, August 12, 2019, https://www.nytimes.com/2019/08/12/business/energy-environment/plastics-shell-pennsylvania-plant.html

55. Wang Tao, "Managing China's Petcoke Problem," Carnegie Endowment for International Peace, May 2015, https://carnegieendowment.org/files/petcoke.pdf

56. See OCI+ Web Tool for Canadian oil sands with petcoke production turned on and off: https://dxgordon.github.io/OCIPlus/#oil/canada-athabasca-dc-sco?model=00000100&carbonToggle=off&carbonTax=20.00; and comparing oil sands to Saudi Ghawar: https://dxgordon.github.io/OCIPlus/#compare/saudi-arabia-ghawar/canada-athabasca-dc-sco?model=00000100

57. There are different petcoke grades. The most degraded, fuel-grade petcoke, which is high in sulfur and heavy metals, is produced from oil sands. See Suncor,

"Petcoke," https://www.suncor.com/en-CA/about-us/supply-and-trading/petroleum-coke

58. Associated Press, "Alberta Oil Sands Waste Exported by American Refineries to Pollution-Choked India," December 1, 2017, https://www.cbc.ca/news/canada/edmonton/exporting-pollution-american-refineries-oilsands-india-1.4428065

59. Andrea Thompson, "The Science and History of Oil Spills," Live Science, April 23, 2010, https://www.livescience.com/9885-faq-science-history-oil-spills.html

60. International Gas Union, "Global Gas Trade, 2000–2015," https://www.igu.org/resources-data

61. International Gas Union, "2019 World LNG Report," https://www.igu.org/app/uploads-wp/2019/06/IGU-Annual-Report-2019_23.pdf; and CEDIGAZ, "The Global Gas Market in 2018," https://www.cedigaz.org/the-global-gas-market-in-2018. For LNG conversion units, see https://qp.com.qa/en/Pages/ConversionFactor.aspx

62. International Gas Union, "2019 World LNG Report," https://www.igu.org/app/uploads-wp/2019/06/IGU-Annual-Report-2019_23.pdf

63. For example, see Australia Gorgon with and without LNG transport using the OCI+ Preview, https://dxgordon.github.io/OCIPlus/#oil/australia-gorgon?model=00100000&carbonToggle=off&carbonTax=20.00

64. Mohammad Masnadi and Adam Brandt, "Climate Impacts of Oil Extraction Increase Significantly with Oilfield Age," *Nature Climate Change* 7 (2017): 551–56, https://www.nature.com/articles/nclimate3347

65. James Chen, "Brent Blend," *Investopedia*, August 20, 2018, https://www.investopedia.com/terms/b/brentblend.asp; David Braziel, "Lift Me Up! The Brent Complex, Linkages That Make It Work and Implications for Global Markets, Part 2," *RBN*, June 15, 2020, https://rbnenergy.com/lift-me-up-the-brent-complex-linkages-that-make-it-work-and-implications-for-global-markets-part-2

66. UK Brent plotting total processed oil, NGLs, and gas (1,750 BOE/d) versus only oil (370 BOE/d), the difference of which amounts to 1,380 BOE/d of NGLs and gas, OCI+ Preview Web Tool, https://dxgordon.github.io/OCIPlus/#analysis?xSelect=crudeProductionVolume&ySelect=productionVolume&oiltypeSelect=Depleted%20Gas

67. Compare UK Brent to PA Marcellus using twenty-year GWPs, OCI+ Preview Web Tool, https://dxgordon.github.io/OCIPlus/#oil/uk-brent and https://dxgordon.github.io/OCIPlus/#compare/u.s.-pennsylvania-marcellus/uk-brent

68. Ann Scarborough Bull and Milton Love, "Worldwide Oil and Gas Platform Decommissioning: A Review of Practices and Reefing Options," *Ocean & Coastal Management* 168 (February 1, 2019), https://doi.org/10.1016/j.ocecoaman.2018.10.024

69. GlassPoint is working with oil companies to generate zero-GHG steam through some of the world's largest solar thermal projects. See GlassPoint, https://www.glasspoint.

com/; https://www.glasspoint.com/technology/lowest-cost/; https://www.glasspoint.com/belridgesolar/; and BusinessWire, "Aera Energy and GlassPoint to Build California's Largest Solar Energy Project at Kern County's Belridge Oilfield," November 29, 2017, https://www.businesswire.com/news/home/20171129006025/en/Aera-Energy-GlassPoint-Build-California%E2%80%99s-Largest-Solar

70. Adam Brandt, "Why Are California's Oils Some of the World's Dirtiest: A Panel with the Oil-Climate Index Team," April 15, 2019, https://west.stanford.edu/research/works/why-are-californias-oils-some-worlds-dirtiest-panel-oil-climate-index-team

71. Jingfan Wang et al., "Potential Solar Energy Use in the Global Petroleum Sector," *Energy* 118 (January 1, 2017), https://doi.org/10.1016/j.energy.2016.10.107

72. Minimizing fossil fuel inputs includes more than solar steam, the only option currently modeled. Comparing solar steam at 0 percent and 75 percent, OCI+ Preview Web Tool, https://dxgordon.github.io/OCIPlus/#oil/u.s.-california-midway-sunset?model=00100000&carbonToggle=off&carbonTax=20.00 and https://dxgordon.github.io/OCIPlus/#oil/u.s.-california-midway-sunset?model=00100020&carbonToggle=off&carbonTax=20.00

73. Compare Argentina Vaca Muerta low to high methane, OCI+ Preview Web Tool, https://dxgordon.github.io/OCIPlus/#oil/argentina-vaca-muerta?model=02100002&carbonToggle=off&carbonTax=20.00 and https://dxgordon.github.io/OCIPlus/#oil/argentina-vaca-muerta?model=01100001&carbonToggle=off&carbonTax=20.00

74. John Wihbey, "Pros and Cons of Fracking: 5 Key Issues," Yale Climate Connections, May 27, 2015, https://www.yaleclimateconnections.org/2015/05/pros-and-cons-of-fracking-5-key-issues/?gclid=CjoKCQjw4s7qBRCzARIsAImcAxZHmoHg2b4ltWxdRmEd13bd4cA4Y83K3VjqAg2qJ_dbvhPBnZifRlQaAh7xEALw_wcB

75. Hector Herrera, "The Legal Status of Fracking Worldwide," Global Network for Human Rights and the Environment, January 6, 2020, https://gnhre.org/2020/01/06/the-legal-status-of-fracking-worldwide-an-environmental-law-and-human-rights-perspective/

76. Robert Jackson et al., "The Environmental Costs and Benefits of Fracking," *Annual Review of Environment and Resources* 39 (2014), https://www.annual-reviews.org/doi/full/10.1146/annurev-environ-031113-144051; David Roberts, "Fracking May Be a Bigger Climate Problem Than We Thought," *Vox*, August 16, 2019, https://www.vox.com/energy-and-environment/2019/8/15/20805136/climate-change-fracking-methane-emissions

77. Robert Howarth, "Ideas and Perspectives: Is Shale Gas a Major Driver of Recent Increase in Global Atmospheric Methane?," *Biogeosciences* 16 (August 14, 2019), https://www.biogeosciences.net/16/3033/2019/

78. Rob Jackson et al., "Increasing Anthropogenic Methane Emissions Aris Equally from Agricultural and Fossil Fuel Sources," *Environmental Research Letters* 15, no. 7 (July 2020), https://iopscience.iop.org/article/10.1088/1748-9326/ab9ed2

79. International Energy Agency, "Methane Tracker 2020," March 2020, https://www.iea.org/reports/methane-tracker-2020/methane-abatement-options

80. Carbon capture currently only applies to $CO_2$ and not other GHGs. The "S" can also stand for "sequestration," permanently storing $CO_2$ that is captured.

81. When CCS techniques also utilize the captured carbon, this is known as carbon capture, utilization, and storage.

82. Louise Jeffery et al., "Options for Supporting Carbon Dioxide Removal," New Climate Institute, July 2020, https://www.c2g2.net/wp-content/uploads/Options-for-supporting-Carbon-Dioxide-Removal_July_2020.pdf

83. International Energy Agency, "Carbon Capture, Utilization and Storage," June 16, 2020, https://www.iea.org/fuels-and-technologies/carbon-capture-utilisation-and-storage

84. U.S. Wyoming Natural $CO_2$ EOR compared to the United Arab Emirates man-made $CO_2$ EOR, OCI+ Web Tool Preview, https://dxgordon.github.io/OCIPlus/#compare/u.s.-wyoming-natural-co2-eor-example/uae-man-made-co2-eor-example

85. Saudi Ghawar production only emits an estimated 21 kg $CO_2$e/BOE out of a total 576 kg $CO_2$e/BOE, OCI+ Web Tool Preview, https://dxgordon.github.io/OCIPlus/#oil/saudi-arabia-ghawar

86. "Map of CCS Projects in Europe," International Oil and Gas Producers, https://www.iogp.org/bookstore/product/map-of-ccs-projects-in-europe/

87. Deborah Gordon, "Understanding Climate Engineering," Carnegie Endowment for International Peace, August 21, 2017, https://carnegieendowment.org/files/Understanding_Climate_Engineering_Web.pdf; Also see Daisy Dunne, "Explainer: Six Ideas to Limit Global Warming with Solar Engineering," *CarbonBrief*, May 9, 2018, https://www.carbonbrief.org/explainer-six-ideas-to-limit-global-warming-with-solar-geoengineering

88. William Safire, "Location, Location, Location," *New York Times*, June 26, 2009, https://www.nytimes.com/2009/06/28/magazine/28FOB-onlanguage-t.html?auth=login-email

89. See, for example, Sergey Budovicz et al., "Cryovolcanism on the Earth: Origin of a Spectacular Crater in the Yamal Peninsula (Russia)," *Scientific Reports*, September 10, 2018, https://www.nature.com/articles/s41598-018-31858-9/; and Stephanie Pappas, "Oozing Methane Blasts Holes in Siberian Tundra," LiveScience, July 6, 2017, https://www.livescience.com/59705-oozing-methane-blasts-craters-in-siberian-tundra.html

90. Jason G. Goldman, "Ecuador Has Begun Drilling for Oil in the World's Richest Rainforest," *Vox*, January 14, 2017, https://www.vox.com/energy-and-environment/2017/1/14/14265958/ecuador-drilling-oil-rainforest

91. Kelly Swing, "Day of Reckoning for Ecuador's Biodiversity," *Nature* 469 (January 2011), https://www.researchgate.net/publication/49768588_Day_of_reckoning_for_Ecuador's_biodiversity

92. According to the Oil Production Greenhouse Gas Emissions Estimator (OPGEE), baseline drilling and development combustion GHGs as low as 0.03 g $CO_{2e}$/MJ are estimated in conventional ecosystems in Azerbaijan or offshore.

93. OPGEE estimates baseline drilling and development land use and combustion GHGs much larger in fragile ecosystems than other ecosystems (where GHGs hover at zero). For the Indonesia Duri and Minas oil fields (which are located in the Sumatran rainforest), drilling and development emissions are estimated at 6 g $CO_{2e}$/MJ, and for Alaska North Slope (located in the Arctic permafrost) emissions are estimated at 5 g $CO_{2e}$/MJ. Compare these GHGs to Norway Ekofisk (an off-shore field) with drilling and development emissions estimated at 0.03 g $CO_{2e}$/MJ. See OPGEE3.0a Beta to do these runs, https://github.com/arbrandt/OPGEE/blob/master/model/OPGEE_3.0a_BETA.xlsm

94. Assumes upstream GHGs from Russia Urengoyskoye (located in Siberian per-mafrost) with high venting and fugitive emissions values estimated at 268 kg $CO_{2e}$/BOE, using the OCI+ Preview Web Tool, https://dxgordon.github.io/OCIPlus/#oil/russia-urengoyskoye?model=02100002&carbonToggle=off&carbonTax=20.00

95. Note that a gigatonne equals 1 billion tonnes. For reference, Saudi Ghawar, which supplies about 5 percent of the world's oil, has total lifecycle GHGs estimated at just under 1 Gt $CO_{2e}$ a year. Adam Brandt et al., "Climate-Wise Choices in a World of Oil Abundance," *Environmental Research Letters* 13 (April 5, 2018), https://iop-science.iop.org/article/10.1088/1748-9326/aaae76; https://dxgordon.github.io/OCIPlus/#total-emissions

96. Note: Using a twenty-year GWP, estimated GHG reductions in refining of 35 Gt$CO_2$e are assumed to increase to 50 Gt. See Liang Jing et al., "Carbon Intensity of Global Crude Oil Refining and Mitigation Potential," *Nature Climate Change* 10 (2020), https://www.nature.com/articles/s41558-020-0775-3?proof=t

97. Jeff Tollefson, "IPCC Says Limiting Global Warming to 1.5 °C Will Require Drastic Action," *Nature*, October 8, 2018, https://www.nature.com/articles/d41586-018-06876-2

98. Intergovernmental Panel on Climate Change, "Global Warming of 1.5 °C," 2018, https://www.ipcc.ch/site/assets/uploads/sites/2/2018/07/SR15_SPM_version_stand_alone_LR.pdf. Note that the Intergovernmental Panel on Climate Change's 770 Gt $CO_2$ emission limit applies to $CO_2$ only and is one of many estimates that continue to change based on different models according to a range of probabilities of staying below the 1.5 °C threshold. Also see AirClim, "How Much More Can Be Omitted?," https://www.airclim.org/how-much-more-can-be-emitted; UN Environment, "Emissions Gap Report 2018," November 2018, http://wedocs.unep.org/bitstream/handle/20.500.11822/26895/EGR2018_FullReport_EN.pdf

99. International Energy Agency, *World Energy Outlook 2018*, Chapter 11, https://www.iea.org/weo2018/

100. For assumptions regarding this estimate, see comments and sources in Table 4.3.

101. International Energy Agency, *World Energy Outlook 2018*, Appendix A.3, https://www.iea.org/weo2018/

102. Calculated based on oil and gas sector emission savings of 8.6 Gt compared to International Energy Agency current policies estimate of 24.5 Gt in 2040. See ibid.

103. Mohammad Masnadi et al., "Global Carbon Intensity of Crude Oil Production," *Science* 361 (August 2018), https://science.sciencemag.org/content/361/6405/851.summary; Liang Jing et al., "Carbon Intensity of Global Crude Oil Refining and Mitigation Potential," *Nature Climate Change* 10 (2020)

104. Jan C. Minx et al., "Negative Emissions—Part 1: Research Landscape and Synthesis," *Environmental Research Letters* 13, no. 6 (May 22, 2018), https://iopscience.iop.org/article/10.1088/1748-9326/aabf9b; US Department of Energy, "Demand: Effects of Travel Reduction and Efficient Driving on Transportation: Energy Use and Greenhouse Gas Emissions," March 2013, https://www.nrel.gov/docs/fy13osti/55635.pdf

105. Benjamin Storrow, E&E News, "Can a Big Oil Company Go Carbon-Free?," *Scientific American*, December 6, 2019, https://www.scientificamerican.com/article/can-a-big-oil-company-go-carbon-free/

106. For example, the US Greenhouse Gas Reporting Program covers only those sources over 25,000 metric tonnes $CO_2$e per year; see https://www.epa.gov/ghgreporting/ghgrp-and-oil-and-gas-industry. See also Transparency Pathway Initiative, http://www.lse.ac.uk/GranthamInstitute/tpi/; Carbon Disclosure Project, https://www.cdp.net/en; and Carbon Tracker, https://www.carbon-tracker.org/

107. Rachel Adams-Heard, "What Happens When an Oil Giant Walks Away," Bloomberg Green, April 15, 2021, https://www.bloomberg.com/graphics/2021-tracking-carbon-emissions-BP-hilcorp/

CHAPTER 5

1. Office of Congressional Ethics, "David Skaggs, Chairman," https://oce.house.gov/about/board-and-staff/david-skaggs-chairman

2. James Hansen et al., "Climate Impact of Increasing Atmospheric Carbon Dioxide," *Science* 213, no. 4511 (August 28, 1981), https://pubs.giss.nasa.gov/docs/1981/1981_Hansen_ha04600x.pdf

3. "Congressional Testimony of Dr. James Hansen," before the US Senate Committee on Energy and Natural Resources, June 23, 1988, https://www.sealevel.info/1988_Hansen_Senate_Testimony.html#

4. *Ozone-Forming Potential of Reformulated Gasoline* (Washington, DC: National Academies of Press, 1999), https://www.nap.edu/download/9461

5. ARCO was formed in 1966 by the merger of Atlantic Refining and the Richfield Oil Corporation when they bought Sinclair Oil Corporation. Through the 1980s, ARCO was one of the largest global oil companies. ARCO was bought by BP in 2000. In 2013, BP sold ARCO and its other Southern California assets to Tesoro (which was renamed Andeavor in 2017). Marathon Petroleum acquired the entire holding (including ARCO) in 2018. https://www.britannica.com/topic/Atlantic-Richfield-Company

6. Matthew Wald, "ARCO Offers New Gasoline to Cut Up to 15% of Old Cars' Pollution," *New York Times*, August 16, 1989, https://timesmachine.nytimes.com/timesmachine/1989/08/16/001489.html?pageNumber=1

7. Rachel Adams-Heard, "What Happens When an Oil Giant Walks Away?" *Bloomberg Green*, April 15, 2021, https://www.bloomberg.com/graphics/2021-tracking-carbon-emissions-BP-hilcorp/

8. ExxonMobil, "Our History," https://www.exxon.com/en/history

9. Note that the OCI+ considers oil refining as a midstream operation to disaggregate these GHGs from downstream marketing activities that determine all end use emissions.

10. Natural Resources Governance Institute, "National Oil Company Database," December 2019, https://www.nationaloilcompanydata.org/indicator

11. Qatar Petroleum, for example, operates in joint ventures with Eni, Shell, and others worldwide. See https://qp.com.qa/en/MediaCentre/Lists/QPPublications/Attachments/29/QP%202018%20Sustainability%20Report.pdf; Saudi Aramco is also considered an INOC. For more information on INOCs, see Peter Ramsay, "The Rise of the Mid-East 'INOC,'" *Petroleum Economist*, March 6, 2020, https://www.petroleum-economist.com/articles/politics-economics/middle-east/2020/the-rise-of-the-mid-east-inoc

12. Author's calculations using individual company annual financial reports, along with other databases, including *Oil & Gas Journal*, OGJ150, September 3, 2018, 24 and 41, https://digital.ogj.com/ogjournal/20180903/MobilePagedReplica.action?pm=2&folio=6#pg8; c&en, Top 50 Chemical Companies or 2018, https://cen.acs.org/business/finance/CENs-Global-Top-50-chemical/97/i30. It is reported that NOCs collectively have oil and gas assets valued at over $3 trillion; see Joshua Olufemi, "National Oil Companies with $3.1 Trillion in Assets Dangerously Under-Scrutinised—NRGI," *Premium Times*, July 31, 2019, https://www.premiumtimesng.com/business/327048-national-oil-companies-with-3-1-trillion-in-assets-dangerously-under-scrutinised-nrgi.html

13. An indicator of changing valuation in the oil and gas industry according to the OGJ150 share values follows: $2,021 per share on October 9, 2018, $707 on March 19, 2020 (a 65% drop in overall value), and $1,225 on January 14, 2021 (a 73% rise in overall value). See OGJ150, Financial Content Services, https://markets.financial-content.com/stocks/quote/historical?Symbol=CIX%3AOGJ200&Month=10&Range=1&Year=2021

14. Independent petroleum companies (known as wildcatters) are common in the United States and are not listed in the table. Examples include Devon, EOG, Apache, Noble, Pioneer, Continental, Chesapeake, and Southwestern, all of which are located in the United States and have asset values ranging from $10 to $30 billion. Several chemical companies are independent of IOCs, including DowDuPont, BASF, Formosa, Mitsubishi, LG, Ineos, and LyondelleBasell, with asset values from $22 to $192 billion. About 9,000 independent producers drill over 90 percent of oil and gas wells in the United States, accounting for 54 percent of America's oil production and 85 percent of its natural gas and natural gas liquid production. See Independent Petroleum Association of America, "About IPAA," https://www.ipaa.org/independent-producers/. Note that the US Internal Revenue Service even publishes a tax definition for independents that produce fewer than 75,000 barrels per day of oil or have retail sales less than $5 million a year. See US Internal Revenue Service, "4.41.1 Oil and Gas Handbook," December 3, 2013, https://www.irs.gov/irm/part4/irm_04-041-001

15. *Oil & Gas Journal*, OGJ150, September 3, 2018; *Oil & Gas Journal*, "O&GJ Refining Survey 2018"; OGJ100, September 3, 2018. For the above referenced OGJ publications see https://digital.ogj.com/ogjournal/20180903/MobilePagedReplica.action?pm=2&folio=Cover#pg1; NRGI National Oil Company Database, ttps://resourcegovernance.org/analysis-tools/publications/national-oil-company-database; Platts Top250 Rankings, https://www.spglobal.com/platts/top250

16. Visualcapitalist.com, "Chart of the Week: Big Oil" (2015 data), https://oilprice.com/Energy/Crude-Oil/The-17-Trillion-Oil-Industry-Isnt-Going-Anywhere.html

17. Investopedia, "What Percentage of the Global Economy Consists of the Oil and Gas Drilling Sector?," September 10, 2018, https://www.investopedia.com/ask/answers/030915/what-percentage-global-economy-comprised-oil-gas-drilling-sector.asp

18. *Oil & Gas Journal*, "OGJ150: Leading Oil and Gas Companies in the US" and "OGJ100: Leading Oil and Gas Companies Outside the US," September 3, 2018, 34–45.

19. BP, Oil and Natural Gas Prices, https://www.bp.com/en/global/corporate/energy-economics/statistical-review-of-world-energy/oil.html#oil-prices and https://www.bp.com/en/global/corporate/energy-economics/statistical-review-of-world-energy/natural-gas.html#natural-gas-prices

20. For OCI+ calculations between min and max values, for oil and gas producers see OCI+ Preview webtool, https://dxgordon.github.io/OCIPlus/#supply-chain?model=00000000&carbonToggle=off&carbonTax=20.00&stepSelect=upstream&sortSelect=true

21. Ibid.

22. For example, compare China Sulige shipping with and without LNG transport, OCI+ Preview web tool, https://dxgordon.github.io/OCIPlus/#supply-chain?m

odel=00000000&carbonToggle=off&carbonTax=20.00&stepSelect=midstream
&sortSelect=true; https://dxgordon.github.io/OCIPlus/#supply-chain?model=
00002000&carbonToggle=off&carbonTax=20.00&stepSelect=midstream&sort
Select=true

23. See, for example, IPIECA, "Sustainability Reporting Guidance for the Oil and Gas
Industry," March 2020, https://www.ipieca.org/media/5110/ipieca_sustainability-
guide_2020_mod3-cce.pdf; American Petroleum Institute, "Compendium of
Greenhouse Gas Emissions Methodologies for the Oil and Gas Industry," 2011,
https://www.api.org/~/media/Files/EHS/climate-change/GHG_industry-
guidelines-IPIECA.pdf; World Resources Institute and World Business Council
for Sustainable Development, "GHG Accounting Protocol," https://ghgpro-
tocol.org/sites/default/files/standards/ghg-protocol-revised.pdf;     and     ISO,
"Specification with Guidance at the Organizational Level for Quantification and
Reporting of Greenhouse Gas Emissions and Removals," 2006.

24. World Resources Institute and World Business Council for Sustainable
Development, "GHG Accounting Protocol," https://ghgprotocol.org/sites/
default/files/standards/ghg-protocol-revised.pdf

25. Efforts are underway to align corporate commitments that are reported on a different
basis. See Simon Dietz et al., "Carbon Performance of European Integrated Oil and
Gas Companies: Briefing Paper," Transition Pathway Initiative, May 2020, https://
www.transitionpathwayinitiative.org/publications/58.pdf?type=Publication

26. BP, "BP Sets Ambition for Net Zero by 2050, Fundamentally Changing Organization
to Deliver," February 12, 2020, https://www.bp.com/en/global/corporate/news-
and-insights/press-releases/bernard-looney-announces-new-ambition-for-bp.html

27. BP, "From International Oil Company to Integrated Energy Company: BP Sets
Out Strategy for Decade of Delivery towards Net Zero Ambition," August 4, 2020,
https://www.bp.com/en/global/corporate/news-and-insights/press-releases/
from-international-oil-company-to-integrated-energy-company-bp-sets-out-
strategy-for-decade-of-delivery-towards-net-zero-ambition.html

28. Will Kennedy, Laura Hurst, and Kevin Crowley, "BP Walks Away from the
Oil Supermajor Model It Helped Crete," *Bloomberg Green*, August 4, 2020,
https://www.bloomberg.com/news/articles/2020-08-04/bp-walks-away-from-
the-oil-supermajor-model-it-helped-create

29. The EU nations (which excludes the United Kingdom, Norway, and Russia) are
entirely dependent on imported oil. This holds for gas and petroleum products
as well, as discussed in chapter 6. See "Oil and Petroleum Products - A Statistical
Overview,"    Eurostat,    June    2020,    https://ec.europa.eu/eurostat/statistics-
explained/index.php?title=Oil_and_petroleum_products_-_a_statistical_
overview&oldid=315177#

30. Corbin Hiar, "BP, Facing Scrutiny, Renews Bold Climate Goals," *ClimateWire*,
June 4, 2020, https://www.eenews.net/stories/1063312133

31. "Petroleum Refining & Marketing," Reliance Industries, https://www.ril.com/OurBusinesses/PetroleumRefiningAndMarketing.aspx. Assumes 83 million bpd crude oil portion of refinery throughput in 2019, see N. Sonnichsen, "Oil Refinery Capacity Worldwide by Leaking Country, 2010–2019," Statista, August 17, 2020,https://www.statista.com/statistics/273579/countries-with-the-largest-oil-refinery-capacity/#

32. BP, "Reliance and BP to Create Major World-Class Fuels Partnership for India's Fast-Growing Market," August 6, 2019, https://www.bp.com/en/global/corporate/news-and-insights/press-releases/reliance-and-bp-to-create-major-world-class-fuels-partnership-for-indias-fast-growing-market.html

33. Author's calculations based on Jamnagar's 1.24 mbpd refining capacity with estimated average petroleum product Scope 3 emissions of 450 kilograms of $CO_2e$ per BOE. For petroleum product emission factors, see Deborah Gordon et al., "Know Your Oil," Carnegie Endowment for International Peace, March 11, 2015, https://carnegieendowment.org/2015/03/11/know-your-oil-creating-global-oil-climate-index-pub-59285

34. Chevron, "2019 Performance Data," https://www.chevron.com/-/media/shared-media/documents/2019-sustainabilty-performance-data.pdf; https://www.chevron.com/corporate-responsibility/climate-change/greenhouse-gas-management; Shell, "Greenhouse Gas Emissions," https://www.shell.com/sustainability/sustainability-reporting-and-performance-data/performance-data/greenhouse-gas-emissions.html

35. Ibid.

36. CDP, "Beyond the Cycle: Which Oil and Gas Companies Are Ready for the Low-Carbon Transition?," www.cdp.net/en/investor/sector-research/oil-and-gas-report

37. Dietz et al., "Carbon Performance of European Integrated Oil and Gas Companies"; for detailed information on Shell and Eni, see https://www.transitionpathwayinitiative.org/tpi/publications/59.pdf?type=Publication; and for an assessment of the climate competencies of fifty oil and gas companies, see https://www.transitionpathwayinitiative.org/tpi/sectors/oil-gas

38. Norman Wisely and Munir Hassan, "Energy Transition: Evolution or Revolution?," CMS, January 2020, https://cms.law/en/int/publication/energy-transition-evolution-or-revolution

39. Shell, "What Are Shell Scenarios?," https://www.shell.com/energy-and-innovation/the-energy-future/scenarios/what-are-scenarios.html

40. Shell, "Management Day 2019," June 4–5, 2019, slides 43 and 81, https://www.shell.com/investors/news-and-media-releases/investor-presentations/2019-investor-presentations/management-day-2019/_jcr_content/par/textimage_af3b.stream/1572529480536/e35f4f28f2cda0b5769605468243b3538b1a3ec8/shell-management-day2019-webcast-presentation-slides-updated.pdf

41. Ibid., slides 43, 57, and 66.

42. "PBF Energy to Buy Shell's Martinez, California Refinery for up to $1Billion," *Reuters*, June 11, 2019, https://www.reuters.com/article/us-pbfenergy-shell-martinez/pbf-energy-to-buy-shells-martinez-california-oil-refinery-for-up-to-1-billion-idUSKCN1TC2M7

43. Jessica Resnick-alt, et. al., "Shell Plans to Exit California with ExxonMobil," *Reuters*, July 1, 2021, https://www.reuters.com/business/energy/exclusive-shell-plans-exit-california-joint-venture-with-exxon-mobil-sources-2021-07-01/

44. Shell, "A Better Life with a Healthy Planet: Pathways to Net-Zero Emissions," May 2016, https://www.shell.com/energy-and-innovation/the-energy-future/scenarios/a-better-life-with-a-healthy-planet.html

45. Note that Shell assigns the lowest GHG reduction potential to its own operating emissions. See "Lowering Our Net Carbon Footprint," Shell Sustainability Report 2019, https://reports.shell.com/sustainability-report/2019/sustainable-energy-future/net-carbon-footprint.html

46. "Repsol Will Be a Net Zero Emissions Company by 2050," December 2, 2019, https://www.repsol.com/en/press-room/press-releases/2019/repsol-will-be-a-net-zero-emissions-company-by-2050.cshtml

47. Katherine Dunn, "Shell Becomes the Largest Global Energy Company to Commit to a Net-Zero Emissions Goal by 2050," *Fortune*, April 16, 2020, https://fortune.com/2020/04/16/net-zero-emissions-shell-oil-industry-gas/

48. "Universite Total: Talking about the Energy Transition," https://www.total.com/media/video/universite-total-talking-about-energy-transition

49. Philippe Roos, "Q&A: Mathieu Soulas on Total's Transition Strategy," *EI New Energy*, July 16, 2020, http://www.energyintel.com/pages/eig_article.aspx?DocId=1077894

50. "From Net Zero Ambition to Total Strategy," Total, September 30, 2020, https://www.total.com/media/news/press-releases/2020-strategy-outlook-presentation?source=content_type:react|first_level_url:news|section:main_content|button:body_link

51. Francois de Beaupuy, "Total to Replace Oil Refinery with Clean Fuel and Plastic Plants," September 24, 2020, https://www.bloomberg.com/news/articles/2020-09-24/total-to-replace-oil-refinery-with-clean-fuel-and-plastic-plants

52. Bjorn Otto Sverdrup, "My Oil Firm Wants Results from COP21. But Not the Kind You Might Expect," *The Guardian*, November 27, 2015, https://www.theguardian.com/commentisfree/2015/nov/27/cop-21-paris-climate-talks-low-carbon-oil-statoil

53. Harald Eraker et al., "Pure Norwegian Oil?," *NRK*, October 3, 2016, https://www.nrk.no/dokumentar/xl/ren_-norsk-olje_-1.13150883

54. For more information on this particular case study for Ekofisk, see Julia Benz, "Leading Emission Reduction Opportunities and Renewable Technologies for Oil and Gas Production," Senior thesis, Brown University, March 26, 2021.

55. The initial Phase 1 and 2 OCI+ assessments of Norwegian oil fields can be found at Carnegie Endowment for International Peace, http://oci.carnegieendowment.org/#supply-chain. Note that in a recent analysis Norway places sixth lowest in its volume-weighted oil production GHG intensity. But this only considers Norwegian production, not lifecycle GHGs or Equinor operations beyond Norway. See Mohammad Masnadi et al., "Global Carbon Intensity of Crude Oil Production," *Science* 361, no. 6405 (August 31, 2018). In the updated OCI+, Norway Ekofisk is estimated at 19 percent lower in lifecycle GHGs than Saudi Ghawar; see https://dxgordon.github.io/OCIPlus/#compare/norway-ekofisk/saudi-arabia-ghawar

56. See NRK documentary, minute 43: https://tv.nrk.no/serie/brennpunkt/2016/MDDP11001016/avspiller

57. Norwegian Petroleum, "Exports of Oil and Gas," https://www.norskpetroleum.no/en/production-and-exports/exports-of-oil-and-gas/; and Chevron, "2018 Supplement to the Annual Report," https://www.chevron.com/investors

58. Ariel Cohen, "Saudi Aramco IPO Hits $2 Trillion Mark amid Guarded Forecast," *Forbes*, December 18, 2019, https://www.forbes.com/sites/arielcohen/2019/12/18/saudi-aramco-ipo-hits-2-trillion-mark-but-forecast-still-guarded/#7ea04d6b42e6. Note that Aramco was traded on the Riyadh stock exchange. In 2020, Apple passed Aramco, with a market valuation of $1.84 trillion compared to Aramco's 1.76 trillion. Seehttps://www.theverge.com/2020/7/31/21350154/apple-worlds-most-valuable-company-saudi-aramco

59. "Saudi Aramco: The Oil Colossus," *Reuters*, November 3, 2019, https://www.reuters.com/article/us-saudi-aramco-ipo-factbox/saudi-aramco-the-oil-colossus-idUSKBN1XD03T

60. Michael Ives, "'Real Impact' – Across the Kingdom and around the Globe – Aramco's First Branding Campaign," *Aramco ExPats*, October 30, 2019, https://www.aramcoexpats.com/articles/real-impact-across-the-kingdom-and-around-the-globe-aramcos-first-branding-campaign/ Also see Matthew Taylor, "Saudi Aramco Removes 'Sustainable' Oil Adverts after Complaints," *The Guardian*, April 29, 2020, https://www.theguardian.com/business/2020/apr/29/saudi-aramco-removes-sustainable-oil-adverts-after-complaints

61. Masnadi et al., "Global Carbon Intensity of Crude Oil Production."

62. The speed at which Aramco can act relates to its direct ties to the monarchy with absolute control. The ease at which it can control oil volumes relates to the fact that Aramco has fewer huge oil fields that require relatively fewer inputs to produce compared to other global assets.

63. Ariel Cohen, "Too Little Too Late? Russia and Saudi Arabia Reach Truce in Oil Price War," *Forbes*, April 10, 2020, https://www.forbes.com/sites/arielcohen/2020/04/10/too-little-too-late-russia-and-saudi-arabia-reach-truce-in-oil-price-war/#5bc5c0c56ff0

64. "Was BP's 'Beyond Petroleum' Campaign Greenwashing?," Climate one, November 14, 2019, https://www.youtube.com/watch?v=91fnsnIOYJg

65. John Kenney, "Beyond Propaganda," *New York Times*, August 14, 2006, https://www.nytimes.com/2006/08/14/opinion/14kenney.html

66. Dietz, "Carbon Performance of European Integrated Oil and Gas Companies."

67. *Oil & Gas Journal*, OGJ150, September 3, 2018.

68. Darius Snieckus, "Investment giant BlackRock backs calls for oil supermajor BP to 'accelerate' climate plans," *Recharge,* May 31, 2021, https://www.rechargenews.com/energy-transition/investment-giant-blackrock-backs-calls-for-oil-supermajor-bp-to-accelerate-climate-plans/2-1-1018172

69. For more information on a voluntary methane standard that BP and others are piloting, see Methane Intelligent Quotiet, https://miq.org/

70. There is no online copy of a public brochure produced by Chevron titled "Building Effective Climate Policy Together," which was personally handed to me in February 2020.

71. Government of Western Australia, Department of Mines, Industry Regulation and Safety, "Gorgon Carbon Dioxide Injections Project," https://www.dmp.wa.gov.au/Petroleum/Gorgon-CO2-injection-project-1600.aspx

72. See OCI+ model input data for Australia Gorgon and Pennsylvania Marcellus, https://dxgordon.github.io/OCIPlus/#methodology;

73. Emissions estimated using the OPGEE3.0a Beta model; see https://github.com/arbrandt/OPGEE. For more information on this particular case study, see Benz, "Leading Emission Reduction Opportunities and Renewable Technologies for Oil and Gas Production."

74. The project plans to inject between 3.4 and 4 million tonnes of reservoir $CO_2$ each year. See Chevron, "Gorgon Project: An Australian Icon," https://australia.chevron.com/our-businesses/gorgon-project; and Chevron, "Gorgon Carbon Dioxide Injection Project," 2019, https://australia.chevron.com/-/media/australia/publications/documents/gorgon-co2-injection-project.pdf

75. Between project start-up in March 2016 and CCS commencement in August 2019, Gorgon emitted an estimated total of 31 million metric tonnes of GHGs. In 2017 and 2018, the Australian government reported that Gorgon's GHG emissions totaled 18 million tonnes. An additional estimated 13 million tonnes were emitted in 2016 and 2019. See Australian Government Clean Energy Regulator, "Safeguard Facility Reported Emissions," National Greenhouse Gas and Energy Reporting, March 26, 2020; for 2017–2018 records, see http://www.cleanenergyregulator.gov.au/NGER/National%20greenhouse%20and%20energy%20reporting%20data/safeguard-facility-reported-emissions/safeguard-facility-emissions-2017-18 (see records 61–91); and for 2018–2019 records, seehttp://www.cleanenergyregulator.gov.au/NGER/National%20greenhouse%20and%20energy%20reporting%20data/safeguard-facility-reported-emissions/

safeguard-facility-emissions-2018-19?Paged=TRUE&p_Title=Gibson%20 Island&p_ID=63&View=%7b2E72BD28%2d8DE5%2d4AA9%2d9795%2dD9A 6B430FBBD%7d&PageFirstRow=61 (see records 61–91).

76. Lisa Cox, "Western Australia LNG Plant Faces Calls to Shut Down Until Faulty Carbon Capture System is Fixed," *The Guardian*, January 14, 2021, https://www. theguardian.com/environment/2021/jan/15/western-australia-lng-plant-faces- calls-to-shut-down-until-faulty-carbon-capture-system-is-fixed

77. Chevron, "Gorgon Carbon Dioxide Injection Project," 2019, https://australia. chevron.com/-/media/australia/publications/documents/gorgon-co2-injection- project.pdf

78. IHS Markit, "Leadership Dialogue with Mike Wirth, Chairman & CEO, Chevron," CERAWeek, May 14, 2020, https://www.youtube.com/watch?re load=9&v=jQVNRPdCtvs&fbclid=IwAR1ow2hU4g_uC8h5FbtGEfJO- 4nkndbK9Mh1SSUlHKw6JpPGVJPINdI7OqM&app=desktop

79. Steve Coll, *Private Empire: ExxonMobil and American Power* (New York: Penguin Press, 2012).

80. Mark Hulbert, "Exxon's Getting Booted from the Dow Jones Industrial Average May Be a Blessing in Disguise for Its Investors," *Market Watch*, August 30, 2020, https://www.marketwatch.com/story/exxon-getting-booted-from-the- dow-jones-industrial-average-may-be-a-blessing-in-disguise-for-its-investors- 2020-08-25

81. Matt Phillips, "Exxon's Board Defeat Signals the Rise of Social-Good Activists," *New York Times*, June 9, 2021, https://www.nytimes.com/2021/06/09/business/ exxon-mobil-engine-no1-activist.html

82. Kevin Crowley and Bryan Gruley, "The Humbling of Exxon," *Bloomberg Business Week*, May 4, 2020, https://www.bloomberg.com/features/2020-exxonmobil- coronavirus-oil-demand/

83. Erik Larson, "ExxonMobil Beats New York's Climate-Change Accounting Case," Bloomberg, December 10, 2019, https://www.bloomberg.com/news/ articles/2019-12-10/exxon-prevails-over-n-y-in-climate-change-accounting- case?sref=kDVWiOVx

84. In 2020, China opened its doors for the first time to non-state-owned enterprises (domestic and foreign) to explore and produce oil and gas. See Oceana Zhou and Daisy Xu, "China's 2019 Oil Product Exports Jump 20% on Year," S&P Global, January 24, 2020, https://www.spglobal.com/platts/en/market-insights/latest- news/oil/012420-china-data-2019-oil-product-exports-jumps-20-on-year-99-of- export-quotas-utilized

85. Knoema, "Total Petroleum Consumption," https://knoema.com/atlas/topics/ Energy/Oil/Petroleum-consumption

86. International Energy Agency, "Oil Information 2019," August 2019, https://www. iea.org/reports/oil-information-2019

87. Zhou and Xu, "China's 2019 Oil Product Exports Jump 20% on Year,"

88. China is adding significant coking refinery capacity and is setting up to be the world's largest exporter of fuel-grade petcoke. See "China to Dominate Asian Refinery Coking Units' Capacity Growth by 2024," *Energy Northern Perspective*, August 10, 2020, https://energynorthern.com/2020/08/10/china-to-dominate-asian-refinery-coking-units-capacity-growth-by-2024/

89. For background on Southwestern, see SWN, https://www.swn.com/about/

90. See Table 3.2.

91. Oxy, "Enhanced Oil Recovery," https://www.oxy.com/OurBusinesses/OilandGas/Technology/Enhanced-Oil-Recovery/Pages/default.aspx

92. OCI+ Gas Preview Beta Web Tool Under Development, https://dxgordon.github.io/OCIPlus/#analysis?xSelect=years&ySelect=ghgTotal&oiltypeSelect=CO2%20EOR%20Oil

93. Notable exceptions to date include Rockefeller Brothers Endowment; Norges Bank Investment Management (Norway Sovereign Wealth Fund); Zurich Insurance; Government of Ireland; New York City Public Pension; and HSBC, which has divested from coal, oil sands, and/or arctic drilling, but not petroleum more broadly.

94. Doug Dannemiller and Sean Collins, "2020 Investment Management Outlook," Deloitte, December 3, 2019, https://www2.deloitte.com/us/en/insights/industry/financial-services/financial-services-industry-outlooks/investment-management-industry-outlook.html

95. BlackRock and Vanguard are the biggest shareholders in ExxonMobil (owning 13% of the company's stock). It was reported that State Street (another major financial firm) also joined ranks. See Steven Mufson, "Financial Firms Lead Shareholder Rebellion against ExxonMobil Climate Change Policies," *Washington Post*, May 31, 2017, https://www.washingtonpost.com/news/energy-environment/wp/2017/05/31/exxonmobil-is-trying-to-fend-off-a-shareholder-rebellion-over-climate-change/

96. Eric Rosenbaum, "Activists Thought BlackRock, Vanguard Found Religion on Climate Change. Not Anymore," *CNBC*, October 13, 2019, https://www.cnbc.com/2019/10/13/blackrock-vanguard-found-religion-on-climate-doubts-are-growing.html

97. Transition Pathway Initiative, 2019; see http://www.climateaction100.org/

98. Task Force on Climate-Related Financial Disclosures, "About the Task Force," https://www.fsb-tcfd.org/about/#

99. Climate-Related Financial Disclosures, "Task Force on Climate-Related Financial Disclosures Overview," March 2020, https://www.fsb-tcfd.org/wp-content/uploads/2020/03/TCFD_Booklet_FNL_Digital_March-2020.pdf

100. Climate Action 100+, "Global Investors Driving Business Transition," http://www.climateaction100.org/

101. Climate Action 100+, "2019 Progress Report," September 2019, http://www.climateaction100.org/

102. Damian Carrington, "Fossil Fuel Divestment Funds Rise to $6tn," *The Guardian*, September 10, 2018, https://www.theguardian.com/environment/2018/sep/10/fossil-fuel-divestment-funds-rise-to-6tn; Richard Mattison, "Accounting for Climate: The Next Frontier in ESG," *S&P Global*, October 11, 2019, https://www.spglobal.com/en/research-insights/featured/accounting-for-climate-the-next-frontier-in-esg

103. Jillian Ambrose and Jon Henley, "European Investment Bank to Phase Out Fossil Fuel Financing," *The Guardian*, November 15, 2019, https://www.theguardian.com/environment/2019/nov/15/european-investment-bank-to-phase-out-fossil-fuels-financing

104. Investor oversight on best practices is very helpful. See RMI Center for Climate Aligned Finance, https://climatealignment.org/

105. Jeremy Grantham, "Ten Quick Topics to Ruin Your Summer," *GMO Quarterly Newsletter*, 2Q2015, https://advisoranalyst.com/2015/08/05/jeremy-grantham-10-quick-topics-to-ruin-your-summer.html/

106. GMO website, https://www.gmo.com/americas/

107. For a 2019 video of Jeremy Grantham discussing climate change see https://www.youtube.com/watch?v=xVVbCyrBQVg. For information on the United Nations Principle for Responsible Investment (PRI) initiative see https://www.unpri.org/pri

108. Grantham, "Ten Quick Topics to Ruin Your Summer."

109. Industry advisers include Halliburton and Schlumberger, for example, which also provide field, installation, and equipment services. Other advisers specialize in oil and gas, such as Gaffney Cline (part of Baker Hughes), Wood Mackenzie, IHS CERA, and Rystad. And other consultants, such as McKinsey, Boston Consulting Group, and Bain, also work closely with the oil industry. See individual websites for more information; http://gaffney-cline-focus.com/files/Brochures/GCA_Carbon_Management_Brochure_July_2019.pdf; https://www.woodmac.com/news/feature/can-the-energy-industry-rise-to-the-challenge-of-climate-change/; https://cdn.ihsmarkit.com/www/pdf/1219/Energy-and-Climate-scenarios.pdf; https://www.rystadenergy.com/newsevents/news/press-releases/top-5-energy-policy-predictions-for-2020/; https://www.mckinsey.com/~/media/McKinsey/dotcom/client_service/Sustainability/cost%20curve%20PDFs/Climate_Change_Business_final_report.ashx; https://www.bcg.com/capabilities/social-impact/climate/center-for-climate-action.aspx; https://www.bain.com/consulting-services/sustainability-corporate-responsibility/

110. For their part, Google announced plans to close down its oil and gas unit for climate reasons in 2020. See Brian Merchant, "Google Says It Will Not Build Custom A.I. for

Oil and Gas Extraction," *Medium*, May 19, 2020, https://onezero.medium.com/google-says-it-will-not-build-custom-a-i-for-oil-and-gas-extraction-72d1f71f42c8

111. Bain & Company, "Oil & Gas," https://www.bain.com/industry-expertise/oil-gas/

112. Matthew Farmer, "What Does the Future Hold for the Oil and Gas Workforce?," *Offshore Technology*, May 1, 2020, https://www.offshore-technology.com/features/workforce-offshore-oil-gas-shale-employment-investment/

113. David Sheppard and Neil Hume, "Vitol's Account Show Potential Value of up to $20bn," *Financial Times*, October 5, 2017, https://www.ft.com/content/283a9606-a77a-11e7-93c5-648314d2c72c#targetText=Based%20on%20listed%20commodity%20traders,the%20figure%20was%20%241.4bn

114. For more information about these companies see https://gunvorgroup.com/news/gunvor-group-2017-results/; https://www.glencore.com/dam/jcr:62bed41c-1627-4bf5-bc43-cf5518ba1193/glen-2017-annual-report.pdf; https://www.glencore.com/sustainability/climate-change; https://www.trafigura.com/financials/2018-the-year-in-review/; https://www.trafigura.com/responsibility/responsible-operation/environment-and-climate-change/; https://www.forbes.com/pictures/5980cf5b4bbe6f3eb80126fa/2-koch-industries/#5664c0c31907; http://www.mercuria.com; https://www.mercuria.com/about-ushttps://www.washingtonpost.com/news/energy-environment/wp/2016/06/06/what-charles-koch-really-thinks-about-climate-change/

115. Julia Payne and Dmitry Zhdannikov, "Probes and Squeezed Profits Change the Oil Trading Game," *Reuters*, June 20, 2019, https://www.reuters.com/article/us-oil-traders/probes-and-squeezed-profits-change-the-oil-trading-game-idUSKCN1TL1IE

116. Lloyds Register, "Assurance Statement Related to the Royal Dutch Shell plc Greenhouse Gas Assertion for the Operational Control Greenhouse Gas Inventory for Calendar Year Ending December 31, 2019," February 25, 2020, https://www.shell.com/sustainability/sustainability-reporting-and-performance-data/performance-data/greenhouse-gas-emissions/_jcr_content/par/tabbedcontent/tab_210321073/textimage.stream/1586188454581/4b0bb77ec597689dec2f2b1ad-3677bf34b1aea78/sov-direct-and-energy-indirect-ghg-assurance-statement.pdf. The California Air Resource Board maintains a list of verification bodies; see https://ww2.arb.ca.gov/verification

117. United Nations Framework Convention on Climate Change, "QA/QC and Verification Reporting Requirements and Their Application in Review of National Greenhouse Gas Inventories," March 13, 2019, https://unfccc.int/sites/default/files/resource/QAQC%20and%20verification_refresher%20seminar%202019_0.pdf

118. For example, see Ernst & Young report for Chevron Corporation, December 13, 2016: https://www.chevron.com/-/media/chevron/sustainability/documents/GHG-inventory-2016.pdf

119. California Air Resources Board, "Mandatory GHG Reporting – Verification," https://ww2.arb.ca.gov/verification

120. California Air Resources Board, "Regulation for the Mandatory Reporting of Greenhouse Gas Emissions," https://ww3.arb.ca.gov/cc/reporting/ghg-rep/regulation/mrr-2018-unofficial-2019-4-3.pdf?_ga=2.155010173.490630161.157825179 1-642710237.1576431103

121. California Air Resources Board, "Annual Summary of GHG Mandatory Reporting: Calendar Year 2018," November 4, 2019, https://ww2.arb.ca.gov/mrr-data. See link to 2019 GHG Facility and Entity Emissions and sort for oil and gas entities.

122. Note that, in addition to the industry associations discussed in this section, a host of smaller industry groups also exist that have taken steps to acknowledge climate risks. But, like the API, they generally offer too little (and often outdated materials) in terms of actionable solutions. For example, see International Association of Oil and Gas Producers, "Climate Change: IOGP Position Paper," https://www.iogp.org/blog/position-statements/climate-change-iogp-position-paper/. For more on energy coalitions, se: Christian Downie, "Ad Hoc Coalitions in the U.S. Energy Sector: Case Studies in the Gas, Oil, and Coal Industries," *Business and Politics* 20, no. 4 (2018), https://www.cambridge.org/core/journals/business-and-politics/article/ad-hoc-coalitions-in-the-us-energy-sector-case-studies-in-the-gas-oil-and-coal-industries/57FE3703B0BB 443B6F04B7F4629F23DC

123. American Petroleum Institute, "Members," https://www.api.org/membership/members

124. Sharon Kelly, "'Time Is Running Out,' American Petroleum Institute Chief Said in 1965 Speech on Climate Change," *Desmog*, November 20, 2018, https://www.desmogblog.com/2018/11/20/american-petroleum-institute-1965-speech-climate-change-oil-gas

125. American Petroleum Institute, "Climate Change," https://www.api.org/oil-and-natural-gas/environment/climate-change

126. Note that a range of geoengineering techniques (large-scale technological interventions to counteract climate change) are under various stages of development. Not only does their safe application need to be carefully vetted, but also none is a substitute for continued successful climate mitigation efforts. In other words, geoengineering alone cannot solve our climate problems and the industry should not attempt to cover its climate footprints using engineering solutions. For background on climate engineering/geoengineering, see https://carnegieendowment.org/files/Understanding_Climate_Engineering_Web.pdf; https://carnegieendowment.org/files/Gordon_Advancing_Public_Disclosure_On_Climate_Engineering_June_2018.pdf

127. Oil and Gas Climate Initiative, "Climate Investments," https://oilandgasclimateinitiative.com/climate-investments/

128. Oil and Gas Climate Initiative, "At Work: Committed to Climate Action," September 2018, http://vklb72qn0p747zkmy18w0m8g.wpengine.netdna-cdn. com/wp-content/uploads/2018/09/OGCI_Report_2018.pdf

129. Climate Action 100+, "2019 Progress Report," September 2019, http://www.climateaction100.org/

130. ALEC is a conservative political group. See https://www.alec.org

131. American Fuels and Petrochemicals Manufacturers, "Member Directory," https://www.afpm.org/membership-directory

132. Deborah Gordon and Stephen D. Ziman, "Petroleum Companies Need a Credible Climate Plan," Carnegie Endowment for International Peace, November 2019, https://carnegieendowment.org/files/Gordon_Petro_Companies_Need_Climate_Plan_Nov2018.pdf

133. Sarah Kent, "Shell to Link Carbon Emission Targets to Executive Pay," *Wall Street Journal*, December 3, 2018, https://www.wsj.com/articles/ shell-to-link-carbon-emissions-targets-to-executives-pay-1543843441

134. For more information on science-based targets that help chart the pathway to net-zero emissions, see World Resources Institute, "What Is a Science-Based Target?," https://sciencebasedtargets.org/; Carbon Disclosure Project, World Resources Institute, United Nations Global Compact, World Wildlife Fund, "Science Based Targets," https://sciencebasedtargets.org/

135. Nicole Winfield and Frank Jordans, "Major Oil Companies Commit to Climate Pricing at Vatican," *AP News*, June 14, 2019, https://apnews.com/ 3460d18f3d414f65b9a70575a3080832#

136. "In-Situ Molecular Manipulation," Society of Petroleum Engineers, https:// www.spe.org/en/industry/in-situ-molecular-manipulation/; https://www.spe. org/industry/globalchallenges.php

137. Sami Alnuaim, "Circular Economy: A Sustainability Innovation and Solution for Oil, Gas, and Petrochemical Industries," *Journal of Petroleum Technology*, May 1, 2019, https://pubs.spe.org/en/jpt/jpt-article-detail/?art=5340#

138. Sabic, "Circular Economy," https://www.sabic.com/en/reports/sustainability-2019/innovation/circular-economy

139. Business Roundtable, "Business Roundtable Redefines the Purpose of a Corporation to Promote 'an Economy That Serves All Americans,'" August 19, 2019, https://opportunity.businessroundtable.org/ourcommitment/

140. The following nine oil and gas industry actors signed the Business Roundtable statement: BP, Chevron, ConocoPhillips, Exxon Mobil, Marathon Oil, Marathon Petroleum, Noble Energy, Phillips 66, and TC Energy. Two other industry-related petrochemical firms, Dow and LyondellBasell, also signed on even though they do not themselves produce, refine, or transport oil and gas. As such, only 6 percent (11 out of 184) signatories are from the oil and gas sector.

141. Milton Friedman, "The Social Responsibility of Business Is to Increase Its Profits," *New York Times Magazine*, September 13, 1970, https://www.

nytimes.com/1970/09/13/archives/a-friedman-doctrine-the-social-responsibility-of-business-is-to.html

142. Leonardo Maugeri, *The Age of Oil* (Westport, CT: Praeger Publishers, 2006).

143. The Economist, "Standard Ogre," December 23, 1999, https://www.economist.com/business/1999/12/23/standard-ogre

144. "The Climate Denial Machine: How the Fossil Fuel Industry Blocks Climate Action," Climate Reality Project, September 5, 2019, https://www.climaterealityproject.org/blog/climate-denial-machine-how-fossil-fuel-industry-blocks-climate-action; Shannon Hall, "Exxon Knew about Climate Change Almost 40 Years Ago," *Scientific American*, October 26, 2015, https://www.scientificamerican.com/article/exxon-knew-about-climate-change-almost-40-years-ago/; Also see https://exxonknew.org/

145. Noah Kaufman, Twitter posting, April 7, 2020, https://twitter.com/noahqk/status/1247841052154908673

146. Adele Peters, "Big Oil Is Cheap: Should the Government Take It Over?," *Fast Company*, March 27, 2020, https://www.fastcompany.com/90482315/big-oil-is-cheap-should-the-government-take-it-over

## CHAPTER 6

1. United Nations Climate Change, "Nationally Determined Contributions (NDCs)," https://unfccc.int/nationally-determined-contributions-ndcs; United Nations Framework Convention on Climate Change,"NDC Registry (Interim)," https://www4.unfccc.int/sites/NDCStaging/Pages/All.aspx

2. Note that all references to NDC percentages in this chapter are cited from the following article: Lewis C. King and Jeroen C. J. M. van den Bergh, "Normalisation of Paris Agreement NDCs to Enhance Transparency and Ambition," *Research Letters*, July 26, 2019, https://iopscience.iop.org/article/10.1088/1748-9326/ab1146. See Appendix D for assessment of whether each nation's GHGs are projected to increase or decrease between 2015 and 2030 under their submitted NDC.

3. The UN Framework Convention on Climate Change posts historic 2015 NDCs and 2020 updates on the NDC Registry; see https://www4.unfccc.int/sites/NDCStaging/Pages/All.aspx. To track which nations have updated their 2020 NDC submissions, see Climate Action Tracker, https://climateactiontracker.org/climate-target-update-tracker/

4. United Nations, "The Heat Is On: NDC Global Outlook Report 2019," September 2019, https://reliefweb.int/sites/reliefweb.int/files/resources/NDC_Outlook_Report_2019.pdf

5. World Meteorological Organization, "Carbon Dioxide Levels Continue at Record Levels, despite COVID-10 Lockdown," November 23, 2020, https://public.wmo.int/en/media/press-release/carbon-dioxide-levels-continue-record-levels-despite-covid-19-lockdown

6. Deborah Gordon, "Not All Oil Is Equal. As Economies Recover, Which Oils Should Stay in the Ground?," *Brink*, June 21, 2020, https://www.brinknews.com/not-all-oil-is-equal-which-types-should-oil-companies-turn-back-on-as-the-economy-recovers-paris-agreement-climate-change-post-covid/

7. Global Carbon Project, https://www.globalcarbonproject.org/index.htm

8. William Nordhaus, "Projections and Uncertainties about Climate Change in an Era of Minimal Climate Policies," *American Economic Journal: Economic Policy* 10, no. 3 (2018): 333–60, https://pubs.aeaweb.org/doi/pdfplus/10.1257/pol.20170046

9. Barry Rabe, ed., *Greenhouse Governance: Addressing Climate Change in America* (Washington, DC: Brookings Institution Press, 2010).

10. United Nations, "What Is the United Nations Framework Convention on Climate Change?," https://unfccc.int/process-and-meetings/the-convention/what-is-the-united-nations-framework-convention-on-climate-change

11. As of 2015, there were 197 parties (196 states and the European Union) to the UN Framework Convention on Climate Change. See https://unfccc.int/process-and-meetings/the-convention/status-of-ratification/status-of-ratification-of-the-convention. As the source of the largest share in past and present emissions, the onus is put on developed countries to lead the way. Industrialized countries fall into three categories: Annex I Parties that have committed to adopting national policies; Annex II Parties that are obliged to provide financial resources; and Annex B Parties that set emission reduction targets (2008–2012). These parties are joined by Non-Annex I parties that largely represent developing countries. See J. M. Allwood et al., Glossary. in: *Climate Change 2014: Mitigation of Climate Change, Contribution of Working Group III to the Fifth Assessment Report of the Intergovernmental Panel on Climate Change*, ed. O. Edenhofer et al. (Cambridge: Cambridge University Press, 2014); and United Nations Framework Convention on Climate Change, "Parties & Observers," https://unfccc.int/parties-observers. Numerous nongovernmental parties are admitted into the UNFCCC as observers. See United Nations Framework Convention on Climate Change, "Overview," https://unfccc.int/process-and-meetings/parties-non-party-stakeholders/non-party-stakeholders/overview

12. One hundred and ninety-seven parties have signed the UNFCCC, including all UN member states (193); the UN General Assembly observes the State of Palestine, UN nonmember states of Niue and the Cook Islands, and the European Union. United Nations, "Treaty Collection, United Nations Framework Convention on Climate Change," status as of January 19, 2020, https://treaties.un.org/Pages/ViewDetailsIII.aspx?src=IND&mtdsg_no=XXVII-7&chapter=27&Temp=mtdsg3&clang=_en. Note that the UN considers Taiwan to be represented by China. See US Office of the Historian, "All Countries," https://history.state.gov/countries/all

13. United Nations Framework Convention on Climate Change, "NDC Registry (Interim)," https://www4.unfccc.int/sites/NDCStaging/Pages/All.aspx

14. See https://www4.unfccc.int/sites/ndcstaging/Pages/Home.aspx

15. King and van den Bergh, "Normalisation of Paris Agreement NDCs to Enhance Transparency and Ambition."

16. "List of Non Updating Countries," CAT Climate Target Update Tracker, https://climateactiontracker.org/climate-target-update-tracker/list-non-updating-countries/; also see UNFCCC, "Latest Submissions," https://www4.unfccc.int/sites/NDCStaging/Pages/LatestSubmissions.aspx

17. Ibid.; King and van den Bergh, "Normalisation of Paris Agreement NDCs to Enhance Transparency and Ambition."

18. "2030 Climate Target Plan," European Commission, April 2021, https://ec.europa.eu/clima/policies/eu-climate-action/2030_ctp_en

19. Noemie Liprince-Ringuet, "Which Countries Will Step Up Climate Commitments in 2020? What We Know Now," World Resources Institute, November 21, 2019, https://www.wri.org/blog/2019/11/which-countries-will-step-climate-commitments-2020-what-we-know-now; European Union, "Committing to Climate-Neutrality by 2050: Commission Proposes European Climate Law and Consults on European Climate Pact," March 4, 2020,https://ec.europa.eu/commission/presscorner/detail/en/ip_20_335

20. The United Kingdom withdrew its membership from the European Union in January 2020. For a current list of EU members, see https://europa.eu/european-union/about-eu/countries_en

21. International Energy Agency, "Climate Change," https://www.iea.org/topics/climate-change

22. International Energy Agency, "The Oil and Gas Industry in Energy Transitions: Insights from IEA Analysis," January 2020, https://www.iea.org/reports/the-oil-and-gas-industry-in-energy-transitions

23. Equatorial Guinea and Republic of Congo have much smaller baseline emissions than the other OPEC nations.

24. King and van den Bergh, "Normalisation of Paris Agreement NDCs to Enhance Transparency and Ambition."

25. Ibid. Note that Chile has a relatively small 2015 GHG baseline of 3.6 gigatonnes of $CO_2$e.

26. The nonfounding member nations in OPEC (and the year they joined) include Libya (1962), the United Arab Emirates (1967), Algeria (1969), Nigeria (1971), Ecuador (joined 1973, suspended 1992, rejoined 2007), Gabon (joined 1975, terminated 1995, rejoined 2016), Angola (2007), Equatorial Guinea (2017), Congo (2018), Qatar (joined 1961, terminated 2019), and Indonesia (joined 1962, suspended 2009, reactivated then resuspended 2016). See Organization of Petroleum Exporting Countries, "OPEC Member Countries," https://www.opec.org/opec_web/en/about_us/25.htm; and Organization of Petroleum Exporting Countries, "Our Mission," https://www.opec.org/opec_web/en/about_us/23.htm

27. The other OPEC-plus nations include Azerbaijan, Bahrain, Brunei, Kazakhstan, Malaysia, Mexico, Oman, South Sudan, and Sudan. In 2020, together with the OPEC nations, they attempted but failed to freeze production to prop up falling oil prices. See Brian Wingfield et al., "New Decade, New OPEC Oil Curbs, Same Mixed Results," *Bloomberg*, February 24, 2020, https://www.bloomberg.com/graphics/opec-production-targets/

28. "OPEC: Energy, Climate Change and Sustainable Development," *OPEC Bulletin*, April 2019, https://www.opec.org/opec_web/static_files_project/media/downloads/publications/OB042019.pdf

29. "Member States of the APEC Asia-Pacific Economic Cooperation," WorldData.info, https://www.worlddata.info/alliances/apec-asia-pacific-economic-cooperation.php

30. APEC also includes Hong Kong and Taiwan, which both joined in 1991 and are not considered as countries under UNFCCC. See "APEC Members," https://www.apec.org/About-Us/About-APEC/Member-Economies

31. Arthur D. Little, "Dead End Approaches for High-Sulfur Fuels," April 2018, https://www.adlittle.com/sites/default/files/viewpoints/adl_dead_end_approaches_for_high-sulfur_fuels.compressed.pdf

32. Extractive Industries Transparency Initiative, "Global Factsheet," August 2020, https://eiti.org/files/documents/eiti_factsheet_en_08.2020.pdf

33. Extractive Industries Transparency Initiative, "EITI Launches 2019 EITI Standard," June 18, 2020, https://eiti.org/news/eiti-launches-2019-eiti-standard#

34. It is unclear whether the omission of oil and gas supply-side emission reductions may also be an indication of chronic GHG undercounting in national inventories.

35. Cleo Verkuijl, Natalie Jones, and Michael Lazarus, "Untapped Ambition: Addressing Fossil Fuel Production through NDCs and LEDS," Stockholm Environmental Institute (SEI), June 2019, https://www.sei.org/wp-content/uploads/2019/06/untapped-ambition-addressing-fossil-fuel-production-through-ndcs-and-leds.pdf (see Tables 2 and 3).

36. In its updated NDC submitted on April 21, 2021, the United States mentions "natural gas" one time: "To address methane, the United States will update standards and invest in plugging leaks from wells and mines and across the natural gas distribution infrastructure." See https://www4.unfccc.int/sites/ndcstaging/PublishedDocuments/United%20States%20of%20America%20First/United%20States%20NDC%20April%202021%202021%20Final.pdf

37. Author's calculations using data cited in Figures 6.2 and 6.3.

38. Methane is the main component of natural gas, making it difficult to increase gas use without increasing methane emissions, which accelerates global warming when gas leaks into the atmosphere.

39. As of September 2020, 170 nations have not updated their NDC targets. Many are not planning to do so. To track which countries have provided updates, see

https://climateactiontracker.org/climate-target-update-tracker/ and https://www4.unfccc.int/sites/ndcstaging/Pages/LatestSubmissions.aspx

40. International Energy Agency, *World Energy Outlook 2018*, Table 3.8, November 2018.
41. Organization for Economic Co-operation and Development, "Investing in Climate, Investing in Growth," May 23, 2017, https://www.oecd.org/env/investing-in-climate-investing-in-growth-9789264273528-en.htm
42. Author's calculations using sources listed in Tables 6.1 to 6.5.
43. Rabe, ed., *Greenhouse Governance*.
44. Government of Canada, "Canada's Action on Climate Change," https://www.canada.ca/en/services/environment/weather/climatechange/climate-action.html
45. Government of Canada, "Regulations Respecting the Reduction in the Release of Methane and Certain Volatile Organic Compounds (Upstream Oil and Gas Sector)," https://pollution-waste.canada.ca/environmental-protection-registry/regulations/view?Id=146
46. Rattan Mall, "Positioning Canada as Supplier of World's Cleanest Gas," *Voice Online*, August 29, 2019, https://www.voiceonline.com/positioning-canada-as-supplier-of-worlds-cleanest-natural-gas/
47. David Coglon, "How Nanotechnology Could Reduce Oil Sands Carbon Footprint," *Energy Examined*, Canadian Oil and Natural Gas Producers, https://context.capp.ca/energy-matters/2018/dyk_university-team-and-nanotechnology
48. Norskolje&gass, "Norway's Petroleum History," October 19, 2017, https://www.norskoljeoggass.no/en/about-us/oljehistorien/
49. "Norway Supports Efforts to Eliminate Gas Flaring by 2030," Government.no, April 17, 2015, https://www.regjeringen.no/en/aktuelt/eliminate_flaring/id2407055/#
50. Richard Milne, "Oil and the Battle for Norway's Soul," *Financial Times*, July 27, 2017, https://www.ft.com/content/c2dad93c-7192-11e7-aca6-c6bd07df1a3c
51. In 2001, Statoil became listed on the Oslo and New York stock exchanges, partly privatizing the company. In 2018, Statoil rebranded itself as Equinor, https://www.equinor.com/
52. In 2018, nearly one-third of Equinor's revenues came from its oil and gas operations in Europe, Africa, and North America. See Equinor, "Annual Report 2018," https://www.equinor.com/en/investors.html#annual-reports
53. Moves are afoot to sell some $6 billion (USD) in oil and gas holdings in Norway's sovereign wealth fund, although it may retain shares in integrated majors, Shell, Exxon-Mobil, BP, and Total. See Terje Solsvik, "Norway Sovereign Wealth Fund to Divest Oil Explorers, Keep Refiners," *Reuters*, October 1, 2019, https://www.reuters.com/article/us-norway-swf-oil/norway-sovereign-wealth-fund-to-divest-oil-explorers-keep-refiners-idUSKBN1WG4R9
54. Harry Cockburn, "Norway Refuses to Drill for Billions of Barrels of Oil in Arctic, Leaving 'Whole Industry Surprised and Disappointed,'" April 9, 2019,

https://www.independent.co.uk/environment/norway-oil-drilling-arctic-ban-labor-party-unions-a8861171.html

55. *Canberra Times*, "Australia Becomes the Largest Liquefied Natural Gas Exporter in the World," January 7, 2020, https://www.canberratimes.com.au/story/6568957/australia-now-the-worlds-largest-natural-gas-exporter/?cs=14231

56. Somini Sengupta, "Climate Change Policy Toppled Australia's Leader. Here's What it Means for Others," *New York Times*, August 24, 2018, https://www.nytimes.com/2018/08/24/climate/australia-climate-change.html; King and van den Bergh, "Normalisation of Paris Agreement NDCs to Enhance Transparency and Ambition."

57. Chris Barrie, "Climate Change Poses a 'Direct Threat' to Australia's National Security. It Must Be a Political Priority," *The Conversation*, October 7, 2019, http://theconversation.com/climate-change-poses-a-direct-threat-to-australias-national-security-it-must-be-a-political-priority-123264

58. Judith Lewis Mernit, "Gavin Newsom and Big Oil: It's Complicated," *Salon*, October 6, 2020, https://www.salon.com/2020/10/06/gavin-newsom-and-big-oil--its-complicated_partner/

59. California Air Resources Board, Zero-Emission Vehicle Program, https://ww2.arb.ca.gov/our-work/programs/zero-emission-vehicle-program

60. California Air Resources Board, "Low Carbon Fuel Standard," https://ww2.arb.ca.gov/our-work/programs/low-carbon-fuel-standard

61. Brad Plumer and Jill Cowan, "A Scorched California Moves to Halt Sales of New Gas-Powered Cars by 2035," *New York Times*, September 23, 2020, https://www.nytimes.com/2020/09/23/climate/california-ban-gas-cars.html

62. "California's Oil Refineries," California Energy Commission, https://www.energy.ca.gov/data-reports/energy-almanac/californias-petroleum-market/californias-oil-refineries

63. Cal Matters, "California's 'Hydrogen Highway' Never Happened. Could 2020 Change That?," https://calmatters.org/environment/2020/01/why-california-hydrogen-cars-2020/

64. The US oil and gas production and refining trifecta holds since 2014 and extends annually through 2019. For times series data on oil (petroleum and other liquids) production, see https://www.eia.gov/international/data/world/petroleum-and-other-liquids/annual-petroleum-and-other-liquids-production; for data on gas production, see https://www.eia.gov/international/data/world/petroleum-and-other-liquids/annual-petroleum-and-other-liquids-production; for benchmarking refined oil volumes refer to refined petroleum products production, https://www.eia.gov/international/data/world/petroleum-and-other-liquids/annual-petroleum-and-other-liquids-production

65. US Energy Information Administration, "Petroleum & Other Liquids," https://www.eia.gov/dnav/pet/pet_sum_sndw_dcus_nus_w.htm. Note that while

volumes of US oil imports and exports vary from month to month, the relative reliance on oil imports remained relatively constant in 2019, dipping only slightly to 37 percent in end September 2020 during the coronavirus pandemic, while the share of oil exports rose to 33 percent in end September 2020, likely due to a reduction in oil volumes refined.

66. US Energy Information Administration, "Natural Gas Monthly," https://www.eia.gov/ naturalgas/monthly/. Most of the excess gas that the United States produces but does not consume is exported, but consequential volumes are also disposed of (flared and vented). Note that the shares of gas the United States produced but did not consume as well as the share it exported remained relatively constant in 2020 compared to 2019.

67. For more on why a meaningful conversation is lacking on US climate policymaking, see Rabe, ed., *Greenhouse Governance*.

68. National Aeronautics and Space Administration, "Carbon Monitoring System," https://carbon.nasa.gov/; "Biden Signs Bill to Restore Regulations on Climate-Warming Methane Emissions," National Public Radio, June 30, 2021, https://www.npr.org/2021/04/28/991635101/congress-votes-to-restore-regulations-on-climate-warming-methane-emissions

69. Hiroko Tabuchi, "Oil and Gas May Be a Far Bigger Climate Threat Than We Knew," *New York Times*, February 19, 2020, https://www.nytimes.com/2020/02/19/climate/methane-flaring-oil-emissions.html

70. "Your Texas Economy," Federal Reserve Bank of Dallas, January 31, 2020, https://www.dallasfed.org/-/media/Documents/research/econdata/texaseconomy.pdf

71. New Mexico Executive Order 2019-003, "Addressing Climate Change and Energy Waste Prevention," January 29, 2019, https://www.governor.state.nm.us/wp-content/uploads/2019/01/EO_2019-003.pdf

72. Judith Kohler, "Oil-Gas Law Is a Paradigm Shift, Co-Sponsor Tells Those Writing New Rules," *Denver Post*, September 25, 2019, https://www.denverpost.com/2019/09/25/colorado-oil-gas-commission-181-climate-change/#; Dan Grossman, "Colorado's Landmark Methane Rules Raise Bar for Federal Climate Action," Environmental Defense Fund, March 18, 2021, https://blogs.edf.org/energyexchange/2021/03/18/colorados-landmark-methane-rules-raise-bar-for-federal-climate-action/

73. David Taylor et al., "Black to Green: A BIG Opportunity for North Dakota's Oil and Gas Producers," *Science* 242 (May 13, 2019), https://www.sciencedirect.com/science/article/pii/S0306261919305665

74. Arturo Balderas Torres, Priscila Lazaro Vargas, and Jouni Paavola, "The Systemic and Governmental Agendas in Presidential Attention to Climate Change in Mexico 1994–2018," *Nature Communications* 11 (January 23, 2020): 455, https://www.nature.com/articles/s41467-019-14048-7

75. Mauricio Savarese, "Brazil's Role Questioned after UN Global Warming Meeting," *AP News*, December 16, 2019, https://apnews.com/62ed107987486bf96537b88a88edf6ef

76. See data sources in Figure 6.3.

77. Deloitte, "Brazilian Pre-Salt: Petrobras Is Betting Its Future on Pre-Salt Assets, Promising to Rise Its Production," https://www2.deloitte.com/br/en/pages/energy-and-resources/upstream-guide/articles/pre-salt-brazil.html

78. S&P Platts, "Brazilian Biodiesel, Ethanol Output Advance on Year in October," December 3, 2019, https://www.spglobal.com/platts/en/market-insights/latest-news/agriculture/120319-brazilian-biodiesel-ethanol-output-advance-on-year-in-october

79. Falih Hassan and Elian Peltier, "Scorching Temperatures Bake Middle East Amid Eid al-Adha Celebrations," *New York Times*, July 31, 2020, https://www.nytimes.com/2020/07/31/world/middleeast/Middle-East-heat-wave.html

80. National Aeronautics and Space Administration, "Earth Observatory," https://earthobservatory.nasa.gov/world-of-change/global-temperatures

81. Marc Lynch and Laurie Brand, "Refugees and Displacement in the Middle East," Carnegie Endowment for International Peace, March 29, 2017, https://carnegieendowment.org/2017/03/29/refugees-and-displacement-in-middle-east-pub-68479

82. Lee Fang and Sharon Lerner, "Saudi Arabia Denies Its Key Role in Climate Change Even as It Prepares for the Worst," *The Intercept*, September 18, 2019 https://theintercept.com/2019/09/18/saudi-arabia-aramco-oil-climate-change/

83. Climate Change Performance Index, "Saudi Arabia," https://www.climate-change-performance-index.org/country/saudi-arabia; Note: the Climate Change Performance Index is an NGO collaboration by GermanWatch, https://www.germanwatch.org/en/about; New Climate Institute, https://newclimate.org/; and Climate Action Network, http://www.climatenetwork.org/

84. King and van den Bergh, "Normalisation of Paris Agreement NDCs to Enhance Transparency and Ambition."

85. Chloe Farand, "UN Report on 1.5C Blocked from Climate Talks after Saudi Arabia Disputes Science," *Climate Home News*, June 27, 2019, https://www.climatechangenews.com/2019/06/27/un-report-1-5c-blocked-climate-talks-saudi-arabia-disputes-science/

86. Mohammad Masnadi et al., "Global Carbon Intensity of Crude Oil Production," *Science*, August 31, 2018, https://science.sciencemag.org/content/361/6405/851.summary; King and van den Bergh, "Normalisation of Paris Agreement NDCs to Enhance Transparency and Ambition." Note that the same issues affect the United States, Russia, and other major oil and gas producers whose sheer production volumes translate into large GHG emissions even when each barrel produced has a lower GHG intensity.

87. Saudi Aramco, "Addressing the Climate Challenge," https://www.saudiaramco.com/en/making-a-difference/planet/climateinitiative

88. Frank Kane, "Aramco CEO in Call for 'Ultra Clean Energy,'" *Arab News*, September 10, 2019, https://www.arabnews.com/node/1552781/business-economy

89. King and van den Bergh, "Normalisation of Paris Agreement NDCs to Enhance Transparency and Ambition."

90. Steven Mufson, "Facing Unbearable Heat, Qatar Has Begun to Air-Condition the Outdoors," *Washington Post*, October 16, 2019, https://www.washingtonpost.com/graphics/2019/world/climate-environment/climate-change-qatar-air-conditioning-outdoors/

91. Tsvetana Paraskova, "Qatar Builds the Biggest Carbon Capture Plant in the Middle East," Oilprice. com, October 8, 2019, https://oilprice.com/Latest-Energy-News/World-News/Qatar-Builds-The-Biggest-Carbon-Capture-Plant-In-The-Middle-East.html

92. Sania Aziz Rahman, "Is UAE Leading the Way for Concentrated Solar Power in the Gulf?," *Eco-Business*, May 10, 2019, https://www.eco-business.com/news/is-uae-leading-the-way-for-concentrated-solar-power-in-the-gulf/

93. Anmar Frangoul, "Oil Field in Oman Set to Get a Solar Plant to Aid Its Crude Production," *CNBC*, November 14, 2018, https://www.cnbc.com/2018/11/14/oilfield-in-oman-set-to-get-a-solar-plant-to-aid-its-crude-production.html

94. Nima Khorrami Assl, "Oman's Economic Ambitions," Sada Carnegie Endowment for International Peace, December 14, 2018, https://carnegieendowment.org/sada/77972

95. Nick Butler, "Kuwait's Efforts on Green Energy Fails to Match Its Best Intentions," *Financial Times*, September 9, 2019, https://www.ft.com/content/bdcee818-c35f-11e9-ae6e-a26d1d0455f4

96. Clifford Krauss, "Canada Oil-Sands Plan Collapses Over Politics and Economics," *New York Times*, February 24, 2020, https://www.nytimes.com/2020/02/24/business/energy-environment/frontier-oil-sands-canada.html

97. Matthew Johnson et al., "Comparisons of Airborne Methane Measurements and Inventory Estimates of Methane Emissions in the Alberta Upstream Oil and Gas Sector," *Environment Science and Technology* 51, no. 21 (October 17, 2017), https://pubs.acs.org/doi/10.1021/acs.est.7b03525

98. Janet French, "Alberta, Ottawa Reach Preliminary Deal on Methane Emissions Regulation," *CBC*, May 13, 2020, https://www.cbc.ca/news/canada/edmonton/alberta-ottawa-reach-preliminary-deal-on-methane-emissions-regulation-1.5568924

99. Benjamin Israel et al., "The Oilsands in a Carbon-Constrained Canada," Pembina Institute, February 19, 2020, https://www.pembina.org/pub/oilsands-carbon-constrained-canada

100. See data sources in Figure 6.3.

101. In 2019, I met with representatives from the Japan Research Network, which is working with Japan's Ministry of Economy, Trade, and Industry (METI) and the Mitsubishi Research Institute to study GHG emission reduction measures in the oil and gas sector.

102. Verkuijl et al., "Untapped Ambition: Addressing Fossil Fuel Production through NDCs and LEDS."

103. Peter Hansen, "Nigeria News Brief and Action Alert," Climate Scorecard, December 20, 2020, https://www.climatescorecard.org/2020/12/nigeria-has-experienced-a-271-increase-in-greenhouse-gas-emissions-since-1990/

104. See data sources in Figure 6.3.

105. "Russia's New Arctic Project Will Be Biggest in Global Oil - Rosneft," *Barents Observer*, February 14, 2020, https://www.themoscowtimes.com/2020/02/14/russias-new-arctic-project-will-be-biggest-in-global-oil-rosneft-a69294

106. Neven Valev, "Political Stability – Country Ratings," theglobaleconomy.com, https://www.theglobaleconomy.com/rankings/wb_political_stability/

107. King and van den Bergh, "Normalisation of Paris Agreement NDCs to Enhance Transparency and Ambition."

108. "Eyes on Nigeria: Technical Report," American Association for the Advancement of Science, 2011, https://www.aaas.org/sites/default/files/AAAS_EyesOnNigeria_Technical_Report.pdf

109. Berkeley Earth, "List of States and Provinces," http://berkeleyearth.lbl.gov/state-list/. Note that Canada's Nunavut and Northwest Territories (which abut Alaska's Prudhoe Bay oil fields) also have the dubious distinction of temperatures that have risen over 4.5 degrees Celsius.

110. "Russia's New Arctic Project Will Be Biggest in Global Oil – Rosneft," *Barents Observer*, February 14, 2020, https://www.themoscowtimes.com/2020/02/14/russias-new-arctic-project-will-be-biggest-in-global-oil-rosneft-a69294

111. For more information on Vostok's plans see Caepla, "Russia Planning Largest Oil Project in World History: $155B Investment & 100,000 Jobs," February 28, 2020, https://www.caepla.org/russia_planning_largest_oil_project_in_world_history_155b_investment_100_000_jobs; Olga Yagova and Aaron Sheldrick, "Russia's Rosneft Seeks Japanese Investors for Giant Vostok Oil Development," *Reuters*, December 11, 2019, https://www.reuters.com/article/russia-oil-japan/russias-rosneft-seeks-japanese-investors-for-giant-vostok-oil-development-idUSL4N28L1QF

112. See data sources in Figure 6.3.

113. King and van den Bergh, "Normalisation of Paris Agreement NDCs to Enhance Transparency and Ambition."

114. Steven Lee Myers, "Xi Sets 40-Year Target in Ambitious Pledge on Carbon Neutrality," *New York Times*, September 23, 2020, https://www.nytimes.com/2020/09/23/world/asia/china-climate-change.html

115. Joanna Slater, "Can India Chart a Low-Carbon Future? The World Might Depend on It," *Washington Post*, June 12, 2020, https://www.washingtonpost.com/climate-solutions/2020/06/12/india-emissions-climate/?arc404=true

116. Daniel Kaufmann and Rebecca Iwerks, "Close to Home: The Critical Importance of Subnational Governance in Oil, Gas, and Mining," Brookings Institute,

September 27, 2016, https://www.brookings.edu/blog/future-development/ 2016/09/27/close-to-home-the-critical-importance-of-subnational-governance- in-oil-gas-and-mining/

117. C40 Dashboard, https://www.c40.org/other/gpc-dashboard

118. Somini Sengupta, "Earth Is Overheating. Millions Are Already Feeling the Pain," *New York Times*, August 7, 2020, https://www.nytimes.com/interactive/2020/ 08/06/climate/climate-change-inequality-heat.html

119. United Nations, "Polar Vortex Responsible for Texas Deep Freeze, Warm Arctic Temperatures," *United Nations News*, March 9, 2021, https://news.un.org/ en/story/2021/03/1086752; Chris Stipes, "Harris County Took the Brunt of February Freeze," University of Houston, April 8, 2021, https://uh.edu/news- events/stories/2021/april-2021/04082021-hobby-winter-hc.php

120. National Weather Service, "Houston Climate Data," https://www.weather. gov/hgx/climate_iah. Note that reported air temperatures do not account for Houston's high humidity that factors into the heat index, driving Houston's heat well over 100 degrees Fahrenheit on a typical summer day. For calculation see https://www.wpc.ncep.noaa.gov/html/heatindex.shtml

121. John Nova Lomax, "The Evolution of the Energy Capital of the World," *Texas Monthly*, February 14, 2017, https://www.texasmonthly.com/energy/ evolution-energy-capital-world/

122. "Top 30 U.S. Ports," Logistics Management, https://www.logisticsmgmt.com/ article/top_30_u.s._ports_trade_tensions_determine_where_cargo_goes_next

123. Katie Watkins, "Report: Oil & Gas Industry Set to Release an Extra 220 Million Tons of Greenhouse Gases by 2025," *Houston Public Media*, January 8, 2020, https://www.houstonpublicmedia.org/articles/news/energy-environment/ 2020/01/08/356799/report-oil-gas-industry-set-to-release-an-extra-220-million- tons-of-greenhouse-gases-by-2025/

124. In a 1981 article, *Texas Monthly* dubbed nearby Port Neches the "Cancer Belt." See https://www.texasmonthly.com/articles/the-cancer-belt/

125. A Google search on October 8, 2020, for "chemical plants in Beaumont Tx" listed twenty-four facilities, and a Yellow Pages search on March 6, 2020, returned simi- lar results; see https://www.yellowpages.com/beaumont-tx/chemical-plants

126. Environmental Integrity Project, "31 New or Expanded Petrochemical Plants Approved in Hurricane Zone along TX and LA Coast," September 26, 2018, https:// environmentalintegrity.org/news/31-new-or-expanded-petrochemical-plants/

127. Houston Climate Action Plan, Draft, July 29, 2019, http://greenhoustontx.gov/ climateactionplan/2019-DRAFT-CAP.pdf

128. ConocoPhillips, "Climate Change Position," http://www.conocophillips.com/ sustainability/integrating-sustainability/sustainable-development-governance/ policies-positions/climate-change-position/

129. Ben Steverman, "The Energy Elite Have Started Listening to Their Enemy No. 1 in Houston," *Bloomberg Green*, March 3, 2020, https://www.bloomberg.com/

news/features/2020-03-03/texas-s-energy-elite-are-listening-to-climate-lawyer-blackburn?sref=dcitGkK3&utm_source=url_link

130. Stateside Staff, "A Three-Story Pile of Pet Coke Next to the Detroit River, Where Will It Go?," *Michigan Public Radio*, May 22, 2013, https://www.michiganradio.org/post/three-story-pile-pet-coke-next-detroit-river-where-will-it-go

131. Chad Livengood, "Detroit Denies Marathon's Request to Keep Pet Coke Piles Uncovered," *Crain's Detroit Business*, April 22, 2019, https://www.crainsdetroit.com/energy/detroit-denies-marathons-request-keep-pet-coke-piles-uncovered

132. Keith Matheny, "Marathon Wants to Keep Storing Pet Coke Uncovered near Detroit River," *Detroit Free Press*, February 15, 2019, https://www.freep.com/story/news/local/michigan/wayne/2019/02/15/marathon-pet-coke-petroleum-detroit-river-fugitive-dust/2881924002/

133. Relinde Van Loo, "The Rise and Fall of the Dutch Groningen Gas Field," E&C Consultants for the *EU Blog*, August 15, 2018, https://www.eecc.eu/blog/groningen-gas-field

134. Stanley Reed, "Earthquakes Are Jolting the Netherlands. Gas Drilling Is to Blame," *New York Times*, October 24, 2019, https://www.nytimes.com/2019/10/24/business/energy-environment/netherlands-gas-earthquakes.html#

135. Benjamin Storrow, @bstorrow Tweet, Twitter, May 28, 2020, 8:21 AM, https://twitter.com/bstorrow/status/1265981622316302337

136. "US Government Role in Shale Gas Fracking History: A Response to Our Critics," Breakthrough Institute, March 1, 2012, https://thebreakthrough.org/issues/energy/us-government-role-in-shale-gas-fracking-history-a-response-to-our-critics

137. Researchers in government from Norway, Chile, Canada, New Zealand, Germany, India, and others are also involved. For examples of government entities involved see National Energy Technology Laboratory, https://netl.doe.gov/oil-gas/gas-hydrates; US Department of Energy, Office of Oil and Gas, https://www.energy.gov/fe/science-innovation/oil-gas-research; US Geological Survey,https://www.usgs.gov/centers/whcmsc/science/us-geological-survey-gas-hydrates-project?qt-science_center_objects=0#qt-science_center_objects; and Hawaii Natural Energy Institute/US Naval Research Laboratory,https://www.hnei.hawaii.edu/research/alternative-fuels/methane-hydrates/overview

138. Pacific Northwest National Laboratory, "Algae to Crude Oil: Million-Year Natural Process Takes Minutes in the Lab," December 17, 2013, https://www.pnnl.gov/news/release.aspx?id=1029

139. For examples see National Aeronautics and Space Administration Carbon Monitoring System, https://carbon.nasa.gov/objectives.html?; Japan GOSAT Project, http://www.gosat.nies.go.jp/en/; the Netherlands TROPOMI, http://www.tropomi.eu/

140. The local government of the German state of North Rhine-Westphalia is also involved. See David Weston, "German-Dutch Hydrogen Feasibility Study Launched," January 30, 2020, https://www.windpowermonthly.com/article/1672522/german-dutch-hydrogen-feasibility-study-launched

141. Sarah E. Baker et al., "Getting to Neutral: Options for Negative Carbon Emissions in California," January, 2020, Lawrence Livermore National Laboratory, LLNL-TR-796100, https://www-gs.llnl.gov/content/assets/docs/energy/Getting_to_Neutral.pdf

142. Joe Biden, https://joebiden.com/racial-economic-equity/

143. "Global Research and Development Expenditures Fact Sheet," Congressional Research Service, April 29, 2020, https://fas.org/sgp/crs/misc/R44283.pdf. See Table 3 for R&D as a share of purchasing power parity (PPP).

144. James Pethokoukis, "US Federal Research Spending Is at a 60-Year Low. Should We Be Concerned?," American Enterprise Institute, May 11, 2020, https://www.aei.org/economics/us-federal-research-spending-is-at-a-60-year-low-should-we-be-concerned/. See Exhibit 1.

145. US Department of Energy, "National Laboratories," https://www.energy.gov/national-laboratories

146. Adrienne Blume, "IRPC '19: Keynote Speakers Tackle Shifts in HPI Landscape," Hydrocarbon Processing, September 25, 2019, https://www.hydrocarbonprocessing.com/conference-news/2019/09/irpc-19-keynote-speakers-tackle-shifts-in-hpi-landscape. In particular, see Madhav Acharya's (ARPA-e) comments.

147. See President Joe Biden's climate plan: https://joebiden.com/climate-plan/#

148. International Energy Agency, "World Energy Investment 2020," May 2020, https://www.iea.org/reports/world-energy-investment-2020/rd-and-technology-innovation

149. Francis Bator, "The Anatomy of Market Failure," *Quarterly Journal of Economics* 72 (August 1958), https://academic.oup.com/qje/article-abstract/72/3/351/1859737?redirectedFrom=fulltext

### CHAPTER 7

1. The Bill Lane Center for the American West, https://west.stanford.edu/events/why-are-california-s-oils-some-world-s-dirtiest-panel-oil-climate-index-team

2. Judith Lewis Mernit, "Gavin Newsom and Big Oil: It's Complicated," *Salon*, October 6, 2020, https://www.salon.com/2020/10/06/gavin-newsom-and-big-oil--its-complicated_partner/

3. California Legislative Information, "AB-3217: Greenhouse Gases: Crude Oil," http://leginfo.legislature.ca.gov/faces/billTextClient.xhtml?bill_id=201920200AB3217; Ann Alexander, "Know Your Oil Bill: Ignorance about CA Crude Is Not Bliss," February 25, 2020, https://www.nrdc.org/experts/ann-alexander/know-your-oil-bill-ignorance-about-ca-crude-not-bliss

4. Note that the California legislation follows from similar federal bills: H.R. 6082 and H.R. 3286, introduced by Democratic Representative Jared Huffman (California). See Jared Huffman, "Rep. Huffman Introduces Energy Transparency Legislation," July 19, 2017, https://huffman.house.gov/media-center/press-releases/rep-huffman-introduces-energy-transparency-legislation-know-your-oil-act-and-carbon-pollution-transparency-act; H.R. 6082, 114th Congress (2015–2016), https://www.govtrack.us/congress/bills/114/hr6082; H.R. 3285, 115th Congress (2017–2018), https://www.congress.gov/bill/115th-congress/house-bill/3286?s=1&r=82

5. Even those who currently work in (or are retired from) government and industry also qualify as part of civil society in their personal lives. For a discussion on civil society actors, see Paul Hawken, "The Movement with No Name," *Tricycle*, Fall 2008, https://tricycle.org/magazine/movement-no-name/

6. Stephen Campbell, UMI Fund, personal communication in Zoom meeting, April 28, 2020; see https://www.umi-center.com/about

7. "America's Energy Choices: Investing in a Strong Economy and a Clean Environment," Alliance to Save Energy, American Council for an Energy Efficient Economy, Natural Resources Defense Council, and the Union of Concerned Scientists, 1991.

8. International Energy Agency, "The Oil and Gas Industry in Energy Transitions," 2020, https://webstore.iea.org/download/direct/2935

9. The first wave of environmentalism reportedly dates back to the early 1900s and the creation of national parks, while the second was set off by Earth Day in 1970. See Noel Perrin, "'Think Globally, Act Locally,'" *New York Times*, January 24, 1993, https://timesmachine.nytimes.com/timesmachine/1993/01/24/issue.html

10. David Schlosberg and Roman Coles, "The New Environmentalism of Everyday Life: Sustainability, Material Flows and Movements," *Contemporary Political Theory* 15 (2016): 160–81, https://doi.org/10.1057/cpt.2015.34

11. See data sources in Table 7.1.

12. David Brancaccio and Ali Oshinskie, "Ralph Nader Wants You to Wear a Seat Belt," *NPR Marketplace*, September 28, 2018, https://www.marketplace.org/2018/09/28/ralph-nader-wants-you-wear-your-seat-belt/

13. Patrick McGee, "How VW's Cheating on Emissions Was Exposed," *Financial Times*, January 11, 2017, https://www.ft.com/content/103dbe6a-d7a6-11e6-944b-e7eb37a6aa8e

14. For examples of other NGOs involved in utility sector work, see US Environmental Protection Agency, "Green Power Partnership," https://www.epa.gov/green-power/green-power-partnership-related-programs-organizations

15. Rocky Mountain Institute, "Reinventing Fire: Industry," 2011, https://rmi.org/insight/reinventing-fire-industry/

16. The IEA is an "autonomous agency" composed of member countries and associated with the Organisation for Economic Co-operation and Development. But it is not a

government entity. In assessing countries in chapter 6, IEA membership is sorted and discussed. It is covered in this chapter as a quasi-NGO in terms of the function the Energy Information Administration (EIA) serves informing civil society.

17. Frank Newport, "Americans Want Government to Do More on Environment," March 29, 2018, https://news.gallup.com/poll/232007/americans-want-government-more-environment.aspx

18. For OCI+ outreach examples see Oil Climate Project website, https://www.oil-climateproject.com/presentations-and-seminars. And previous outreach efforts in the press and through events are archived at the Carnegie Endowment for International Peace, "Oil Initiative," https://carnegieendowment.org/specialprojects/carnegieoilinitiative/all/1253?lang=en&pageOn=1

19. Coursehero; see https://www.coursehero.com/file/14653096/unconventional-oil/. Also see Wikipedia, "Unconventional Oil," https://en.wikipedia.org/wiki/Unconventional_oil

20. For event recording and presentation materials, see "Oil and Climate Change in the Twenty-First Century," Panel presentation at COP21, Paris, France, December 11, 2015, https://carnegieendowment.org/2015/12/11/oil-and-climate-change-in-twenty-first-century-event-5054

21. Reddit AMAs with hundreds of Q&A interactions were conducted in December 2015 and December 2016; see https://www.reddit.com/r/geopolitics/comments/3wsblg/we_are_four_climate_and_energy_experts_here_to/; https://www.reddit.com/r/science/comments/5glqrw/science_ama_series_were_a_team_of_researchers/

22. Oil Climate Project, https://www.oilclimateproject.com/

23. The IPCC was created in 1988 by the United Nations and World Meteorological Organization. See https://www.ipcc.ch/

24. Intergovernmental Panel on Climate Change, "Global Warming of 1.5 °C," October 6, 2018, https://www.ipcc.ch/sr15/. There are hundreds of stories; for examples see World Resources Institute, "8 Things You Need to Know about the IPCC 1.5°C Report," October 7, 2018, https://www.wri.org/blog/2018/10/8-things-you-need-know-about-ipcc-15-c-report; Union of Concerned Scientists, "The IPCC Gets Real about the 1.5°C Target," October 8, 2018, https://blog.ucsusa.org/peter-frumhoff/the-ipcc-gets-real-about-the-1-5c-target; and Greenpeace, "IPCC Report Points to Stark Choices Ahead, but Reason for Hope and Action," October 8, 2018, https://www.greenpeace.org/international/press-release/18839/ipcc-report-points-to-stark-choices-ahead-but-reason-for-hope-and-action-greenpeace/

25. Intergovernmental Panel on Climate Change, "Characteristics of Four Illustrative Model Pathways," October 2018, https://www.ipcc.ch/sr15/graphics/#cid_6333

26. National Academy of Sciences, "Automotive Fuel Economy: How Far Can WeGo?," 1992, http://www.nap.edu/catalog/1806/automotive-fuel-economy-how-far-can-we-go

27. NAS, "Partnership for a New Generation of Vehicles," 1998, https://www.nap.edu/catalog/6127/review-of-the-research-program-of-the-partnership-for-a-new-generation-of-vehicles

28. National Academy of Sciences, "Gulf Research Program Awards $7.25 Million to Eight Projects Working to Advance Safety Culture in the Offshore Oil and Gas Industry," January 16, 2020, https://www.nationalacademies.org/news/2020/01/gulf-research-program-awards-7-25-million-to-eight-projects-working-to-advance-safety-culture-in-the-offshore-oil-and-gas-industry

29. For congressional testimony transcripts see https://carnegieendowment.org/files/2014-04-02-Debbie_Gordon_testimony.pdf and https://carnegieendowment.org/files/Deborah_Gordon_subcommittee_testimony_-_12_11_12_SS_NP_CG.pdf

30. "Federal Crowdsourcing Webinar Series, Episode 1: Citizen Science," April 2019, https://www.citizenscience.gov/about/#. Also see "Orphan Well Location Survey," https://www.citizenscience.gov/catalog/499/# and https://www.citizenscience.gov/catalog/496/#

31. US House of Representatives, "Crowdsourcing and Citizen Science," 15 USC 3724, 2017; See laws in effect on June 3, 2020, https://uscode.house.gov/view.xhtml?req=granuleid:USC-prelim-title15-section3724)&num=0&edition=prelim

32. Sam Sanders, "Some Google Street View Cars Now Track Pollution Levels," *NPR*, July 29, 2015, https://www.npr.org/sections/alltechconsidered/2015/07/29/427462846/some-google-street-view-cars-now-track-pollution-levels#. For more information about Aclima, https://aclima.io/#mission

33. Motor Vehicle Fuel Efficiency Act of 1990, 101st Congress, 1989–1990, https://www.congress.gov/bill/101st-congress/senate-bill/1224/amendments

34. "Fuel-Efficiency Effort Defeated in Senate," *CQ Almanac*, 1990, https://library.cqpress.com/cqalmanac/document.php?id=cqal90-1112545

35. Jon Miltimore, "What Is Lobbying (and Do We Really Need Lobbyists)?," Foundation for Economic Education, September 26, 2019, https://fee.org/articles/what-is-lobbying-and-do-we-really-need-lobbyists/

36. For example see *Reveal* expose interviewing two former ExxonMobil lobbyists, June 30, 2021, https://twitter.com/UE/status/1410300881761882112

37. To block the Trump administrations attempted CAFE standard rollback in 2020, ten environmental NGOs are suing the Trump administration; see https://www.nrdc.org/media/2020/200527. Also see Safe Climate Campaign, http://safeclimatecampaign.org/; Union of Concerned Scientists, https://www.ucsusa.org/resources/brief-history-us-fuel-efficiency

38. See sources in Figure 7.1.

39. Deborah Gordon and Madhav Acharya, "Oil Shake-Up: Refining Transitions in a Low-Carbon Economy," Carnegie Endowment for International Peace, April

2018, https://carnegieendowment.org/files/Gordon_DrivingChange_Article_April2018_final.pdf

40. Examples of analysis produced with former oil industry experts include Gordon and Acharya, "Oil Shake-Up," and "Petroleum Companies Need a Credible Climate Plan," November 2018. See Oil Climate Project for details, https://www.oilclimateproject.com/

41. For CCL founder Marshall Saunders's nerve-wracking story of his first-ever advocacy meeting in Congress, see https://citizensclimatelobby.org/about-ccl/citizens-climate-lobbys-founder/; https://citizensclimatelobby.org/about-ccl/advisory-board/

42. For example, a carbon price tailored to the oil and gas sector would distinguish between heterogeneous resources using the OCI+. See Deborah Gordon and Jessica T. Mathews, "Smart Tax: Pricing Oil for a Safe Climate," Carnegie Endowment for International Peace, June 15, 2016, https://carnegieendowment.org/2016/06/15/smart-tax-pricing-oil-for-safe-climate-pub-63765

43. Reference to lyrics by Steely Dan, "Barrytown," on the album *Pretzel Logic*, February 20, 1974, https://www.azlyrics.com/lyrics/steelydan/barrytown.html

44. "54 Sources for Climate Change News," *GW Public Health Online*, January 6, 2020, https://publichealthonline.gwu.edu/blog/sources-for-climate-news/

45. Carbon Mapper, https://carbonmapper.org/

46. "Methane Detectors Challenge | Environmental Defense Fund," https://www.edf.org/methane-detectors-challenge.

47. See, for example, Al Gore, "We Can Solve the Climate Crisis by Tracing Pollution Back to Its Sources," *Medium*, July 15, 2020, https://medium.com/@algore/we-can-solve-the-climate-crisis-by-tracing-pollution-back-to-its-sources-4f535f91a8dd; and Environmental Defense Fund, "This Is the Next Wave of Environmentalism," https://www.edf.org/approach/fourth-wave

48. United Nations Framework Convention on Climate Change, "Handbook on Measurement, Reporting and Verification," 2014, https://unfccc.int/sites/default/files/non-annex_i_mrv_handbook.pdf

49. The new nonprofit, MiQ, (a collaboration between RMI and SystemIQ) has developed a voluntary certification standard for low-methane natural gas. See https://miq.org/

50. Lewis King and Jeroen C. J. M. van den Bergh, "Normalisation of Paris Agreement NDCs to Enhance Transparency and Ambition," *Environmental Research Letters* 14 (2019), https://iopscience.iop.org/article/10.1088/1748-9326/ab1146/pdf

51. *The Greenhouse Gas Protocol: A Corporate Accounting and Reporting Standard*, rev. ed., World Business Council for Sustainable Development and World Resources Institute, March 2004, https://ghgprotocol.org/sites/default/files/standards/ghg-protocol-revised.pdf

52. Simon Dietz et al., "Carbon Performance Assessment of Oil & Gas Producers: Note on Methodology," Transition Pathway Initiative, October 2020, https://www.transitionpathwayinitiative.org/publications/62.pdf?type=Publication

53. Simon Dietz, et al., "TPI State of Transition Report 2020," Transition Pathway Initiative, May 2020, https://www.transitionpathwayinitiative.org/publications/50.pdf?type=Publication

54. The CCAC is composed of sixty-nine government partners, eighteen intergovernmental organizations, and fifty-eight NGOs. For more information see Climate and Clean Air Coalition, https://ccacoalition.org/en

55. Climate and Clean Air Coalition, "Global Alliance to Significantly Reduce Methane Emissions in the Oil and Gas Sector by 2030," 2019, https://ccacoalition.org/en/activity/global-alliance-significantly-reduce-methane-emissions-oil-and-gas-sector-2030

56. World Bank, "Saudi Aramco Joins World Bank's Initiative: 'Zero Routine Flaring by 2030,'" November 6, 2019, https://www.saudiaramco.com/en/news-media/news/2019/zero-routine-flaring-by-2030-initiative

57. National Oceanic and Atmospheric Administration, "Visible Infrared Imaging Radiometer Suite (VIIRS)," https://ncc.nesdis.noaa.gov/VIIRS/. Christopher Elvidge is the principal investigator; see https://payneinstitute.mines.edu/eog/. Regarding VIIRS accuracy see Adam Brandt, "Accuracy of Satellite-Derived Estimates of Flaring Volume for Offshore Oil and Gas Operations in Nine Countries," *Environmental Research Communications*, May 11, 2020, https://iopscience.iop.org/article/10.1088/2515-7620/ab8e17/pdf

58. Natural Resources Defense Council, "Court Battles," https://www.nrdc.org/court-battles

59. Our Children's Trust (an NGO) has filed numerous lawsuits in US state courts. See https://www.ourchildrenstrust.org/pending-state-actions/ and http://blogs2.law.columbia.edu/climate-change-litigation/wp-content/uploads/sites/16/case-documents/2015/20150812_docket-615-cv-1517_complaint.pdf

60. Umair Irfan, "Pay Attention to the Growing Wave of Climate Change Lawsuits," *Vox*, June 4, 2019, https://www.vox.com/energy-and-environment/2019/2/22/17140166/climate-change-lawsuit-exxon-juliana-liability-kids

61. Reeves Wiedeman, "The Rockefellers vs. the Company That Made Them Rockefellers," *New York Magazine*, January 8, 2018, https://nymag.com/intelligencer/2018/01/the-rockefellers-vs-exxon.html

62. David Hasemyer, "Fossil Fuels on Trial. Where the Major Climate Change Lawsuits Stand Today," *Inside Climate News*, January 17, 2020, https://insideclimatenews.org/news/04042018/climate-change-fossil-fuel-company-lawsuits-timeline-exxon-children-california-cities-attorney-general; Damian Carrington, "Can Climate Litigation Save the World?," *The Guardian*, March 20, 2018, https://www.theguardian.com/environment/2018/mar/20/can-climate-litigation-save-the-world

63. The impetus for this lawsuit against ExxonMobil started with Steve Coll, then-president of New America, when he published *Private Empire: ExxonMobil and American Power* (New York: Penguin Books, 2012). Coll stepped down at the NGO in 2013 to become dean of Journalism at Columbia University. He continued his research, which led to a collaboration with the *Los Angeles Times* and the publication of "What Exxon Knew about the Earth's Melting Arctic," by Sara Jerving et al., October 9, 2015, https://graphics.latimes.com/exxon-arctic/

64. Ron Bousso, et al., "Shell Ordered to Deepen Carbon Cuts in Landmark Dutch Case," *Reuters*, May 26, 2021, https://www.reuters.com/business/sustainable-business/dutch-court-orders-shell-set-tougher-climate-targets-2021-05-26/

65. The First Amendment to the US Constitution states that "Congress shall make no law . . . abridging . . . the right of the people peaceably to assemble," https://www.archives.gov/founding-docs/bill-of-rights-transcript#toc-amendment-i-2

66. Dareh Gregorian, Courtney Kube, and Carol E. Lee, "Trump Says He Will Deploy Military if State Officials Can't Contain Protest Violence," *NBC News*, June 1, 2020, https://www.nbcnews.com/politics/white-house/trump-considering-move-invoke-insurrection-act-n1221326

67. Tal Axelrod, "Three States Push Criminal Penalties for Fossil Fuel Protests amid Coronavirus," March 27, 2020, https://thehill.com/policy/energy-environment/489960-three-states-push-criminal-penalties-for-fossil-fuel-protests-amid

68. For a few of the many recent examples of oil and gas protests, see https://www.energyvoice.com/oilandgas/north-sea/218708/oil-giant-targeted-in-climate-protest/; https://www.vox.com/2020/1/16/21067763/harvard-law-climate-protest-exxon-paul-weiss; https://apnews.com/149c94212b1d4843bcf1ef41f6c94647;

69. Google searches on June 3, June 10, and October 12, 2020, yielded 49.9 to 51.3 million hits in 0.5 to 0.6 seconds.

70. Lisa Johnson and Moira Wyton, "Counter-Protesters Break Down Rail Blockade in West Edmonton as CN Wins Injunction Against Pipeline Opponents," *Edmonton Journal*, February 20, 2020, https://edmontonjournal.com/news/local-news/protesters-block-cn-rail-line-in-west-edmonton; "Greenpeace Activists Board Shell Oil Rigs in Protest against Plans to Leave Behind Oil in the North Sea," October 14, 2019, https://www.greenpeace.org/international/press-release/24826/greenpeace-activists-board-shell-oil-rigs-in-protest-against-plans-to-leave-behind-oil-in-the-north-sea/; Nicholas Kusnetz, "Texas Charges Oil Port Protesters under New Fossil Fuel Protection Law," September 19, 2019, https://insideclimatenews.org/news/16092019/oil-port-protesters-charged-texas-fossil-fuel-infrastructure-law-pipelines-greenpeace-houston

71. Kai Bosworth, "The People Know Best: Situating the Counterexpertise of Populist Pipeline Opposition Movements." *Annals of the American Association of Geographers*, 2019, https://www.tandfonline.com/doi/full/10.1080/24694452.2018.1494538

72. For protest archives, see 350.org, "Stop the Keystone XL Pipeline," https://350. org/stop-keystone-xl/; https://350.org/category/topic/kxl/; https://350.org/ kxl-victory/. For a chronology through 2017, see "A Chronological History of Controversial Keystone XL Pipeline Project," *CBC*, January 24, 2017, https:// www.cbc.ca/news/politics/keystone-xl-pipeline-timeline-1.3950156

73. IHS Cambridge Energy Research Associates, "The Role of Canadian Oil Sands in US Oil Supply," 2010, https://cdn.ihs.com/ihs/cera/The-Role-of-Canadian-Oil-Sands-in-US-Oil-Supply.pdf

74. Canadian Association of Petroleum Producers, "Oil and Natural Gas Pipelines," 2019, https://www.capp.ca/explore/oil-and-natural-gas-pipelines/

75. "Revisiting the Keystone Pipeline," Carnegie Endowment for International Peace, August 1, 2016, https://www.youtube.com/watch?v=e3Vx3Sc3SQE&feature=yo utu.be

76. Alexis Madrigal, "The Great Crude-Oil Fireball Test," *The Atlantic*, February 5, 2018, https://www.theatlantic.com/technology/archive/2018/02/the-great-crude-oil-fireball-test/552029/

77. Jeff Brady, "Builder of Controversial Keystone XL Pipeline Says It's Moving Forward," *NPR*, March 31, 2020, https://www.npr.org/2020/03/31/824445928/ builder-of-controversial-keystone-xl-pipeline-says-its-moving-forward

78. Jariel Arvin, "More Than 20 Republican-Led States Sue Biden for Canceling the Keystone XL Pipeline," *Vox*, March 18, 2021, https://www.vox.com/22306919/ biden-keystone-xl-trudeau-oil-pipeline-climate-change

79. Michael Coren, "Covid-19 Is Closing Canada's Carbon-Intensive Oil Sands For Business," *Quartz*, April 29, 2020, https://qz.com/1846830/ covid-19-is-closing-canadas-carbon-intensive-oil-sands/

80. Cary Krosinsky, "The Seven Tribes of Sustainable Investing," LinkedIn, September 11, 2018, https://www.linkedin.com/pulse/seven-tribes-sustainable-investing-cary-krosinsky

81. Billy Nauman, "Sharp Rise in Number of Investors Dumping Fossil Fuel Stocks," *Financial Times*, September 9, 2019, https://www.ft.com/content/ 4dec2ce0-d0fc-11e9-99a4-b5ded7a7fe3f

82. Tate Williams, "As Top Foundations Resist Divesting from Fossil Fuels, What Might Change Their Minds?," *Inside Philanthropy*, January 13, 2020, https:// www.insidephilanthropy.com/home/2020/1/13/as-top-foundations-resist-divesting-from-fossil-fuels-what-might-change-their-minds; and Tate Williams, "Major Climate Funders Are Still Invested in Fossil Fuels. Why is That?," *Inside Philanthropy*, December 19, 2019, https://www.insidephilanthropy.com/home/ 2019/12/19/major-climate-funders-are-still-invested-in-fossil-fuels-why-is-that

83. Ellen Burstein and Camille Caldera, "Harvard Endowment Commits to Net-Zero Greenhouse Gas Emissions by 2050," *Harvard Crimson*, April 22, 2020, https:// www.thecrimson.com/article/2020/4/22/endowment-ghg-neutral-2050/

84. Cara Lombardo, Miriam Gottfried, and Rebecca Elliott, "Marathon Petroleum Revives Talks," *Wall Street Journal*, June 19, 2020, https://www.wsj.com/articles/marathon-petroleum-in-talks-with-potential-buyers-of-speedway-gas-station-unit-11592515818

85. For examples, see Tim Faulkner, "Providence Divestment from Fossil Fuels Moves Slowly," *EcoRI News*, February 22, 2016, https://www.ecori.org/government/2016/2/22/providence-divestment-from-fossil-fuels-moving-slowly; Obi Ezekoye and Jannick Thomsen, "Going, Going, Gone: A Quicker Way to Divest Assets," McKinsey & Company, 2018, https://www.mckinsey.com/~/media/McKinsey/Business%20Functions/Strategy%20and%20Corporate%20Finance/Our%20Insights/Going%20going%20gone%20A%20quicker%20way%20to%20divest%20assets/Going-going-gone-a-quicker-way-to-divest-assets.pdf

86. For examples, see Dow Jones, https://www.indexologyblog.com/2019/11/25/should-green-benchmarks-include-fossil-fuel-stocks/; Bloomberg New Energy Finance, https://data.bloomberglp.com/bnef/sites/4/2014/08/BNEF_DOC_2014-08-25-Fossil-Fuel-Divestment.pdf; MSCI, https://www.msci.com/www/research-paper/foundations-of-esg-investing/0795306949; Sustainalytics, https://www.sustainalytics.com/esg-blog/esgarp-scores-in-search-of-reasonably-priced-low-esg-risk-stocks/

87. Sophie Yeo, "Here's How Shareholders Are Pressuring Oil Companies to Act on Climate Change," *Desmog Blog*, May 13, 2020, https://www.desmogblog.com/2020/05/13/here-s-how-shareholders-are-pressuring-oil-companies-act-climate

88. See, for example, https://www.upcounsel.com/shareholder-resolution; Paul Keirnan, "SEC Seeks to Curb Shareholder Resolutions," *Wall Street Journal*, February 9, 2020, https://www.wsj.com/articles/sec-seeks-to-curb-shareholder-resolutions-11581264001

89. "Oil Majors Face Shareholder Resolutions on Climate Change," *The Economist*, May 30, 2019, https://www.economist.com/business/2019/05/30/oil-majors-face-shareholder-resolutions-on-climate-change

90. Climate Action 100+, "Global Investors Driving Business Transition," http://www.climateaction100.org/

91. Ceres, "Shareholder Resolutions Database," https://www.ceres.org/shareholder-resolutions-database

92. "Hedge Fund TCI Vows to Punish Directors over Climate Change," ICAST Marketing, February 20, 2020, https://www.icastusa.org/tcitopunishdirectors/

93. Clifford Kraus, "Exxon Board to Get a Third Activist Pushing Cleaner Energy, " New York Times, June 2, 2021,https://www.nytimes.com/2021/06/02/business/exxon-board-clean-energy.html

94. David Wethe, "Chevron Shareholders Want Evidence of Climate Change Performance," *World Oil*, May 28, 2020, https://www. worldoil.com/news/2020/5/28/chevron-shareholders-want-evidence-of-climate-change-performance

95. Inma Galvez-Robles, "19 Youth Climate Activists You Should Be Following on Social Media," EarthDay.org, June 14, 2019, https://www.earthday.org/19-youth-climate-activists-you-should-follow-on-social-media/

96. Ernst & Young, "How Do We Regenerate This Generation's View of Oil and Gas?," 2017, https://assets.ey.com/content/dam/ey-sites/ey-com/en_gl/topics/oil-and-gas/ey-how-do-we-regenerate-this-generations-view-of-oil-and-gas.pdf

97. "Greta Thunberg: What Does the Teenage Climate Change Activist Want?," *BBC*, February 28, 2020, https://www.bbc.com/news/world-europe-49918719

98. Ben Webster, "Professor at St John's College, Oxford, Turns Oil Row into a Heated Debate," *The Times*, January 31, 2020, https://www.thetimes.co.uk/article/professor-at-st-johns-college-oxford-turns-oil-row-into-a-heated-debate-0zr2wpmb5

99. "A Heated Oxford Education," Editorial Board, *Wall Street Journal*, February 2, 2020, https://www.wsj.com/articles/a-heated-oxford-education-11580680180

100. David Gorton, "Top 5 Largest University Endowments," *Investopedia*, February 6, 2020, https://www.investopedia.com/articles/markets/081616/top-5-largest-university-endowments.asp

101. Harvard Management Company, "Harvard Endowment Net-Zero Portfolio Commitment," April 2020, http://www.hmc.harvard.edu/content/uploads/2020/04/Net-Zero-Commitment.pdf

102. John McCormack, Carol Propper, and Sarah Smith, "Herding Cats? Management and University Performance," *Economic Journal*, 124, no. 578 (October 2013), https://www.researchgate.net/publication/259548344_Herding_Cats_Management_and_University_Performance

103. Laura Henry et al., "NGO Participation in Global Governance Institutions: International and Domestic Drivers of Engagement," *Interest Groups and Advocacy* 8, no. 3 (August 2019): 291–332, https://www.researchgate.net/publication/335846215_NGO_participation_in_global_governance_institutions_international_and_domestic_drivers_of_engagement

104. Michael Edwards, "Herding Cats? Civil Society and Global Governance," *New Economy* 9, no. 2 (May 2002): 71–76, https://www.researchgate.net/publication/247643064_Herding_cats_Civil_society_and_global_governance

105. This phenomenon is not new. According to Adam Werbach, past president of the Sierra Club, "Every time environmentalists step outside the confines of the environmental discourse to articulate a more expansive, more inclusive and more compelling vision for the future, they cease being environmentalists and start becoming American progressives." See Adam Werbach, "Where the Environmental Movement Can and Should Go from Here," *Grist*, January 13, 2005, https://grist.org/article/werbach-reprint/

106. The IPCC defined the co-benefits of climate mitigation in 2007. See https://web.archive.org/web/20160525042147/http://www.ipcc.ch/publications_and_data/ar4/wg3/en/ch4s4-5-3.html

107. Drew Shindell, et al., "Global Methane Assessment," United Nations Environment Programme and Climate and Clean Air Coalition, May 2021, https://www.unep.org/resources/report/global-methane-assessment-benefits-and-costs-mitigating-methane-emissions

108. Emily Atkin, "The Climate Movement's Silence," May 31, 2020, https://heated.world/p/the-climate-movements-silence

109. Emma Foehringer Merchant, "How the Environmental Movement Has Changed," *New Republic*, November 29, 2015, https://newrepublic.com/article/124647/environmental-movement-changed

110. Frederik Heylen, Evelien Willems, and Jan Beyers, "Do Professionals Take Over? Professionalisation and Membership Influence in Civil Society Organizations," *Voluntas*, March 24, 2020, https://link.springer.com/article/10.1007/s11266-020-00214-9

111. Heather Houser, *Infowhelm* (New York: Columbia University Press, 2020).

112. Carbon Tracker, "Carbon Bubble," August 23, 2017, https://carbontracker.org/terms/carbon-bubble/

113. Ozge Zihnioglu, "The Prospects of Civic Alliance: New Civic Activists Acting Together with Civil Society Organizations," *International Journal of Voluntary and Nonprofit Organizations* 30, https://link.springer.com/article/10.1007/s11266-018-0032-9

CHAPTER 8

1. US White House, "Fact Sheet: The White House Announces Commitments to the American Business Act on Climate Pledge," October 19, 2015, https://obamawhitehouse.archives.gov/the-press-office/2015/10/19/fact-sheet-white-house-announces-commitments-american-business-act

2. US Environmental Protection Agency, "Greenhouse Gas Equivalencies Calculator," https://www.epa.gov/energy/greenhouse-gases-equivalencies-calculator-calculations-and-references. Note: To convert natural gas emissions that are reported in metric tons $CO_2$ per thousand cubic feet (scf), it is assumed that there are 5,800 scf per BOE.

3. Dana Nuccitelli, "New Report Finds Costs of Climate Change Impacts Often Underestimated," *Yale Climate Connections*, November 18, 2019, https://yale-climateconnections.org/2019/11/new-report-finds-costs-of-climate-change-impacts-often-underestimated/

4. John Horowitz et al., "Methodology for Analyzing a Carbon Tax," US Department of the Treasury, Office of Tax Analysis, Working Paper 115, January 2017, https://www.treasury.gov/resource-center/tax-policy/tax-analysis/Documents/WP-115.pdf

5. Jonathan Koomey et al., "Getting Smart about Oil in a Warming World," Carnegie Endowment for International Peace, October 5, 2016, https://carnegieendowment.org/files/Gordon-Oil_in_a_warming_world1.pdf

6. Deborah Gordon and Jessica Mathews, "Smart Tax: Pricing Oil for a Safe Climate," Carnegie Endowment for International Peace, June 15, 2016, https://carnegieendowment.org/2016/06/15/smart-tax-pricing-oil-for-safe-climate-pub-63765

7. US National Academy of Sciences, "National Inventories of Greenhouse Gas Emissions," in *Verifying Greenhouse Gas Emissions: Methods to Support International Climate Agreements* (Washington, DC: National Academies Press, 2010), 53-66, http://www.nap.edu/read/12883/chapter/4

8. For examples, see Renee Cho, "More Plastic Is on the Way: What It Means for Climate Change," Earth Institute, Columbia University, February 20, 2020, https://blogs.ei.columbia.edu/2020/02/20/plastic-production-climate-change/; Duke University, "Heavy Oils and Petroleum Coke Raising Vanadium Emissions," December 15, 2017, Physics.org, https://phys.org/news/2017-12-heavy-oils-petroleum-coke-vanadium.html

9. Deborah Gordon and Frances Reuland, "Mapping, Measuring, and Managing Methane: The Critical Role of a Potent Climate Pollutant," Watson Institute for International and Public Affairs, November 2019, https://watson.brown.edu/files/watson/imce/news/ResearchBriefs/2019/Methane%20Report-6%20November%202019.pdf

10. Intergovernmental Panel on Climate Change, "Global Warming of 1.5 °C," October 6, 2018, https://ipcc.ch/sr15/

11. See Scenario P1: Intergovernmental Panel on Climate Change, "Characteristics of Four Illustrative Model Pathways," 1.5 °C Special Report, Summary for Policymakers, Figure SPM.3b, http://www.ipcc.ch/sr15/chapter/spm

12. Paul Davies and Michael Green, "EU Issues New Sustainable Investment Disclosure Rules," Latham & Watkins, April 5, 2019, https://www.globalelr.com/2019/04/eu-issues-new-sustainable-investment-disclosure-rules/

13. UN Climate Change, "Reporting and Review under the Paris Agreement," https://unfccc.int/process-and-meetings/transparency-and-reporting/reporting-and-review-under-the-paris-agreement

14. Studies indicate US oil and gas inventories underestimate methane emissions by as much as 60 percent. See Ramon Alverez et al., "Assessment of Methane Emissions from U.S. Oil and Gas Supply Chain," *Science* 361, no. 6398 (July 2018), http://science.sciencemag.org/content/361/6398/186. The European Union's GHG inventory (EDGAR) uses uniform oil and gas volumes as the indicator of supply-side emissions. See Monica Crippa et al., "High Resolution Temporal Profiles in the Emissions Database for Global Atmospheric Research," *Scientific Data* 7 (April 2020): Article 121, https://www.nature.com/articles/s41597-020-0462-2; for detailed table see https://www.nature.com/articles/s41597-020-0462-2/tables/4

15. US Environmental Protection Agency, Greenhouse Gas Reporting Program (GHGRP), "GHGRP Refineries," July 15, 2021, https://www.epa.gov/ghgreporting/ghgrp-refineries; Dean Foreman, "The U.S. Refining Industry—Striving

to Maintain Global Leadership," American Petroleum Institute, October 2, 2019, https://www.api.org/news-policy-and-issues/blog/2019/10/02/us-refining-industry-striving-to-maintain-global-leadership

16. Eric Kort et al., "Four corners: The largest US methane anomaly viewed from space," *Geophysical Research Letters* (41) 19, 6898-6903, October 16, 2014, https://agupubs.onlinelibrary.wiley.com/doi/full/10.1002/2014GL061503

17. Intergovernmental Panel on Climate Change, "2019 Refinement to the 2006 IPCC Guidelines for National Greenhouse Gas Inventories," https://www.ipcc.ch/report/2019-refinement-to-the-2006-ipcc-guidelines-for-national-greenhouse-gas-inventories/

18. Deborah Gordon and David Livingston, "Petroleum Coke Use in India and South Asia: Recent Trends and Energy Policy Options," in *Connectivity and Trading in Power & Energy: A Regional and International Dimension*, Syed Munir Khasru, ed. 17-30, (Bangladesh: IPAG, 2017), https://carnegieendowment.org/files/PetcokeUseIPAG.pdf. For national flows see Chatham House, "ResourceTrade.earth," https://resourcetrade.earth/data?year=2018&category=141&units=value

19. Wang Tao, "Managing China's Petcoke Problem," Carnegie-Tsinghua Center for Global Policy, https://carnegieendowment.org/files/petcoke.pdf

20. For these nations' (and others') petcoke import/export statistics see http://data.un.org/Data.aspx?q=petroleum&d=EDATA&f=cmID%3aPK. The air quality in Delhi is so bad that the High Court compared the city to "living in a gas chamber." Smog obscured sunlight and visibility and smelled of woodsmoke. See "Death by Pollution: Delhi's Fight for Clean Air," *CNN Money*, February 17, 2016, http://money.cnn.com/2016/02/17/news/economy/india-new-delhi-air-pollution/index.html; For petcoke bans, see "India Bans Pet Coke Imports for Use as a Fuel," *Hindu Business Line*, August 17, 2018, https://www.thehindubusinessline.com/markets/commodities/india-bans-pet-coke-import-for-use-as-fuel/article24716341.ece

21. "China Limits Higher-Sulfur Coke Use in Key Regions," *Argus Media*, July 16, 2019, https://www.argusmedia.com/en/news/1940947-china-limits-highersulphur-coke-use-in-key-regions

22. International Maritime Organization, "Sulphur 2020 – Cutting Sulphur Dioxide Emissions," http://www.imo.org/en/MediaCentre/HotTopics/Pages/Sulphur-2020.aspx

23. India is the only country that is currently banning petcoke. See PTI, "Ensure Ban on Petcoke and Furnace Oil, NGT Directs CPCB," *The Hindu*, July 17, 2020, https://www.thehindu.com/news/national/ensure-ban-on-petcoke-and-furnace-oil-ngt-directs-cpcb/article32112214.ece; for plastic bans see Laura Parker, "Plastic Bag Bans Are Spreading. But Are They Truly Effective?," *National Geographic*, April 17, 2019, https://www.nationalgeographic.com/environment/2019/04/plastic-bag-bans-kenya-to-us-reduce-pollution/

24. Pratim Ranjan Bose, "Petcoke Imports Double on Plunging Global Price, Absence of Carbon Tax," *Hindu Business Line*, June 27, 2016, updated January 20, 2018, https://www.thehindubusinessline.com/economy/policy/petcoke-imports-double-on-plunging-global-prices-absence-of-carbon-tax/article8780325.ece#

25. For details on Governor Newsom's managed oil decline see https://www.sacbee.com/opinion/california-forum/article239537598.html. For details on using the OCI+ to determine which assets to turn on after the COVID pandemic see https://www.brinknews.com/not-all-oil-is-equal-which-types-should-oil-companies-turn-back-on-as-the-economy-recovers-paris-agreement-climate-change-post-covid/

26. Jared Huffman, "Rep. Huffman Introduces Energy Transparency Legislation," July 19, 2017, https://huffman.house.gov/media-center/press-releases/rep-huffman-introduces-energy-transparency-legislation-know-your-oil-act-and-carbon-pollution-transparency-act. For text see H.R. 6082, 114th Congress (2015–2016), https://www.govtrack.us/congress/bills/114/hr6082 and H.R. 3285, 115th Congress (2017–2018), https://www.congress.gov/bill/115th-congress/house-bill/3286?s=1&r=82

27. California Legislative Information, "AB-3217: Greenhouse Gases: Crude Oil and Emissions," February 21, 2020, http://leginfo.legislature.ca.gov/faces/billTextClient.xhtml?bill_id=201920200AB3217

28. Personal communication with Dr. Daniel Jacob, Harvard University, May 7, 2020, following the release of private sector error-prone data from a government satellite as reported here: https://scitechdaily.com/mapping-methane-emissions-on-a-global-scale/

29. Kristina Makansi, "Carbon Expert Brings Mission-Focused, Team Approach Addressing Climate Change," UA@Work, January 28, 2020, https://uaatwork.arizona.edu/lqp/carbon-expert-brings-mission-focused-team-approach-addressing-climate-change; Carbon Mapper, https://carbonmapper.org/

30. Daniel Cusworth et al., "Multisatellite Imaging of a Gas Well Blowout Enables Quantification of Total Methane Emissions," *Geophysical Research Letters* 48, no. 2 (January 28, 2021), https://agupubs.onlinelibrary.wiley.com/doi/full/10.1029/2020GL090864

31. United Nations Climate Change, "Nationally Determined Contributions (NDCs)," https://unfccc.int/process/the-paris-agreement/nationally-determined-contributions/ndc-registry#eq-1

32. Mohammad Masnadi et al., "Global Carbon Intensity of Crude Oil Production," *Science* 361, no. 6405 (August 31, 2018), https://science.sciencemag.org/content/361/6405/851; Liang Jing et al., "Carbon Intensity of Global Crude Oil Refining and Mitigation Potential," *Nature Climate Change* 10 (June 2020),https://www.nature.com/articles/s41558-020-0775-3

33. Quirin Schiermeier, "Global Methane Levels Soar to Record High," *Nature*, July 14, 2020, https://www.nature.com/articles/d41586-020-02116-8

34. Katherine Konschnik and Frances Reuland, "Canadian Oil and Gas Methane Rules to Kick Off in 2020—Others Likely to Follow," International Energy Agency, February 17, 2020, https://www.iea.org/commentaries/canada-steps-up-its-efforts-to-reduce-methane-emissions

35. Katherine Konschnik and Sarah Marie Jordaan, "Reducing Fugitive Methane Emissions from the North American Oil and Gas Sector: A Proposed Science-Policy Framework," *Climate Policy* 18, no. 9 (2018), https://www.tandfonline.com/doi/abs/10.1080/14693062.2018.1427538?journalCode=tcp020

36. Alissa Rubin and Clifford Krauss, "Toxic Twilight as Iraq Burns Precious Gas," *New York Times*, July 17, 2020, https://www.nytimes.com/2020/07/16/world/middleeast/iraq-gas-flaring-cancer-environment.html

37. World Health Organization, "Air Pollution," https://www.who.int/health-topics/air-pollution#tab=tab_1

38. Brad Plumer, "Europe's Love Affair with Diesel Cars Has Been a Disaster," *Vox*, October 16, 2015, https://www.vox.com/2015/10/15/9541789/volkswagen-europe-diesel-pollution

39. Deborah Gordon and Jessica Mathews, "Smart Tax: Pricing Oil for a Safe Climate," Carnegie Endowment for International Peace, 2016, https://carnegieendowment.org/files/GordonMathews_CarbonTax_Final.pdf

40. See "Whitehouse, Booker, Schatz Unveil Methane Fee to Clamp Down on Potent Drive or Climate Change," March 9, 2021, with the linked text of the Methane Emissions Reduction Act, https://www.whitehouse.senate.gov/news/release/whitehouse-booker-schatz-unveil-methane-fee-to-clamp-down-on-potent-driver-of-climate-change

41. Leo Levinson and Deborah Gordon, "DRIVE+: Promoting Cleaner and More Fuel Efficient Motor Vehicles through a Self-Financing System of State Sales Tax Incentives," *Journal of Policy Analysis and Management* 9, no. 3 (Summer 1990), https://www.jstor.org/stable/3325286

42. Inger Sunde et al., "Ren, norsk olje?," *NRK*, December 12, 2015, https://www.nrk.no/dokumentar/xl/ren_-norsk-olje_-1.13150883

43. Thomas Kühn et al., "Effects of Black Carbon Mitigation on Arctic Climate," *Atmospheric Chemistry and Physics* 20 (May 2020), https://doi.org/10.5194/acp-20-5527-2020

44. Merritt Turetsky et al., "Carbon Release through Abrupt Permafrost Thaw," *Nature Geoscience* 13 (February 2020), https://www.nature.com/articles/s41561-019-0526-0.epdf

45. Thomas Lovejoy and Carlos Nobre, "Amazon Tipping Point: Last Chance for Action," *Science Advances* 5, no. 12 (December 2019), https://advances.sciencemag.org/content/5/12/eaba2949

46. J. T. Houghton, G. J. Jenkins, and J. J. Ephraums, eds., *The IPCC Scientific Assessment* (New York: Cambridge University Press, 1990).

47. See Table 3.2 (and text following) for more details.

48. Michelle Cain et al., "Improved Calculation of Warming-Equivalent Emissions for Short-Lived Climate Pollutants," *Nature Climate and Atmospheric Science*, 2019, https://www.researchgate.net/publication/335608530_Improved_calculation_of_warming-equivalent_emissions_for_short-lived_climate_pollutants

49. James Pethokoukis, "US Federal Research Spending Is at a 60-Year Low," American Enterprise Institute, May 11, 2020, https://www.aei.org/economics/us-federal-research-spending-is-at-a-60-year-low-should-we-be-concerned/; International Energy Agency, "Global Energy Investment Stabilized above USD 1.8 Trillion in 2018, but Security and Sustainability Concerns Are Growing," May 14, 2019, https://www.iea.org/newsroom/news/2019/may/global-energy-investment-stabilised-above-usd-18-trillion-in-2018-but-security-.html

50. David Iaconangelo, "ARPA-C? Biden's 100% clean energy plan questioned," *E&E News*, May 28, 2020, https://www.eenews.net/stories/1063251621

51. "Million-Mile Car Batteries Are Coming," *The Economist*, July 30, 2020, https://www.economist.com/science-and-technology/2020/07/30/what-the-million-mile-battery-means-for-electric-cars

52. Robert Service, "Can the World Make the Chemicals It Needs without Oil?," *Science Magazine*, September 19, 2019, https://www.sciencemag.org/news/2019/09/can-world-make-chemicals-it-needs-without-oil

53. "Better living through chemistry" was initially branded by DuPont in 1935. DuPont reportedly dropped "through chemistry" in 1982, and in 1999, the tagline was replaced with "the miracles of science." See https://adage.com/article/btob/dupont-replaces-1935-tagline-reflect-corporate-change/247761

54. Akshat Rathi, "A Tiny Tweak in California Law Is Creating a Strange Thing: Carbon-Negative Oil," *Quartz*, July 1, 2019, https://qz.com/1638096/the-story-behind-the-worlds-first-large-direct-air-capture-plant/

55. Jonathan Watts, "US and Saudi Arabia Blocking Regulation of Geoengineering, Sources Say," *The Guardian*, March 18, 2019, https://www.theguardian.com/environment/2019/mar/18/us-and-saudi-arabia-blocking-regulation-of-geoengineering-sources-say

56. Deborah Gordon, "Understanding Climate Engineering," Carnegie Endowment for International Peace, August 21, 2017, https://carnegieendowment.org/files/Understanding_Climate_Engineering_Web.pdf

57. Deborah Gordon et al., "Advancing Public Climate Engineering Disclosure," Carnegie Endowment for International Peace, May 29, 2018, https://carnegieendowment.org/files/Gordon_Advancing_Public_Disclosure_On_Climate_Engineering_June_2018.pdf; Noah Deich, "Everything and the Carbon Sink," November 18, 2014, https://carbonremoval.wordpress.com/2014/11/18/arpa-c-how-an-advanced-research-projects-agency-for-carbon-could-catalyze-development-of-the-cdr-field/

58. Danish Oil and Gas (DONG) transformed its business into a renewable energy business and was renamed Orsted. By 2025 it plans to be carbon neutral. See Adele

Peters, "How This Danish Energy Company Is Transitioning from Oil and Gas to All Renewables," *Fast Company*, February 5, 2020, https://www.fastcompany.com/90459425/how-this-danish-energy-company-is-transitioning-from-oil-and-gas-to-all-renewables#

59. Colette Cohen, "Opportunity to 'Reimagine' North Sea for Net Zero Says OGTC Chief," *Energy Voice*, December 27, 2019, https://www.energyvoice.com/otherenergy/214554/opportunity-to-reimagine-north-sea-for-net-zero-says-ogtc-chief/

60. Aera, "How Belridge Solar Was Shared around the World," November 29, 2017, https://www.aeraenergy.com/how-belridge-solar-was-shared-around-the-world/

61. "Equinor Wins Opportunity to Develop the World's Largest Offshore Wind Farm," September 20, 2019, https://www.equinor.com/en/news/2019-09-19-doggerbank.html

62. John Cox, "Financing Trouble, Design Changes Delay Major Solar Project in West Kern Oilfield," Bakersfield.com, September 30, 2019, https://www.bakersfield.com/financial-trouble-design-changes-delay-major-solar-project-in-west-kern-oilfield/article_6f692be4-e3bb-11e9-92a5-771e6c569123.html

63. "Product Life Cycle Accounting and Reporting Standard," World Resources Institute and World Business Council for Sustainable Development, September 2011.

64. Pippa Stevens, "Goldman Sachs to Spend $750 Billion on Climate Transition Projects and Curb Fossil Fuel Lending," *CNBC*, December 16, 2019, https://www.cnbc.com/2019/12/16/goldman-sachs-targets-750-billion-for-climate-transition-projects.html

65. Jillian Ambrose and Jon Henley, "European Investment Bank to Phase Out Fossil Fuel Financing," *The Guardian*, November 15, 2019, https://www.theguardian.com/environment/2019/nov/15/european-investment-bank-to-phase-out-fossil-fuels-financing

66. "Hedge Fund TCI Vows to Punish Directors over Climate Change," ICAST Marketing, February 10, 2020, https://www.icastusa.org/tcitopunishdirectors/

67. When companies simply sell their dirtiest assets to other companies, this results in emissions leakage that requires ongoing tracking and warranties to eliminate. See Rachel Adams-Heard, "What Happens When an Oil Giant Walks Away," Bloomberg Green, April 15, 2021, https://www.bloomberg.com/graphics/2021-tracking-carbon-emissions-BP-hilcorp/

68. MiQ, a nonprofit collaboration between RMI and SystemIQ, is a voluntary industry standard that is designed for oil and gas buyers and sellers to opt into. For more information on MiQ, see https://miq.org/

69. Rakesh Sharma, "OPEC vs. the US: Who Controls Oil Prices?," *Investopedia*, October 14, 2020, https://www.investopedia.com/articles/investing/081315/opec-vs-us-who-controls-oil-prices.asp

70. Deborah Gordon, "Not All Oil Is Equal. As Economies Recover, Which Oils Should Stay in the Ground?" *Brink*, June 23, 2020, https://www.brinknews.

com/not-all-oil-is-equal-which-types-should-oil-companies-turn-back-on-as-the-economy-recovers-paris-agreement-climate-change-post-covid/

71. Chris Le Fevre, "Methane Emissions: From Blind Spot to Spotlight," Oxford Institute for Energy Studies, July 2017, https://www.oxfordenergy.org/wpcms/wp-content/uploads/2017/07/Methane-Emissions-from-blind-spot-to-spotlight-NG-122.pdf

72. "Nigerian Oil and Gas News Digest – February 2020 Round Up," *Lexology*, February 29, 2020, https://www.lexology.com/library/detail.aspx?g=788b2fb9-e278-468b-8408-5913d750cf00

73. The World Bank, "Seven Countries Account for Two-Thirds of Global Gas Flaring, April 28, 2021, https://www.worldbank.org/en/news/press-release/2021/04/28/seven-countries-account-for-two-thirds-of-global-gas-flaring

74. Benjamin Storrow, "Are Oil Majors Serious about Cutting Emissions?," *E&E News*, March 25, 2019, https://www.eenews.net/stories/1060128099

75. Climate and Clean Air Coalition, "Reducing Methane Emissions across the Natural Gas Value Chain – Guiding Principles," 2017, https://www.ccacoalition.org/en/resources/reducing-methane-emissions-across-natural-gas-value-chain-guiding-principles

76. Methane Guiding Principles (MGP) is a voluntary, international multistakeholder partnership between industry and nonindustry organizations with a focus on priority areas for methane mitigation along the natural gas supply chain. For more information see https://methaneguidingprinciples.org/

77. "Former Shell Oil President John Hofmeister Weights in on Rollback of Emissions Rules," *NPR All Things Considered*, August 29, 2019, https://www.npr.org/2019/08/29/755555482/former-shell-oil-president-john-hofmeister-weighs-in

78. Catalina Jaramillo, "Bankruptcy Judge Clears Away Final Hurdle in PES Refinery Sale," *WHYY Public Radio*, June 25, 2020, https://whyy.org/articles/bankruptcy-judge-clears-away-final-hurdle-in-pes-refinery-sale/

79. For background on these corporate pledges see https://www.bp.com/en/global/corporate/news-and-insights/press-releases/bernard-looney-announces-new-ambition-for-bp.html; https://www.shell.com/energy-and-innovation/the-energy-future/shells-ambition-to-be-a-net-zero-emissions-energy-business.html

80. Deborah Gordon and Stephen Ziman, "Petroleum Companies Need a Credible Climate Plan," Carnegie Endowment for International Peace, November 15, 2018, https://carnegieendowment.org/files/Gordon_Petro_Companies_Need_Climate_Plan_Nov2018.pdf; Daniel Boffey, "Court Orders Royal Dutch Shell to Cut Carbon Emissions by 45% by 2030, " *The Guardian*, May. 27. 2021, https://www.theguardian.com/business/2021/may/26/court-orders-royal-dutch-shell-to-cut-carbon-emissions-by-45-by-2030

81. Companies need to provide additional disclosure to boost the credibility of their ambitions and the ability to deliver on their climate pledges. According to CDP, no

petroleum companies are not on the climate disclosure "A" list (2019) and many no longer submit transparency documentation: https://www.cdp.net/en/companies/companies-scores

82. Note that Lloyds Banking Group is a financial services provider and is not the same entity as Lloyd's of London, which is a British insurance syndicate that insures against risks. See https://www.lloydsbankinggroup.com/our-brands/ and https://www.investopedia.com/terms/l/lloyds-london.asp

83. Rachel Koning Beals, "JP Morgan Chase Will Withdraw Support for Some Fossil Fuels," *MarketWatch*, February 25, 2020, https://www.marketwatch.com/story/jp-morgan-chase-the-oil-industrys-bank-of-choice-to-withdraw-support-for-some-fossil-fuels-2020-02-25

84. BlackRock Investor Communication, January 2020, https://www.blackrock.com/uk/individual/blackrock-client-letter

85. Ivana Kottasova, "BP Will Link Bonuses to 36,000 Workers to Climate Targets," *CNN Business*, February 1, 2019, https://www.cnn.com/2019/02/01/business/bp-shareholders-paris/index.html

86. For a list of ecolabeling organizations focused on reducing carbon emissions, see Ecolabel Index, http://www.ecolabelindex.com/ecolabels/?st=category,carbon

87. Rocky Mountain Institute and SYSTEMIQ, "Global Methane Solutions," https://www.globalmethanesolutions.org/; Resources for the Future, "Greening Gas: Creating a Market for Low-Methane Natural Gas," https://www.rff.org/events/rff-live/greening-gas-creating-market-low-methane-natural-gas/

88. For example, see LeeAnn Baronett, "Freedom Isn't Free . . . of Methane," Rocky Mountain Institute, June 5, 2019, https://rmi.org/freedom-isnt-freeof-methane/ and Resources for the Future, "Greening Gas: Creating a Market for Low-Methane Natural Gas," April 24, 2020,https://www.rff.org/events/rff-live/greening-gas-creating-market-low-methane-natural-gas/

89. Experts discuss obstacles to pricing "responsible" gas, including the lack of an established definition, variation in US production basins, pipeline systems not directly connecting producers with buyers, and the need for data through the entire supply chain. See S&P Global Platts, "Looking Ahead: The Energy Transition and Future of Index Development," April 16, 2020, https://plattsinfo.spglobal.com/2020-Webinar-Americas-Natural-Gas

90. RMI is using the OCI+ and other diagnostic tools to provide oil and gas GHG inputs to Climate TRACE. For background on Climate TRACE, see https://www.climatetrace.org/

91. Dave Roberts, "The Entire World's Carbon Emissions Will Finally Be Trackable in Real Time," *Vox*, July 16, 2020, https://www.vox.com/energy-and-environment/2020/7/16/21324662/climate-change-air-pollution-tracking-greenhouse-gas-emissions-trace-coalition. See Watt Time: https://www.watttime.org/; Climate TRACE, https://www.climatetrace.org/

92. Oil and gas sector employees expect to retire before age sixty-five. This has been fostered by the industry's generous retirement contributions coupled with higher wages that cushion individual contributions. See Aon, "Oil and Gas 2018 Retirement Benchmarking," 2018, https://www.aon.com/getmedia/da46390f-9dbf-430c-95f7-2e7e83b80b2c/Aon_2018_OG_Drilling_and_Services_Manufacturing_Report_Final.aspx

93. Stephanie Rogers, "Retirement Wave and Digital Reinvention Prompt Urgent Talent Reassessment," *Rigzone*, December 29, 2017, https://www.rigzone.com/news/wire/retirement_wave_and_digital_reinvention_prompt_urgent_talent_reassessment-29-dec-2017-152964-article/

94. Michael Bloomberg, "Let's Hire Laid-Off Oil and Gas Workers to Fight Climate Change," *Bloomberg Opinion*, August 3, 2020, https://www-bloomberg-com.cdn.ampproject.org/c/s/www.bloomberg.com/amp/opinion/articles/2020-08-03/let-s-hire-laid-off-oil-and-gas-workers-to-fight-climate-change

95. See the introduction for my recent experience working with oil and gas industry retirees on creating corporate climate plans and developing GHG mitigation strategies for this sector.

96. Sandia National Laboratories, "Second Act: Sandia Retirees Band Together to Help Small Businesses with Tech Challenges," January 23, 2019, https://www.newswise.com/articles/second-act-sandia-retirees-band-together-to-help-small-businesses-with-tech-challenges

97. Robert Perkins, "Fossil Fuel Angst Darkens Oil Sector Funding Outlook," S&P Global, June 6, 2019, https://www.spglobal.com/en/research-insights/articles/fossil-fuel-angst-darkens-oil-sector-funding-outlook

98. For more information on Climate Action 100, see http://www.climateaction100.org/; Damian Carrington, "Firms Ignoring Climate Crisis Will Go Bankrupt, Says Mark Carney," *The Guardian*, October 13, 2019, https://www.theguardian.com/environment/2019/oct/13/firms-ignoring-climate-crisis-bankrupt-mark-carney-bank-england-governor

99. For example, see the Center for Climate Aligned Finance at RMI, https://climatealignment.org/

100. Cary Krosinsky, "NY State Common's Climate Plan," Top1000funds.com, April 19, 2019, https://www.top1000funds.com/2019/04/ny-state-commons-climate-plan/

101. Using the OCI+, compare, for example, oil sands lifecycle GHGs to those of a depleted oil field or a depleted gas field, whose emissions can be greater yet they are not slated for divestment. See https://dxgordon.github.io/OCIPlus/#total-emissions?ratioSelect=perBarrel&oiltypeSelect=Depleted%20Oil,Extra-Heavy,Depleted%20Gas

102. "Decarbonization Advisory Panel Beliefs and Recommendations," New York State Common Retirement Fund, April 2019, https://www.osc.state.ny.us/sites/

default/files/reports/documents/pdf/2019-07/decarbonization-advisory-panel-2019.pdf

103. Masnadi et al., "Global Carbon Intensity of Crude Oil Production,"; Jing et al., "Carbon Intensity of Global Crude Oil Refining and Mitigation Potential."

104. Using the OCI+ Preview Web Tool, compare UK Brent to Norway Ekofisk or US Texas Eagle Ford Gas at twenty-year GWP and high methane settings. See OCI+ for more details: https://dxgordon.github.io/OCIPlus/#supply-chain?model=02102002&carbonToggle=off&carbonTax=20.00&stepSelect=ghgTotal&sortSelect=true

105. Decommissioning carries high risks, and assurance agents like Lloyd's Register are involved in the process. See https://www.lr.org/en-us/decommissioning-services/

106. Deborah Gordon and Madhav Acharya, "Oil Shake-Up: Refining Transitions in a Low-Carbon Economy," Carnegie Endowment for International Peace, April 2018, https://carnegieendowment.org/files/Gordon_DrivingChange_Article_April2018_final.pdf

107. US Energy Information Administration, "Gasoline and Diesel Refining Margins That Diverged in March Have Moved Closer Together," August 6, 2020, https://www.eia.gov/todayinenergy/detail.php?id=44656

108. Scott Jenkins, "U.S. Refiners Reckon with Uncertain Energy Future," *Chemical Engineering*, May 2019, https://www.chemengonline.com/u-s-refiners-reckon-uncertain-energy-future/

109. For example, see Yuzhong Zhang, "Quantifying Methane Emissions from the Largest Oil-Producing Basin in the United States from Space," *Science Advances* 6, no. 17 (April 22, 2020), https://advances.sciencemag.org/content/6/17/eaaz5120; World Bank, "Global Gas Flaring Jumps to Levels Last Seen in 2009," July 21, 2020, https://www.worldbank.org/en/news/press-release/2020/07/21/global-gas-flaring-jumps-to-levels-last-seen-in-2009

# Index

*For the benefit of digital users, indexed terms that span two pages (e.g., 52–53) may, on occasion, appear on only one of those pages.*

Tables and figures are indicated by *t* and *f* following the page number.